Programmable Microcontrollers

Programmable Microcontrollers
Applications on the MSP432™ LaunchPad

Cem Ünsalan
Marmara University

H. Deniz Gürhan
Yeditepe University

M. Erkin Yücel
Yeditepe University

New York Chicago San Francisco
Athens London Madrid
Mexico City Milan New Delhi
Singapore Sydney Toronto

Library of Congress Control Number: 2017952606

McGraw-Hill Education books are available at special quantity discounts to use as premiums and sales promotions, or for use in corporate training programs. To contact a representative please visit the Contact Us page at www.mhprofessional.com.

Programmable Microcontrollers:
Applications on the MSP432™ LaunchPad

Copyright © 2018 by McGraw-Hill Education. All rights reserved. Printed in the United States of America. Except as permitted under the United States Copyright Act of 1976, no part of this publication may be reproduced or distributed in any form or by any means, or stored in a data base or retrieval system, without the prior written permission of the publisher.

1 2 3 4 5 6   QVS   21 20 19 18 17

ISBN 978-1-259-83619-0
MHID 1-259-83619-3

The pages within this book were printed on acid-free paper.

Sponsoring Editor
Michael McCabe

Editorial Supervisor
Donna M. Martone

Acquisitions Coordinator
Lauren Rogers

Project Manager
Vivek Khandelwal

Copy Editor
Mohd. Taiyab Khan

Proofreader
Yashoda Rawat

Indexer
Ever A. Olano

Production Supervisor
Lynn M. Messina

Composition
MPS Limited

Art Director, Cover
Jeff Weeks

Information contained in this work has been obtained by McGraw-Hill Education from sources believed to be reliable. However, neither McGraw-Hill Education nor its authors guarantee the accuracy or completeness of any information published herein, and neither McGraw-Hill Education nor its authors shall be responsible for any errors, omissions, or damages arising out of use of this information. This work is published with the understanding that McGraw-Hill Education and its authors are supplying information but are not attempting to render engineering or other professional services. If such services are required, the assistance of an appropriate professional should be sought.

Contents

Preface xv
Acknowledgments xvii

1 Introduction 1
   1.1 Hardware to be Used in the Book 1
   1.2 Architecture of the MSP432 Microcontroller 2
       1.2.1 Central Processing Unit 2
       1.2.2 Memory 2
       1.2.3 Input and Output Ports 4
       1.2.4 Timer Modules 4
       1.2.5 Mixed Signal Processing Modules 4
       1.2.6 Digital Communication Modules 5
       1.2.7 Other Modules 5
   1.3 Software to be Used in the Book 5
   1.4 Topics to be Covered in the Book 6

2 Code Composer Studio and Energia 9
   2.1 Code Composer Studio 9
       2.1.1 Downloading and Installing CCS 9
       2.1.2 Starting CCS 9
       2.1.3 CCS Perspectives 10
       2.1.4 Creating a New Project 10
       2.1.5 Adding DriverLib Functions to the Project 13
       2.1.6 Creating an Assembly Project 13
       2.1.7 Program Execution 13
       2.1.8 Inserting a Breakpoint 14
       2.1.9 Adding a Watch Expression 15
       2.1.10 Observing Hardware 15
       2.1.11 The Terminal Window 16
       2.1.12 Closing the Project 18
   2.2 CCS Cloud 19
       2.2.1 Setup 20
       2.2.2 Creating a Project 20
       2.2.3 Debugging and Executing the Project 20
       2.2.4 Observing Hardware 20
   2.3 Energia 21
       2.3.1 Downloading and Installing Energia 21
       2.3.2 Creating, Building, and Executing an Energia Sketch . . 21
       2.3.3 Energia Sketch with CCS 23
       2.3.4 Energia Sketch with CCS Cloud 23

Contents

- 2.4 Platform Comparison 23
- 2.5 Summary 24
- 2.6 Exercises 25

3 MSP432 Programming with C 27
- 3.1 Memory Management 27
  - 3.1.1 The Code 27
  - 3.1.2 Local and Global Variables 27
- 3.2 C Data Types 29
  - 3.2.1 Representing Integer Numbers 29
  - 3.2.2 Representing Fractional Numbers 31
- 3.3 ASCII Code 32
- 3.4 Examples on Data Type Usage 33
- 3.5 Arithmetic and Logic Operations 34
- 3.6 Control Structures 38
  - 3.6.1 Condition Check 38
  - 3.6.2 Loops 38
- 3.7 Arrays and Pointers 39
- 3.8 Structures 41
- 3.9 Miscellaneous Issues 42
  - 3.9.1 Define Statement and Constant Declaration 42
  - 3.9.2 Casting 43
  - 3.9.3 Reaching a Specific Memory Address 44
  - 3.9.4 Advanced Mathematical Operations 45
- 3.10 Summary 46
- 3.11 Exercises 46

4 MSP432 Programming with Assembly 49
- 4.1 MSP432 Core Registers 49
- 4.2 Anatomy of an Assembly Program 50
- 4.3 The Instruction Set 51
  - 4.3.1 Anatomy of an Instruction 51
  - 4.3.2 General Data Moving and Processing Instructions 52
  - 4.3.3 Memory Access Instructions 53
  - 4.3.4 Instructions for Arithmetic and Logic Operations 54
  - 4.3.5 Branch and Control Instructions 57
  - 4.3.6 Floating-Point Instructions 57
  - 4.3.7 Other Instructions 58
- 4.4 Sample Programs on Instruction Set Usage 60
  - 4.4.1 Data Processing and Memory Access Operations 60
  - 4.4.2 Arithmetic and Logic Operations 61
  - 4.4.3 Branch and Control Structures 62
  - 4.4.4 Usage of Other Instructions 63
- 4.5 Inline Assembly 64
- 4.6 Summary 66
- 4.7 Exercises 66

5 Digital Input and Output ... 69
- 5.1 Pin Layout for Digital I/O ... 69
- 5.2 Digital I/O Registers ... 70
  - 5.2.1 Direction Register ... 71
  - 5.2.2 Input Register ... 71
  - 5.2.3 Output Register ... 71
  - 5.2.4 Pull-Up/Down Resistor Register ... 71
  - 5.2.5 Select Register ... 72
- 5.3 Digital I/O via DriverLib Functions ... 72
- 5.4 Digital I/O via Energia Functions ... 73
- 5.5 Digital I/O Hardware Issues ... 74
  - 5.5.1 Active High/Low Input ... 74
  - 5.5.2 Switch Bouncing ... 74
- 5.6 Coding Practices for Digital I/O ... 75
  - 5.6.1 Register-Level Examples ... 75
  - 5.6.2 DriverLib Examples ... 76
  - 5.6.3 Energia Examples ... 78
- 5.7 Application: Home Alarm System ... 79
  - 5.7.1 Equipment List ... 79
  - 5.7.2 Layout ... 80
  - 5.7.3 System Design Specifications ... 80
- 5.8 Summary ... 81
- 5.9 Exercises ... 81

6 Exceptions and Interrupts ... 83
- 6.1 Interrupt as an Exception ... 83
- 6.2 Interrupts in Our Daily Lives ... 84
- 6.3 The Nested Vectored Interrupt Controller ... 85
  - 6.3.1 Serving a Single Interrupt Request ... 85
  - 6.3.2 Serving Multiple Interrupt Requests ... 86
  - 6.3.3 NVIC Registers ... 87
  - 6.3.4 Interrupts via DriverLib Functions ... 88
- 6.4 Port Interrupts ... 90
  - 6.4.1 Port Interrupt Registers ... 90
  - 6.4.2 Port Interrupts via DriverLib Functions ... 90
  - 6.4.3 Port Interrupts via Energia Functions ... 91
- 6.5 Coding Practices for Port Interrupts ... 92
  - 6.5.1 Register-Level Examples ... 92
  - 6.5.2 DriverLib Examples ... 94
  - 6.5.3 Energia Examples ... 98
- 6.6 Application: Home Entrance System ... 99
  - 6.6.1 Equipment List ... 99
  - 6.6.2 Layout ... 100
  - 6.6.3 System Design Specifications ... 102
- 6.7 Summary ... 104
- 6.8 Exercises ... 105

Contents

7 Power Management and Timing Operations 107
- 7.1 Power Supply System 107
- 7.2 Clock System 108
  - 7.2.1 Clock Sources 108
  - 7.2.2 Clocks 108
  - 7.2.3 Clock System Faults 109
  - 7.2.4 Clock System Registers 109
  - 7.2.5 Clock System Usage via DriverLib Functions 111
- 7.3 Power Control Module 113
  - 7.3.1 Power Modes 113
  - 7.3.2 Power Control Module Registers 115
  - 7.3.3 Power Control Module Usage via DriverLib Functions 117
- 7.4 Watchdog Timer 118
  - 7.4.1 Watchdog Timer Registers 118
  - 7.4.2 Watchdog Timer Usage via DriverLib Functions 119
  - 7.4.3 Coding Practices for the Watchdog Timer Module 120
- 7.5 System Timer 122
  - 7.5.1 System Timer Registers 123
  - 7.5.2 System Timer Usage via DriverLib Functions 124
  - 7.5.3 Coding Practices for the System Timer Module 125
- 7.6 Timer32 125
  - 7.6.1 Timer32 Registers 126
  - 7.6.2 Timer32 Usage via DriverLib Functions 128
  - 7.6.3 Coding Practices for the Timer32 Module 129
- 7.7 Timer_A 129
  - 7.7.1 Timer_A Registers 131
  - 7.7.2 Timer_A Usage via DriverLib Functions 135
  - 7.7.3 Coding Practices for the Timer_A Module 138
- 7.8 Real-Time Clock 146
  - 7.8.1 Real-Time Clock Usage via DriverLib Functions 146
  - 7.8.2 Coding Practices for the Real-Time Clock Module 147
- 7.9 Application: Energy Saver System 148
  - 7.9.1 Equipment List 148
  - 7.9.2 Layout 148
  - 7.9.3 System Design Specifications 148
- 7.10 Summary 149
- 7.11 Exercises 149

8 Mixed Signal Systems 151
- 8.1 Analog and Digital Signals 151
- 8.2 The Reference Module 152
  - 8.2.1 Reference Module Registers 153
  - 8.2.2 Reference Module Usage via DriverLib Functions 153
  - 8.2.3 Coding Practices for the Reference Module 154
- 8.3 The Comparator Module 155
  - 8.3.1 Comparator Module Registers 157

	8.3.2	Comparator Module Usage via DriverLib Functions	161
	8.3.3	Coding Practices for the Comparator Module	163
8.4	Analog-to-Digital Conversion		166
	8.4.1	Successive Approximation Register Converter	167
	8.4.2	The ADC14 Module	167
	8.4.3	The Window Comparator	170
	8.4.4	ADC14 Module Registers	170
	8.4.5	ADC14 Module Usage via DriverLib Functions	175
	8.4.6	ADC14 Module Usage via Energia Function	180
	8.4.7	Coding Practices for the ADC14 Module	180
8.5	Digital-to-Analog Conversion		188
	8.5.1	PWM Registers	189
	8.5.2	PWM Usage via DriverLib Function	191
	8.5.3	PWM via Energia Function	192
	8.5.4	Coding Practices for the PWM Module	192
8.6	Application: Leakage Control System		192
	8.6.1	Equipment List	193
	8.6.2	Layout	195
	8.6.3	System Design Specifications	195
8.7	Application: Blinds Control System		196
	8.7.1	Equipment List	196
	8.7.2	Layout	196
	8.7.3	System Design Specifications	197
8.8	Summary		197
8.9	Exercises		197

9 Digital Communication 199

9.1	Enhanced Universal Serial Communication Interface		199
	9.1.1	eUSCI Registers	199
	9.1.2	eUSCI Clocks	200
	9.1.3	Common Properties	201
	9.1.4	Pin Layout for eUSCI	201
9.2	Universal Asynchronous Receiver/Transmitter		201
	9.2.1	UART Registers	204
	9.2.2	eUSCI Module in UART Mode via DriverLib Functions	208
	9.2.3	eUSCI Module in UART Mode via Energia Functions	210
	9.2.4	Coding Practices for the UART Mode	210
9.3	Serial Peripheral Interface		220
	9.3.1	SPI Registers	222
	9.3.2	eUSCI Module in SPI Mode via DriverLib Functions	226
	9.3.3	Coding Practices for the SPI Mode	227
9.4	Inter-Integrated Circuit		233
	9.4.1	I^2C Registers	234
	9.4.2	eUSCI Module in I^2C Mode via DriverLib Functions	239
	9.4.3	Coding Practices for the I^2C Mode	243

x   Contents

       9.5   Application: Flame Detector   248
             9.5.1   Equipment List   248
             9.5.2   Layout   251
             9.5.3   System Design Specifications   251
       9.6   Summary   253
       9.7   Exercises   253

10   **Wireless Communication** **255**
       10.1   Background on Wireless Communication   255
       10.2   Wireless Local Area Network   256
              10.2.1   Definition   256
              10.2.2   WLAN Applications   257
       10.3   Bluetooth   264
              10.3.1   Definition   265
              10.3.2   Bluetooth Applications   266
       10.4   ZigBee   274
              10.4.1   Definition   275
              10.4.2   ZigBee Application   275
       10.5   RFID   278
              10.5.1   Definition   278
              10.5.2   RFID Application   279
       10.6   GSM   281
              10.6.1   Definition   281
              10.6.2   GSM Applications   281
       10.7   GPS   285
              10.7.1   Definition   286
              10.7.2   GPS Application   286
       10.8   Summary   287
       10.9   Exercises   288

11   **Flash Memory and RAM Operations** **289**
       11.1   Flash Memory and Controller   289
              11.1.1   Flash Controller Usage via DriverLib
                       Functions   290
              11.1.2   Coding Practices for the Flash Controller   292
       11.2   Memory Protection Unit   292
              11.2.1   Memory Protection Unit Usage via DriverLib
                       Functions   294
              11.2.2   Coding Practices for the Memory Protection Unit . . .   296
       11.3   Advanced Encryption Standard Accelerator Module   296
              11.3.1   AES256 Module Usage via DriverLib Functions   296
              11.3.2   Coding Practices for the AES256 Module   297
       11.4   Application: Advanced Home Entrance System   298
       11.5   Application: Code Updater   299
       11.6   Summary   299
       11.7   Exercises   299

12 Direct Memory Access ... 301
12.1 The DMA Module ... 301
- 12.1.1 Transfer Types ... 301
- 12.1.2 Fundamental Properties ... 301
- 12.1.3 Cycle Operating Modes ... 303
- 12.1.4 DMA Interrupts ... 304
12.2 DMA Module Usage via DriverLib Functions ... 304
12.3 Coding Practices for the DMA Module ... 307
- 12.3.1 Memory-to-Memory Transfer ... 307
- 12.3.2 Transfer Between Memory and Peripheral Devices ... 307
- 12.3.3 Peripheral to Memory Transfer ... 310
12.4 Application: Baby Monitor System ... 313
- 12.4.1 Equipment List ... 313
- 12.4.2 Layout ... 313
- 12.4.3 System Design Specifications ... 313
12.5 Summary ... 314
12.6 Exercises ... 314

13 Real-time Operating System ... 315
13.1 What Is RTOS? ... 315
- 13.1.1 RTOS Fundamentals ... 315
- 13.1.2 Advantages and Disadvantages of RTOS ... 316
- 13.1.3 RTOS Types ... 316
13.2 TI-RTOS ... 316
- 13.2.1 Kernel ... 316
- 13.2.2 Middleware and Drivers ... 318
- 13.2.3 Instrumentation ... 318
- 13.2.4 Creating a TI-RTOS-Based CCS Project ... 318
13.3 TI-RTOS Kernel ... 319
- 13.3.1 Threads ... 320
- 13.3.2 Timing ... 333
- 13.3.3 Synchronization ... 343
- 13.3.4 Memory Management ... 358
13.4 TI-RTOS Drivers ... 366
- 13.4.1 GPIO ... 366
- 13.4.2 ADC ... 367
- 13.4.3 PWM ... 368
- 13.4.4 UART ... 369
- 13.4.5 SPI ... 371
- 13.4.6 I²C ... 372
13.5 TI-RTOS Instrumentation ... 373
- 13.5.1 Adding the CPU Load Option to the TI-RTOS Project ... 373
- 13.5.2 RTOS Object Viewer ... 373
- 13.5.3 Runtime Object Viewer ... 374
- 13.5.4 RTOS Analyzer ... 374

		13.6	Energia and TI-RTOS	380

13.6 Energia and TI-RTOS 380
13.7 Applications 380
13.8 Summary 381
13.9 Exercises 381

14 Advanced Applications 383
14.1 Garage Gate Control System 383
    14.1.1 Equipment List and Layout 383
    14.1.2 System Design Specifications 384
14.2 Vending Machine 385
    14.2.1 Equipment List and Layout 385
    14.2.2 System Design Specifications 385
14.3 Digital Clock 386
    14.3.1 Equipment List and Layout 386
    14.3.2 System Design Specifications 387
14.4 Audio Spectrum Analyzer 388
    14.4.1 Equipment List and Layout 388
    14.4.2 System Design Specifications 388
14.5 Air Freshener Dispenser 389
    14.5.1 Equipment List and Layout 389
    14.5.2 System Design Specifications 390
14.6 Obstacle Avoiding Tank 390
    14.6.1 Equipment List and Layout 391
    14.6.2 System Design Specifications 391
14.7 Robot Arm Control System 392
    14.7.1 Equipment List and Layout 392
    14.7.2 System Design Specifications 392
14.8 Intelligent Washing Machine 393
    14.8.1 Equipment List and Layout 393
    14.8.2 System Design Specifications 393
14.9 Non-Touch Paper Towel Dispenser 395
    14.9.1 Equipment List and Layout 395
    14.9.2 System Design Specifications 395
14.10 Traffic Lights 397
    14.10.1 Equipment List and Layout 397
    14.10.2 System Design Specifications 398
14.11 Car Parking Sensor System 398
    14.11.1 Equipment List and Layout 398
    14.11.2 System Design Specifications 398
14.12 Body Weight Scale 399
    14.12.1 Equipment List and Layout 399
    14.12.2 System Design Specifications 399
14.13 Intelligent Billboard 400
    14.13.1 Equipment List and Layout 401
    14.13.2 System Design Specifications 401
14.14 Elevator Cabin Control System 401

	14.14.1 Equipment List and Layout	401
	14.14.2 System Design Specifications	402
14.15	Customer Counter System	404
	14.15.1 Equipment List and Layout	404
	14.15.2 System Design Specifications	404
14.16	Frequency Meter	404
	14.16.1 Equipment List and Layout	405
	14.16.2 System Design Specifications	405
14.17	Pedometer	405
	14.17.1 Equipment List and Layout	405
	14.17.2 System Design Specifications	406
14.18	Digital Camera	406
	14.18.1 Equipment List and Layout	407
	14.18.2 System Design Specifications	407
14.19	Irrigation System	408
	14.19.1 Equipment List and Layout	408
	14.19.2 System Design Specifications	408
14.20	Speech Recognizer	409

References 411
Index 413

Preface

The microcontroller usage has two major boosts in recent years. First, the maker movement has promoted the microcontrollers to solve various real-life problems. Second, the Internet of things (IoT) applications have started changing all aspects of our lives, for good. IoT need at least one microcontroller to begin with. Therefore, learning and applying microcontroller concepts are becoming a must for a fresh graduate. For a hobbyist, learning these concepts at a leisure time is fun.

To introduce the working principles of a programmable microcontroller, we have published a book on the MSP430 LaunchPad [11]. There, we aimed to introduce basic microcontroller concepts through real-life applications. The MSP430 microcontroller used in that book had very desirable properties. It could easily be programmed in either C or assembly language. The user could reach all low-level properties of the microcontroller through programming. Therefore, MSP430 was a good candidate to learn basic microcontroller concepts. However, MSP430 is limited by its resources. Therefore, it was not possible to cover advanced topics in our first book.

Texas Instruments (TI) recently introduced the MSP432 microcontroller with the following advantages. First, the architecture of the MSP432 microcontroller is based on ARM® Cortex™-M4F. This architecture is not specific to TI. Therefore, tools introduced for MSP432 can also be used in other microcontrollers with the same architecture as well. Second, resources of the MSP432 microcontroller are far better than the MSP430. This allows the user to apply advanced applications (such as IoT) on this microcontroller. Third, the MSP432 microcontroller can be programmed by TI's standard coding environment Code Compose Studio (CCS) or Energia. CCS is suitable for low-level programming for engineering applications. Energia provides a platform that is suitable for a hobbyist. Therefore, the hobbyist and engineer can benefit from MSP432 through different coding platforms.

This book aims to introduce basic and advanced topics on programmable microcontrollers through MSP432. The emphasis will be on advanced topics. However, basics will also be introduced for the sake of completeness. As in the first book, we will follow an approach based on practical applications in introducing microcontroller concepts. Therefore, we pick the MSP432 LaunchPad that is a compact platform having an MSP432 microcontroller.

We took an undergraduate engineering student and hobbyist as potential readers throughout the book. Therefore, a professional engineer may also benefit from the book. Since we pick the MSP432 LaunchPad, the reader may find a wide variety of applications besides the ones considered in this book. As a result, we expect the reader to become familiar with the basic and advanced microcontroller concepts in action. We believe this will be a valuable asset in today's competitive job market.

Cem Ünsalan, H. Deniz Gürhan, M. Erkin Yücel

Acknowledgments

The authors gratefully acknowledge the support of Texas Instruments in the framing and execution of this work through European University Program. Some of the figures used in this book are the property of TI. They are used here with their permission.

Programmable Microcontrollers

CHAPTER 1
Introduction

The aim of the introductory chapter is to present the layout of the book and the topics covered. As the basic and advanced microcontroller concepts are covered in this book using real-life applications, we will start with the hardware to be used. First we will provide a brief information on the architecture of the MSP432 microcontroller. Then, we will introduce the software to be used in the book. Finally, we will summarize the topics to be covered in the following chapters. This way, we expect the reader to get a feeling on the general structure of the book.

1.1 Hardware to be Used in the Book

There are excellent books on microcontrollers. One group of books explains the theoretical and practical microcontroller concepts on a hypothetical system or microcontroller family (instead of a specific microcontroller). The idea here is to be less specific and be more general. Therefore, the authors aim to explain general concepts and ask the reader to apply them to the specific microcontroller they pick. Since the lifetime of a microcontroller family has a larger lifespan compared to a specific microcontroller, books in this group aim to be used in a longer time period.

The other group of books picks a specific microcontroller and explains detailed concepts on it. By default, concepts explored in these books are specific to the microcontroller at hand. As a result, they are not broad or general. In this book, we follow this approach and pick the Texas Instruments (TI) MSP432 LaunchPad shown in Fig. 1.1. This LaunchPad has an MSP432 microcontroller with the ARM® Cortex™-M4F architecture. To be more specific, the name of the microcontroller on the LaunchPad is MSP432P401R. Throughout the book, we will call it as MSP432.

Selecting a specific microcontroller may seem odd. However, the focus of this book is on explaining the basic and advanced microcontroller concepts through applications. Therefore, it is a must to pick a specific microcontroller and implement all the applications on it. We are aware that the digital electronics industry is dynamic and most probably a microcontroller would be history within 5 or 10 years. However, newer members of the same microcontroller family will be based on the previous ones. Therefore, the current information with applications will be a valuable background for future microcontrollers. Also, other microcontrollers with the ARM® Cortex™-M4F architecture have similar properties with MSP432. Therefore, the topics considered in this book may be applied on them with minor modification.

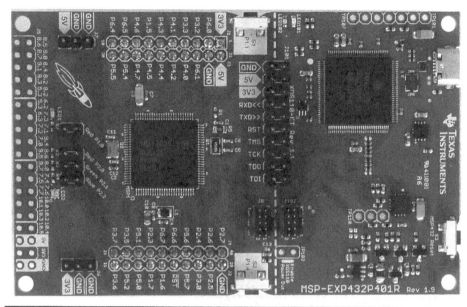

FIGURE 1.1 MSP432 LaunchPad [2].

1.2 Architecture of the MSP432 Microcontroller

The functional block diagram of the MSP432 microcontroller is given in Fig. 1.2. As can be seen in this figure, MSP432 has an ARM® Cortex™-M4F central processing unit (CPU). The CPU can reach clock speeds up to 48 MHz. It has floating-point and memory protection units. MSP432 has 256 kB flash, 32 kB ROM for the driver library, and 64 kB of SRAM including 6 kB backup memory. It has 11 general purpose input and output ports named P1–P10 (with 78 pins) and PJ (with six pins). It has a clock system with four 16-bit and two 32-bit timer modules, real-time clock, and a watchdog timer. MSP432 has a 14-bit analog-to-digital converter (ADC) and two analog comparator modules. It has eight digital communication modules. MSP432 has an eight-channel direct memory access (DMA) module. It has power control manager and power control system modules. MSP432 also has cyclic redundancy check (CRC32), bus control logic, security encryption and description (AES256), reset controller (RSTCTL), system controller (SYSCTL), and reference voltage (REF_A) modules. The MSP432 microcontroller is designed to have low power consumption. If these do not mean much, do not worry. This book is written to explain these concepts.

1.2.1 Central Processing Unit

The MSP432 CPU has a 32-bit ARM® Cortex™-M4F architecture. Hence, its word length is 32 bits. In other words, it can handle data with length 32 bits at once. MSP432 supports the full ARM® instruction set. We will explore this instruction set in detail in Chap. 4.

1.2.2 Memory

The MSP432 microcontroller has three types of memory. These are flash, ROM, and SRAM. Flash is a non-volatile memory. It can keep data when power goes off. Hence, it is

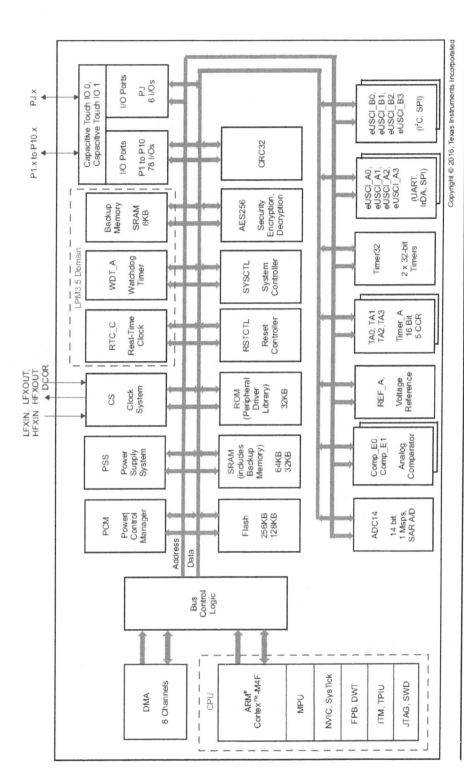

FIGURE 1.2 Functional block diagram of the MSP432 microcontroller [3].

Memory Address	Memory Type
0x00000000-0x003FFFFF	Flash
0x01000000-0x010FFFFF	SRAM
0x02000000-0x020FFFFF	ROM
0x20000000-0x200FFFFF	SRAM
0x22000000-0x23FFFFFF	Bit-banded SRAM
0x40000000-0x400FFFFF	Peripherals (Registers)
0x42000000-0x43FFFFFF	Bit-banded peripherals
0xE0000000-0xFFFFFFFF	Instrumentation, ITM, ...

TABLE 1.1 The Memory Map of MSP432

primarily used to store the code to be executed. ROM is specifically used for storing the driver library that allows executing most predefined functions in a fast manner. SRAM is used for temporary storage. Hence, it is suitable for keeping variables. Within SRAM resides the backup memory. The memory map of MSP432 is presented in Table 1.1.

In Table 1.1, the term bit-banding is mentioned. This method allows reaching a specific bit in a given memory address through memory operations. We will see in Chap. 3 how this can be done.

As can be seen in Table 1.1, the memory map not only represents the flash, ROM, and SRAM, but also represents peripheral registers, peripherals, and instrumentation modules. This is called memory mapped I/O which indicates that peripheral devices (modules on the microcontroller besides the CPU) are handled as if they are memory addresses.

MSP432 has memory protection unit and direct memory access (DMA) module. We will explore the memory protection unit in detail in Chap. 11. In a similar manner, we will introduce the eight-channel DMA module in Chap. 12.

1.2.3 Input and Output Ports

The microcontroller can interact with the outside world through its input and output ports. Here, the processed data can be analog or digital. MSP432 has 84 pins (arranged in 11 ports as P1–P10 and PJ) to be used for input and output. All the pins can be used as input or output. They can also be used for both analog and digital signals. Therefore, they are called general purpose input and output (GPIO). We will consider these starting from Chap. 5.

The pin map of the MSP432 LaunchPad is provided in Fig. 1.3. This figure indicates that not all MSP432 pins are available on the LaunchPad. Therefore, the applications to be developed on it will be limited by the available pins.

1.2.4 Timer Modules

MSP432 has four 16-bit and two 32-bit timer modules, a real-time clock, and a watchdog timer. We will explore their properties in detail in Chap. 7.

1.2.5 Mixed Signal Processing Modules

MSP432 has a 14-bit ADC and two analog comparator modules. The ADC operation is carried out via successive approximation register (SAR) method. We will explore the usage of these modules in Chap. 8.

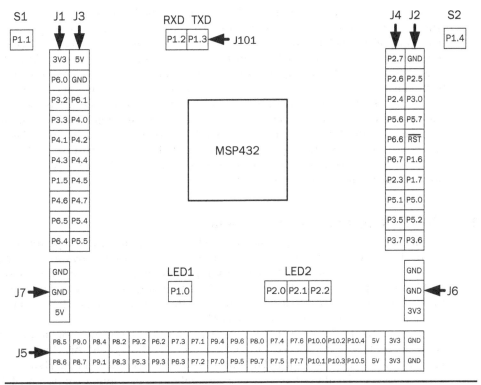

FIGURE 1.3 MSP432 LaunchPad pin map.

1.2.6 Digital Communication Modules

MSP432 has eight digital communication modules called enhanced universal serial communication interface (eUSCI). These modules support universal asynchronous receiver/transmitter (UART), serial peripheral interface (SPI), and inter-integrated circuit (I^2C) communication modes. We will explore these in detail in Chap. 9.

1.2.7 Other Modules

There are also other MSP432 modules that will not be explored in detail in the following chapters. This does not mean that they are redundant. They are just out of the scope of this book. More detail on these modules can be found in [4].

1.3 Software to be Used in the Book

The primary coding environment is Code Composer Studio (CCS). Energia can also be used as the secondary coding platform. CCS is the general environment of all TI devices. CCS is freely distributed by TI (at the time of writing this book). There is also a cloud-based CCS at https://dev.ti.com/, which can be used for programming purposes. Energia is also free and available to the user.

C or assembly language may be preferred for coding. We will use the C language most of the times to explain the concepts in detail throughout the book. While writing the C code, we will provide three different methods. First, we will use the register-level approach. Through it, we will reach the lowest possible hardware level in C language. Although this approach is beneficial for understanding the hardware properties of the microcontroller, it is hard to master. Therefore, we suggest the register-level programming approach to students who want to understand the hardware of MSP432 better. The second approach is using TI's DriverLib functions. These predefined functions can be used in reaching most microcontroller resources with a friendly interface. Therefore, this approach can be used both in advanced and basic levels. However, some hardware properties cannot be reached by DriverLib functions. This is the main limitation for them. The third approach is using Energia. Through it, the user can write C programs similar to Arduino programming. Although this is the easiest method to write a C code, Energia does not cover some hardware properties. Therefore, we suggest only this approach to the hobbyist using MSP432.

1.4 Topics to be Covered in the Book

The first step in understanding a microcontroller is through its CPU and peripheral units. In this chapter, we briefly handled this issue by focusing on the hardware of the MSP432. Therefore, we looked at the CPU, memory, input and output ports, and timer modules, ADC and comparator modules, and the digital communication module of the microcontroller. In other words, this chapter summarized the properties of the MSP432 microcontroller to be considered throughout the book.

The second step in understanding a microcontroller is learning how to program it. To do so, we briefly introduced CCS, CCS Cloud, and Energia in this chapter. We will use these to program the MSP432 microcontroller throughout the book. Hence, we will consider CCS in detail in Chap. 2. CCS is not only a programming environment. Using it, we can observe the status of the hardware components while the program is running. Therefore, it will be of great help in understanding MSP432 in action. We will then introduce Energia to program MSP432 in a simple way. It will be of great help to a hobbyist in the following chapters. In Chap. 3 , we will introduce C programming techniques for MSP432. There we will first consider memory management and data types. Then, we will briefly review basic C concepts. Although the reader may know C language, his or her knowledge of the C language may not be complete from the microcontroller perspective. Therefore, we strongly suggest the reader to revisit this chapter to update his or her C knowledge. Although C language may be sufficient for most applications, learning assembly language is a must to understand the inner working principles of the microcontroller. Therefore, we will focus on assembly language programming for the MSP432 microcontroller in Chap. 4. This will be the only chapter in which we will be using assembly language for programming purposes. The reason for this choice is as follows. Since MSP432 is a 32-bit microcontroller, using assembly language on it is not feasible for advanced applications. Moreover, the reader can always consult books on ARM® assembly language programming for this purpose. Therefore, Chap. 4 only aims to introduce assembly language programming concepts at an introductory level.

The third step in understanding a microcontroller is using it through basic applications. To do so, we should know its related properties in detail. Therefore, we will introduce methods to interact the microcontroller with the outside world in Chap. 5. In Chap. 6, we will focus on the events and interrupt concept that is extremely important in event-driven programming. Therefore, we will consider the occurrence of interrupts as well as the ways to handle them. After interrupts, we will consider time-based operations in Chap. 7. These concepts are also extremely important in real-life applications. In this chapter, we will evaluate the building blocks of the clock system. MSP432 has more than one clock. We explore all these clocks and their usage areas. Another important topic in time-based operations is low power modes. Effectively using them helps power savings in applications. Therefore, we will consider them next. In the same chapter, we will also consider the usage of the watchdog timer, timer, and real-time clock modules. In Chap. 8, we will consider processing of mixed signals. To do so, we will start with the properties of analog and digital signals. Then, we will focus on the analog-to-digital conversion modules in the MSP432 microcontroller. Next, we will focus on the digital-to-analog conversion. Since MSP432 does not have such a module, we will use pulse width modulation instead. In Chap. 9, we will focus on the digital communication module of MSP432. We will consider the UART, SPI, and I^{2}C communication modes under this module.

The fourth step in understanding a microcontroller is using it through advanced applications. Therefore, we will extend digital communication concepts to wireless communication in Chap. 10. This will be a valuable tool in (Internet of Things) IoT applications. In Chaps. 11 and 12, we will explore memory-related concepts as flash memory, RAM operations (including memory protection), and DMA. Memory protection will help us to secure the sensitive data in the microcontroller. DMA will be extensively used in real-time data transfer applications. In Chap. 13, we will introduce the real-time operating system (RTOS) concepts to the reader. This will lead to constructing more complex projects in an effective manner. We will supplement all topics considered up to this chapter by real-life applications. These will be provided at the end of each chapter. The main theme of these applications will be a smart home backed by IoT. Finally, we will provide advanced applications in Chap. 14. These applications will be from a broad perspective. Hence, they can be used to solve diverse real-life problems.

Sample codes in this book are available in the companion website for the reader. Course slides are available in the same website for the reader and instructor. The solution manual is also available in the companion website for only the instructor.

CHAPTER 2
Code Composer Studio and Energia

Code Composer Studio (CCS) is the environment for TI's embedded processors. Although a new version of CCS is introduced every year, the reader should become familiar with at least one CCS version. Therefore, we pick the most recent version of the CCS (version 7.2) in this book [5]. Even if a new version of CCS is introduced in the future, we believe it will not be totally different from this version. In this chapter, we will start with the properties of CCS first. TI recently introduced the cloud version of CCS. Next, we will explore its properties. CCS may be a complex programming environment for a hobbyist. Fortunately, there is a free and community-driven environment called Energia. We will introduce it. Finally, we will compare these coding platforms.

2.1 Code Composer Studio

The first platform we will consider is CCS. Through it, the user can reach almost all properties of the code and hardware it runs on. We start with its setup.

2.1.1 Downloading and Installing CCS

The official version of CCS for MSP432 is freely available on the TI website (at the time of writing this book) [6]. The reader can follow the comments there to download and install CCS.

When the MSP432 LaunchPad is connected to the PC through its USB port, the driver installation starts first. The preprogrammed demo starts automatically after installation. This demo toggles the red, green, and blue colors of LED2 in a sequence. The speed of operation and LED2 color can be changed by pressing buttons S1 and S2. If these operations can be performed, it indicates that the LaunchPad is working properly.

2.1.2 Starting CCS

When the CCS starts for the first time, a window appears such that the location of workspace folder can be configured. Either use the default workspace folder or change the location by clicking the *Browse* button. This workspace folder keeps the project and setting files after CCS is closed. Hence, previous projects and settings will be available when CCS is opened again. Now, CCS is ready to run. A window should appear as in Fig. 2.1.

10 Chapter Two

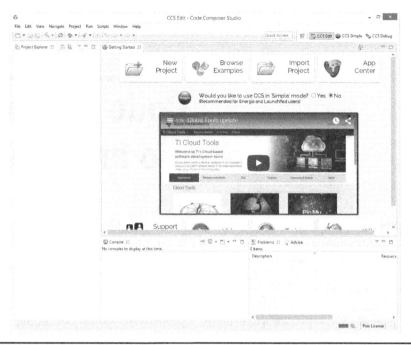

FIGURE 2.1 The CCS window.

2.1.3 CCS Perspectives

There are 10 perspectives for CCS that can be selected from the *Window → Perspective → Open Perspective*. The default ones are CCS Edit and CCS Debug. CCS opens in the Edit perspective every time it starts. This perspective is used for creating projects, building them, and observing errors in them. The CCS Edit perspective is switched to the CCS Debug perspective automatically when the created project is debugged. This perspective is used for debugging projects. It can also be used for observing the hardware (such as registers and memory) or software (such as variables and disassembly) during code execution. The user can switch between these two perspectives using the small icons in the upper right corner of the main CCS window.

2.1.4 Creating a New Project

In this section, we will consider creating a new C or assembly project. CCS generates an executable output file (with extension .out) from these. This file is used by MSP432.

2.1.4.1 A New C Project

To create a new C project under CCS, click *File → New → CCS Project*. A new window will pop up as in Fig. 2.2. In this window, enter a project name and arrange *Target* for the MSP432P401R microcontroller as in Fig. 2.2. Finally, select *Empty Project (with main.c)* and click *Finish*. The project will be created with a source file named main.c after these steps. The generated project should be seen in the *Project Explorer* window.

The compiler optimization runs automatically when the project is created. If the compiler optimization is unnecessary, it can be disabled by the following steps. In the *Project Explorer* window, right-click on the *Project* and select *Properties*. In the pop-up window,

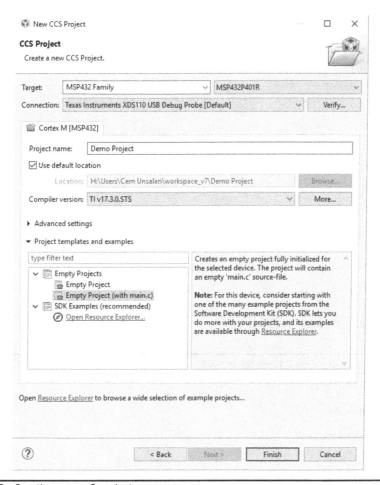

FIGURE 2.2 Creating a new C project.

select *Optimization* and set *Optimization Level* to *off* as in Fig. 2.3. Once the project is created, *The Ultra-Low-Power Advisor* (*ULP Advisor*) provides valuable information on how to use this property at utmost level. The information can be seen from *Infos* in the *Problems* window. The ULP Advisor can be disabled from the *ULP Advisor* window at *Project Properties*.

After writing the C code in main.c, save it by clicking the *Save* button in the upper-left corner of the CCS menu. We will use the code in Listing 2.1. Here, there is an included header file msp.h. Through it, the compiler automatically selects the header file for the correct microcontroller version. The code line WDT_A -> CTL = WDT_A_CTL_PW | WDT_A_CTL_HOLD in Listing 2.1 stops the watchdog timer, which we will see in Chap. 7. Till that chapter, please take this code line as granted.

2.1.4.2 Creating a Header File

A header file may be needed for some projects. To add it to the project, right-click on the project in the *Project Explorer* window and select *New → Header File*; or click *File → New → Header File*. Name the generated header file like header.h and click *Finish*.

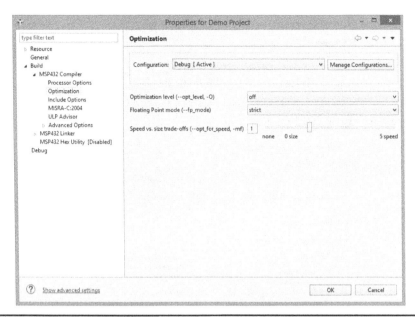

FIGURE 2.3 Disabling optimization.

Listing 2.1 The First C Code for MSP432

```
#include "msp.h"

 int d = 0;

void main(void) {

 WDT_A->CTL = WDT_A_CTL_PW | WDT_A_CTL_HOLD;

 int a = 1;
 float b = -255.25;
 char c = 'c';

 d = d+1;

 while(1);
}
```

An empty window will open for the header file. As this header file is added to the project, do not forget to add the line #include "header.h" in the main C code.

2.1.4.3 Building and Loading the Project

There are two buttons on the horizontal CCS toolbar for code generation. The first one is the *Build* button with the hammer shape. The second one is the *Debug* button with the green bug shape.

The build operation is used for generating the files to be embedded to the target device. Therefore, the main source code is linked with all other source and header files when the build button is clicked. Meanwhile, the map file that keeps the memory address of these files is also created at this step. The running steps can be observed in the *Console* window. The warnings (in yellow), errors (in red), and infos (in blue) can be observed in the *Problems* window. As the code is built, sections with warnings and errors can be reached by double-clicking on them in the *Problems* window.

The debug operation includes the build if the codes are updated. Then, CCS loads the code to target device (here MSP432). The MSP432 LaunchPad should be connected to the host PC to perform this operation. The CCS Edit perspective switches to the CCS Debug perspective after the debug operation is complete. The code is executed till the beginning of the main function.

2.1.5 Adding DriverLib Functions to the Project

We will be using DriverLib functions in the following chapters while working with peripheral devices. The first step in using these functions is adding them to the project. Therefore, download and install the *TI SimpleLink SDK* to your PC first. To do so, click *View → Resource Explorer*. Then, search for *MPS432P401R* using *Select a Device or Board* search box. Select *SimpleLink MSP432 SDK* under *Software* tab. Click *Download and install* button and select *Make Available Offline* to download and install the DriverLib functions to your PC.

Now, a CCS project with DriverLib functions can be created using SimpleLink MSP432 SDK. The recommended way of creating a new project with DriverLib is importing the empty project example from the Resource Explorer and adapting it. To do so, search for *MPS432P401R* using *Select a Device or Board* search box in the *Resource Explorer* window. Select the *empty* project under *Software → SimpleLink MSP432 SDK → Development Tools → MSP-EXP432P401R - Rev 2.x (Red) → DriverLib → empty → No RTOS → CCS Compiler*. Then, click *Import to IDE* icon on the top-right corner of the window. After these steps, the project should be seen in the *Project Explorer* window with the name *empty_MSP_EXP432P401R_nortos_ccs*. You can rename the project as you like by right-clicking on it in the *Project Explorer* window and clicking on the *Rename* button.

2.1.6 Creating an Assembly Project

To create a new assembly project, click on the *File → New → CCS Project* on the main CCS menu. In the pop-up *New CCS Project* window, select *Empty Project* and click *Finish*. The project will be created after these steps. The generated project should be seen in the *Project Explorer* window. Now, a source file should be added to the project. To do so, right-click on the project in the *Project Explorer* window and click on the *File → New → Source File*. In the pop-up *New Source File* window, give a name for the source file, select *None* as file *Template*, and click *Finish*. We will use the sample code in Listing 2.2 as source file. We will see the instructions and directives used here in Chap. 4.

2.1.7 Program Execution

As the main code is debugged, the next step is its execution. Buttons for program execution are placed in the *Debug* window as shown in Fig. 2.4. The name of each button can be observed by moving the cursor over it. These buttons and their functions are explained briefly in the list below.

- **Resume**: Resumes the execution of the code from last location of Program Counter (introduced in Sec. 4.1). When it is pressed, execution continues until a breakpoint or suspend button press.

Listing 2.2 The First Assembly Code for MSP432

```
.text
.align  2
.global main

main:
 MOV R4, #11h
 MOV R5, #00AAh
 AND R3, R4, R5
loop:
 B loop
 .end
```

FIGURE 2.4 Program execution menu.

- **Suspend**: Halts the execution of the code. All windows used to observe software and hardware parts are updated with recent data.
- **Step Into**: Executes the next line of the code. If this line calls a subroutine, the compiler just executes the next line in the subroutine then stops.
- **Step Over**: Executes the next line of the code. If this line calls a subroutine, the compiler executes the subroutine then stops.
- **Assembly Step Into**: Executes the next assembly instruction. If this instruction calls a subroutine, the compiler just executes the next instruction in the subroutine then stops.
- **Assembly Step Over**: Executes the next assembly instruction. If this instruction calls a subroutine, the compiler executes the subroutine then stops.
- **Step Return**: Completes the execution of subroutine.
- **Reset CPU**: Resets the target microcontroller. It works similar to the reset pin. When it is clicked, registers of the device return to their default state.
- **Restart**: Returns the Program Counter to the beginning of the loaded program.

2.1.8 Inserting a Breakpoint

To stop execution of the program in a specific code line, a breakpoint should be added. Right-click on the code line to place the breakpoint and select *Breakpoint (Code Composer Studio)*. There are three breakpoint options under this item as *Breakpoint (Software Breakpoint)*, *Hardware Breakpoint*, and *Watchpoint* as shown in Fig. 2.5.

Software breakpoint is an instruction placed at the breakpoint address to halt the code execution. Hardware breakpoint is an address value that halts the code execution when the program counter matches this value. Watchpoint is actually a special kind of hardware breakpoint based on a specified data value. When the code generates it during execution, the program halts.

Code Composer Studio and Energia   15

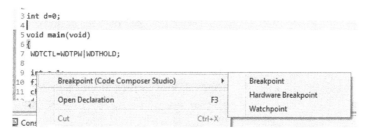

FIGURE 2.5 Adding a breakpoint.

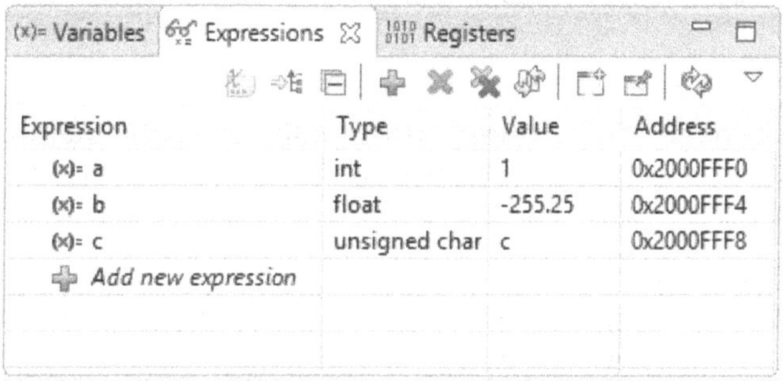

FIGURE 2.6 Adding a watch expression.

There are two ways to alter the inserted breakpoint. First, the breakpoint can be deleted by toggling it. To do so, right-click on the breakpoint. Then, select *Toggle Breakpoint* from the pop-up window. Second, the breakpoint can be disabled. To do so, select *Disable Breakpoint* from the same pop-up window.

2.1.9   Adding a Watch Expression

The *Expressions* window (given in Fig. 2.6) can be used to observe selected variables in CCS. To add a variable to it, select the variable to be observed and right-click on it. Then, click *Add Watch Expression*. Also the *Add new expression* can be clicked in the *Expressions* window and the name of the variable can be entered to the box. Do not forget to halt the execution process to observe changes in selected variables.

Global variables can also be observed in the *Expressions* window either by using the *Add new expression* button or by right-clicking on the *Expressions* window then selecting *Add Global Variables*. However, defining a global variable alone is not enough to arrange a memory location for the compiler. The global variable must be used in the main code.

Local variables can also be observed in the *Variables* window as shown in Fig. 2.7. Both *Expressions* and *Variables* windows are opened automatically after the debugging process. In case of their absence, they can be opened from the *View* menu. We will discuss local and global variables in Sec. 3.1.2.

2.1.10   Observing Hardware

CCS is not only a compiler. Through it, the reader can also observe the hardware status of the microcontroller. We next explore how to observe the key hardware elements of MSP432.

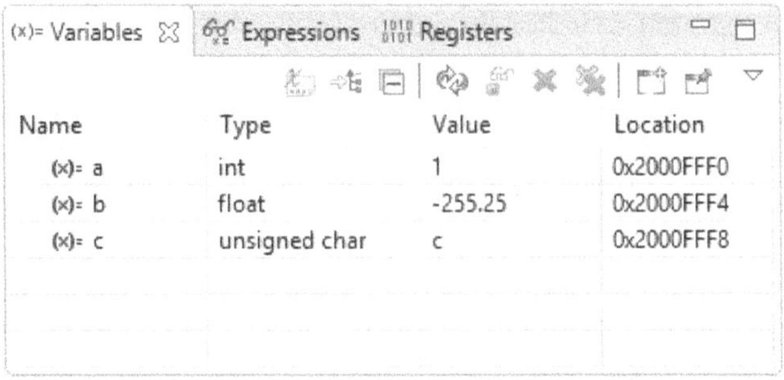

Name	Type	Value	Location
(x)= a	int	1	0x2000FFF0
(x)= b	float	-255.25	0x2000FFF4
(x)= c	unsigned char	c	0x2000FFF8

FIGURE 2.7 Observing local variables in the Variables window.

2.1.10.1 Registers

MSP432 has control microprocessor, peripherals, and special-function registers. The *Registers* window (under the *View* menu) can be used to observe the status of these registers. Here, each register group is listed separately as shown in Fig. 2.8.

2.1.10.2 Memory

To observe the memory contents, click *View → Memory Browser*. Enter the starting address of the memory location to be observed in the empty box as shown in Fig. 2.9. The memory browser can also be used to observe the machine (or assembly) language equivalent of the written C code.

2.1.10.3 Disassembly

CCS provides the assembly code corresponding to the compiled C code. To observe it, the *Disassembly* window should be opened by clicking *View → Disassembly*. The *Disassembly* window will open as presented in Fig. 2.10.

There are four buttons in the vertical column of the *Disassembly* window. These are *Link with Active Debug Context*, *Show Source*, *Assembly Step Into*, and *Assembly Step Over*. *Assembly Step Into* and *Assembly Step Over* buttons have already been explained in Sec. 2.1.7. When the *Link with Active Debug Context* button is pressed, a blue arrow will appear at the left horizontal column of the Disassembly window to follow the assembly code execution. The *Show Source* button may be used to link every C code line with the corresponding assembly line.

2.1.11 The Terminal Window

CCS has an internal terminal program to communicate with the MSP432 LaunchPad through the UART communication mode. We will use this property in Chap. 9. The terminal window can be accessed by clicking on the *View → Other → Terminal*. The terminal window will open as in Fig. 2.11.

Configuration settings should be done first to establish a serial communication link. To do so, the *Terminal Settings* window will be used as shown in Fig. 2.12. The communication type should be selected as "Serial" first. Then the desired baud rate, data bits, stop bits, and parity values should be set. Flow control and timeout settings can be left unchanged.

The port used by the MSP432 LaunchPad can be found under Windows 10. To do so, open the *Device Manager*. Click on the item *Ports (COM&LPT)* in the device list. The port used by MSP432 is named *XDS110 Class Application/User UART*.

FIGURE 2.8 Observing registers.

The link between the terminal program and MSP432 LaunchPad can be established by clicking the green *Connect* mark in the terminal window. The code on the MSP432 microcontroller must be debugged and run to use the terminal window. Let's assume that the code can send and receive data. Data sent from the MSP432 LaunchPad can be observed in the terminal window. It can be cleared anytime by right-clicking and selecting the *Clear Terminal* option. Data can be sent from the terminal in two different ways. In the first option, the user can click anywhere in the terminal window. Then, any character pressed on the keyboard can be sent as soon as the key is pressed. In the second option, the toggle command input field can be used. To do so, right-click on the terminal window and select the *Toggle Command Input Field*. A sub-window will appear below. The text to be sent can be entered here. It will be sent as the enter key is pressed. The link can be terminated by clicking on the red *Disconnect* mark.

18 Chapter Two

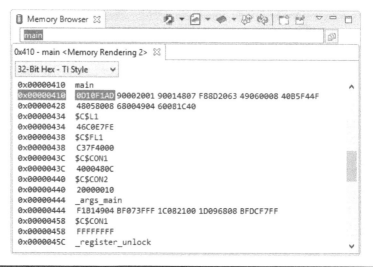

FIGURE 2.9 Observing memory.

```
Disassembly    Memory Browser
                              Enter location here
               main():
♦ 00000410:    F1AD0D10        sub.w      sp, sp, #0x10
      12          WDTCTL = WDTPW | WDTHOLD;           // Stop watchdog timer
  00000414:    4909            ldr        r1, [pc, #0x24]
  00000416:    F44F40B5        mov.w      r0, #0x5a80
  0000041a:    8008            strh       r0, [r1]
      14          int a = 1;
  0000041c:    2001            movs       r0, #1
  0000041e:    9000            str        r0, [sp]
      15          float b = -255.25;
  00000420:    4805            ldr        r0, [pc, #0x14]
  00000422:    9001            str        r0, [sp, #4]
      16          char c = 'c';
  00000424:    2063            movs       r0, #0x63
  00000426:    F88D0008        strb.w     r0, [sp, #8]
      17          d = d+1;
  0000042a:    4805            ldr        r0, [pc, #0x14]
  0000042c:    4904            ldr        r1, [pc, #0x10]
  0000042e:    6800            ldr        r0, [r0]
  00000430:    1C40            adds       r0, r0, #1
  00000432:    6008            str        r0, [r1]
      18          while(1);
```

FIGURE 2.10 The Disassembly window.

2.1.12 Closing the Project

Clicking *Terminate* in the *Debug* window will terminate the active debug session and switch the CCS Debug preference to Edit. Right-click on the project in the *Project Explorer* window and select *Close Project* to close the project. This project can be re-opened by selecting the *Open Project*.

FIGURE 2.11 The terminal window.

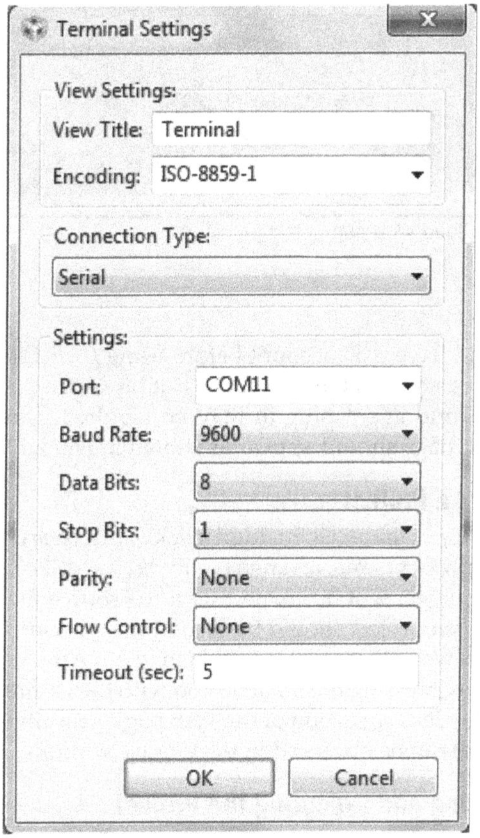

FIGURE 2.12 Terminal settings.

2.2 CCS Cloud

CCS Cloud is a free Web-based integrated development environment (IDE) for TI microcontrollers. It is basically the limited version of the CCS desktop application. Here, we will briefly introduce the properties of CCS Cloud.

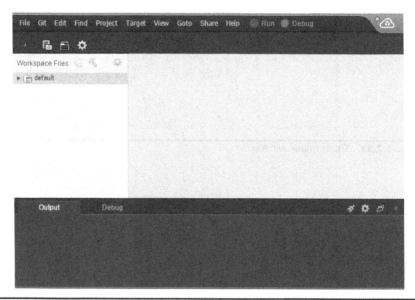

FIGURE 2.13 CCS Cloud Web page.

2.2.1 Setup
The reader should have a TI account before using CCS Cloud. After registration, CCS Cloud can be accessed from the TI website https://dev.ti.com/ide. To work with CCS Cloud, TI Cloud agent plug-in may be required. Download and install it when prompted. A Web page should appear as shown in Fig. 2.13 when CCS Cloud is ready.

2.2.2 Creating a Project
To create a new project under CCS Cloud, click *File → New CCS Project*. On the appearing window, enter a project name, arrange the *Device* as MSP432P401R microcontroller, and click *Finish*. The project will be created with the source file named main.c after these steps. The generated project should be seen in the *Workspace Files* part of the Web page.

In this section, we will use the code in Listing 2.1. After writing the C code in main.c, build the project by pressing the *Build* button with the hammer shape. Running steps can be observed in the *Console* section of the Web page. Warnings (in yellow), errors (in red), and infos (in blue) can be observed in the *Console* window.

2.2.3 Debugging and Executing the Project
The debugging operation can be run by clicking the *Debug* button with the bug shape. When the debug button is clicked, CCS Cloud loads the code to the target device (here MSP432). The MSP432 LaunchPad should be connected to the host computer to perform this operation. After the debug operation is complete, the code is executed till the beginning of the main function. Buttons for program execution are the same as explained in Sec. 2.1.7.

2.2.4 Observing Hardware
In CCS Cloud, execution of the program can be stopped at a specific code line. To do so, a breakpoint can be added by right-clicking on the code line and selecting *Add Breakpoint*

Code Composer Studio and Energia 21

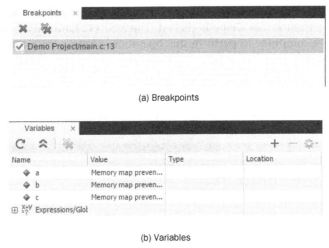

FIGURE 2.14 CCS Cloud breakpoints and variables Web pages.

or left-clicking the left side of the code line number. The *Breakpoints* Web page can be used to observe added breakpoints as appearing in Fig. 2.14a.

The *Variables* Web page (as shown in Fig. 2.14b) can be used to observe selected variables in CCS Cloud. To do so, select the variable to be observed and right-click on it. Then, click *Add Watch Expression*. Also the *Add new expression* can be clicked in the *Variables* Web page and the name of the variable can be entered in to the box. Do not forget to halt the execution of the code to observe changes in selected variables.

2.3 Energia

Energia is the IDE for TI microcontrollers including MSP432. It is basically an abstraction layer similar to the Arduino IDE with almost same preferences, options, and functions. In this section, we will introduce creating and building an Energia sketch using Energia IDE, CCS, or CCS Cloud environment. More detail on Energia can be found on http://energia.nu/.

2.3.1 Downloading and Installing Energia

Energia setup files (in compressed zip form) can be downloaded from http://energia.nu/download/. First, extract the downloaded zip archive to a proper directory. Then, Energia can be started by the executable (energia.exe) file. When Energia is ready, a window should appear as shown in Fig. 2.15.

2.3.2 Creating, Building, and Executing an Energia Sketch

To create a new project under Energia, click *File → New*. Once the project is created, a new window will pop up as shown in Fig. 2.15. Creating the project automatically creates a .ino file with basic required setup and loop functions. The next step is selecting the board, which will be *LaunchPad w/ msp432 EMT (48MHz)* for our case. It should be selected by clicking the *Tools → Board*.

In this section, we will use the code in Listing 2.3. After writing the code, build the project by pressing the *Verify* button with a tick shape. Running steps of the code can be observed in the *Console* window.

22 Chapter Two

Energia loads the code to the target device (here MSP432) when the *Upload* button with a right-arrow shape is clicked. The MSP432 LaunchPad should be connected to the host computer to perform this operation. There is no debugging option under Energia.

FIGURE 2.15 Energia IDE window.

Listing 2.3 The First Energia Sketch for MSP432

```
int a;
float b;
char c;
int d;

void setup(){
  a = 1;
  b = -255.25;
  c = 'c';
  d = 0;
}

void loop(){
  d = d+1;
}
```

2.3.3 Energia Sketch with CCS

Energia can be used jointly with CCS. This way, the programmer can use the advanced hardware debugging options of CCS with Energia. To create a new project with Energia under CCS, click *File → New → Energia Sketch*. A new window will pop up as shown in Fig. 2.16a. Specify the Energia folder that you extracted by clicking *Browse* in this window. By clicking *Next*, a new pop-up window will ask you to write a project name and arrange Device and Board as shown in Fig. 2.16b. Select *MSP432* and *LaunchPad w/ msp432 EMT (48MHz)*. Click *Finish* to create the project. Afterwards, a .ino file with setup and loop functions will be created.

To import an existing Energia sketch to CCS, make sure that the Energia folder is linked to CCS first. To do so, navigate to Window then Preferences under CCS. Select Energia under the preferences window. If the Energia install location is not specified, choose the folder that you extracted Energia by clicking *Browse*. Now, you can import the Energia sketch by navigating to *File → Import → Energia → Existing Energia Sketch*. This will open the *Import* pop-up menu. From here, select the project you would like to import. Then, choose the device and board. Click finish. Building, loading, and execution of an Energia sketch under CCS are the same as discussed in Secs. 2.1.4.3 and 2.1.7.

2.3.4 Energia Sketch with CCS Cloud

An Energia sketch can also be used under CCS Cloud. This way, the programmer can use the advantage of online programming features of CCS with Energia. To create a new project with Energia under CCS Cloud, first open the online IDE. Then, click *File → New Energia Sketch*. From the appearing window, write a project name and arrange Device and Board as *MSP432* and *LaunchPad w/ msp432 EMT (48MHz)*. Select *Empty Sketch* for an empty project. Then, click *Finish* to create the project. Afterwards, a .ino file with basic required setup and loop functions will be created. Building, loading, and execution of an Energia sketch under CCS Cloud are the same as discussed in Secs. 2.2.2 and 2.2.3.

2.4 Platform Comparison

Till now, we have introduced three platforms as CCS, CCS Cloud, and Energia. We tabulate available features for each coding platform in Table 2.1. Hence, the reader can decide on which platform to use in his or her applications.

Feature	CCS	CCS Cloud	Energia
Code editing/building	Yes	Yes	Yes
Support for CCS projects	Yes	Yes	
Support for Energia framework	Yes	Yes	Yes
Run program on local LaunchPad	Yes	Yes	Yes
Breakpoints and observing expressions	Yes	Yes	
View registers and memory content	Yes		
Advanced debug and trace	Yes		
Support for high performance debug probes	Yes		

TABLE 2.1 CCS, CCS Cloud, and Energia Comparison via Available Features [7]

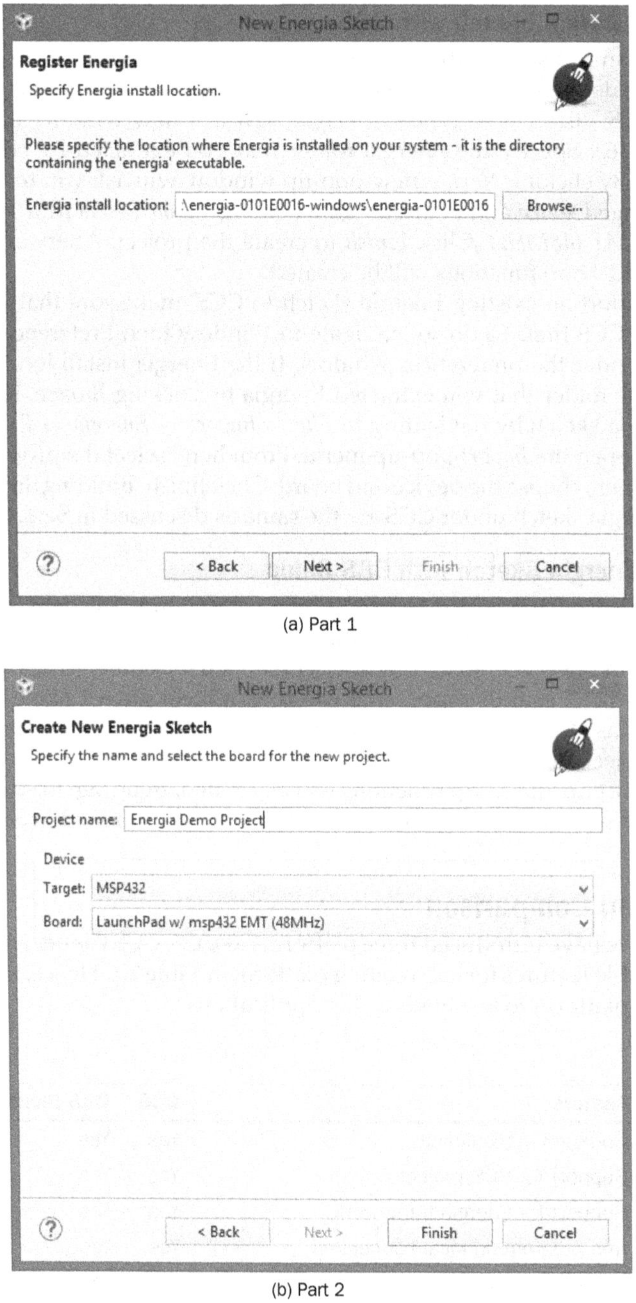

24 Chapter Two

(a) Part 1

(b) Part 2

FIGURE 2.16 Creating a project under Energia.

2.5 Summary

Knowing the hardware of the microcontroller is not enough to use it. The coding environment with all its properties should also be mastered. Therefore, we have introduced

CCS, CCS Cloud, and Energia as the coding environments of the MSP432 microcontroller in this chapter. We started with CCS. Then, we explored the CCS Cloud and Energia environments. Information provided in this chapter will be extensively used in the following chapters. Therefore, we strongly suggest the reader to master them.

2.6 Exercises

2.1 Download the latest version of CCS to your computer and install it.

2.2 Create an empty C project.
  a. Add the code block in Listing 2.1.
  b. Debug the code and run it.
  c. Observe local and global variables.
  d. Add breakpoints and observe their effects.
  e. Obtain the assembly code corresponding to the C code.

2.3 Repeat Exercise 2.2 using CCS Cloud.

2.4 Download the latest version of Energia to your computer and install it.

2.5 Create an empty Energia project.
  a. Add the code block in Listing 2.3.
  b. Debug the code and run it.

2.6 Perform platform migration operations such that the created Energia project in Exercise 2.5 is opened under
  a. CCS.
  b. CCS Cloud.

CHAPTER 3
MSP432 Programming with C

Topics considered in this chapter cover C programming issues for MSP432. Therefore, we strongly suggest the reader to refresh his or her knowledge on C language [8]. Here, we will briefly review basic C concepts from the microcontroller perspective. In other words, we will see how CCS acts while compiling a C code. Therefore, we will first deal with memory management issues. Then, we will consider C data types. Afterwards, we will briefly cover arithmetic and logic operations. Then, we will review control structures. We will also focus on arrays and pointers. Since structures will be extensively used in modifying peripheral devices on the microcontroller, we will review them next. Finally, we will explore miscellaneous issues that we will see in the following chapters.

3.1 Memory Management

Memory location for both the code and data is well defined for MSP432, as can be seen in Table 1.1. Here, we will consider memory issues on actual examples. Since we are using CCS as the compiler, we will see how it manages memory for the code and data.

3.1.1 The Code

After compiling a C code, CCS places the main code block in flash starting from the memory address 0x00000000 to 0x00FFFFFF. Let's consider the sample C code in Listing 3.1. Following the steps in Chap. 2, we can generate a C project under CCS from this code block.

The user can observe how the C code, in Listing 3.1, is placed in memory by using the *Disassembly* window. Therefore, we provide the memory content using the *Disassembly* window from address 0x00000678 to 0x00000690 in Fig. 3.1. As can be seen in this figure, CCS places the code to this memory location.

We can arrange the sample code in Listing 3.1 such that the addition operation is done in a function. The modified code will be as in Listing 3.2. We provide the memory map from address 0x00000678 to 0x0000060A for this code block in Fig. 3.2. As can be seen in this figure, the function is placed after the main code in memory.

3.1.2 Local and Global Variables

We can define a variable either as local or global in C language. As the name implies, the global variable is available to all functions in the code. However, the local variable is

28 Chapter Three

Listing 3.1 The Sample C Program

```c
#include "msp.h"

int a = 1;

void main(void) {

WDT_A->CTL = WDT_A_CTL_PW | WDT_A_CTL_HOLD;

int b = 2;
int c;

c = a+b;

while(1);
}
```

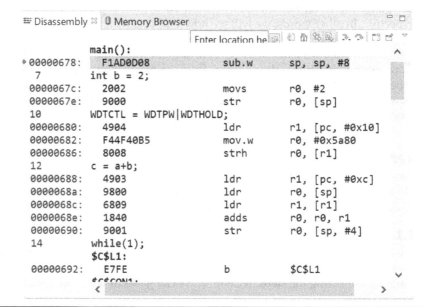

FIGURE 3.1 Memory contents (from address 0x00000678 to 0x00000690) observed by the Disassembly window.

only available to the function it is defined in. CCS keeps local and global variables in two different memory locations. Global variables are kept starting from the lowest possible memory address (0x20000000) in the SRAM. As a new global variable is added, the memory address is incremented and the new variable is saved. On the other hand, local variables are kept in the stack (starting with address 0x20100000). As a reminder, stack is a last-in first-out structure defined to work in the highest memory address available for variables. Based on the working principles of the stack, local variables are saved from top to bottom. Here, it is important to note that the C language takes the main() as a function. Hence, variables defined within it are also treated as local. Therefore, a global variable should be defined before the main().

Listing 3.2 The Sample C Program, with a Function

```c
#include "msp.h"

int sum(int d,int e);

void main(void){
  WDT_A->CTL = WDT_A_CTL_PW | WDT_A_CTL_HOLD;

  int a = 1;
  int b = 2;
  int c;

  c = sum(a,b);

  while(1);
}
int sum(int s1,int s2){
  int s = s1+s2;
  return s;
}
```

We reconsider the sample code in Listing 3.1 to show the difference between local and global variables. Here, the variable a is defined as global. The variables b and c are defined as local. We provide the *Expressions* window in Fig. 3.3. As can be seen in this figure, the global variable a is kept in the memory address 0x20000014 since CCS placed some TI-related variables before it. Local variables b and c are kept in the memory addresses 0x2000FFF8 and 0x2000FFFC (in the stack) as expected.

3.2 C Data Types

We summarize C data types in Table 3.1 with their size, representation, and range. In terms of representation, the first nine data types can be used for integer numbers. The last two data types can be used to represent integer and fractional numbers. Next, let's analyze each group in detail.

3.2.1 Representing Integer Numbers

If a variable does not have fractional part, we can use char, short, int, or long long data type to represent it. These representations have 8, 16, 32, and 64 bits assigned to them, respectively. We should keep in mind that when a number larger than the capacity of its type is assigned to a variable, extra bits could not be saved. Hence, an overflow occurs. In other words, the number is not saved correctly. The user should be aware of this problem.

Data is saved in blocks of 32 bits in memory due to the word length of MSP432. However, the long long data type needs 64 bits. Therefore, the little endian form is used for this purpose. Here, the least significant 32 bits are represented in lower memory address. The most significant 32 bits are represented in the following higher memory address. This way, 64 bits can be saved in memory for the long long data type.

Chapter Three

```
                  main():
00000678:         B50E                    push         {r1, r2, r3, lr}
   7              int a = 1;
0000067a:         2001                    movs         r0, #1
0000067c:         9000                    str          r0, [sp]
   8              int b = 2;
0000067e:         2002                    movs         r0, #2
00000680:         9001                    str          r0, [sp, #4]
  11              WDTCTL = WDTPW|WDTHOLD;
00000682:         490A                    ldr          r1, [pc, #0x28]
00000684:         F44F40B5                mov.w        r0, #0x5a80
00000688:         8008                    strh         r0, [r1]
  13              c = sum(a,b);
0000068a:         9800                    ldr          r0, [sp]
0000068c:         9901                    ldr          r1, [sp, #4]
0000068e:         F000F802                bl           #0x696
00000692:         9002                    str          r0, [sp, #8]
  15              while(1);
                  $C$L1:
00000694:         E7FE                    b            $C$L1
  18              int sum(int s1,int s2){
                  sum():
00000696:         F1AD0D10                sub.w        sp, sp, #0x10
0000069a:         9101                    str          r1, [sp, #4]
0000069c:         9000                    str          r0, [sp]
  19              int s = s1+s2;
0000069e:         9900                    ldr          r1, [sp]
000006a0:         9801                    ldr          r0, [sp, #4]
000006a2:         1840                    adds         r0, r0, r1
000006a4:         9002                    str          r0, [sp, #8]
  20              return s;
```

FIGURE 3.2 Memory contents (from address 0x00000678 to 0x0000060A) observed by the Disassembly window.

Expression	Type	Value	Address
(x)= a	int	1	0x20000014
(x)= b	int	2	0x2000FFF8
(x)= c	int	3	0x2000FFFC
⊕ Add new			

FIGURE 3.3 Observing local and global variables in the Expressions window.

We can represent either positive or positive and negative numbers in each data type. For example, a variable with char data type can only keep an integer number between 0 and 255. On the other hand, a variable with signed char data type can be used to keep an integer number between −128 and 127. Two's complement form is used to represent a negative number. In this form, the negative number is represented as follows. First,

Type	Size (bits)	Representation	Range
unsigned char, char	8	Binary	0 ... 255
signed char	8	Two's complement	−128 ... 127
unsigned short	16	Binary	0 ... 65 535
short, signed short	16	Two's complement	−32 768 ... 32 767
unsigned int	32	Binary	0 ... 4 294 967 295
int, signed int	32	Two's complement	−2 147 483 648 ... 2 147 483 647
unsigned long long	64	Binary	0 ... 18 446 744 073 709 551 615
long long, signed long long	64	Two's complement	−9 223 372 036 854 775 808 ... 9 223 372 036 854 775 807
float	32	IEEE 32-bit	−3.40 282 346e+38 ... 3.40 282 346e+38
double, long double	64	IEEE 64-bit	−1.79 769 313e+308 ... 1.79 769 313e+308

TABLE 3.1 C Data Types

all bit values in the positive number are reversed. Then, one is added to it. Hence, the negative form of the positive number is obtained. Representing a negative number in two's complement form has certain advantages in arithmetic operations [1]. Hence, it is used as a standard form in signed char, short, int, or long long data types.

3.2.2 Representing Fractional Numbers

There are two methods to represent a number with fractional part. These are fixed- and floating-point forms. Unfortunately, C does not have predefined data type for fixed-point number representation. Hence, we will only use floating-point numbers in this book.

As the name implies, the number of bits assigned to the integer and fractional parts are not fixed in floating-point number representation. Instead, the assigned number of bits differs for each number depending on its significant digits. Therefore, a much wider range of values can be handled.

A binary number with fractional part will be formed as $N = (-1)^S \times 2^E \times F$ in floating-point representation. Here, S stands for the sign bit, E represents the exponent value, and F stands for the fractional part. The floating-point number N is saved in memory as $X = SEF$. To represent the floating-point number as $N = (-1)^S \times 2^E \times F$, the number should be normalized such that the integer part will have one digit. The exponent will be biased by $2^{(e-1)} - 1$, where e is the number of bits to be used for E in the given format. Finally, certain number of bits will be assigned to S, E, and F depending on the standard format used for representation. The IEEE 754 standard is used by most digital systems for floating-point numbers. This standard is summarized in Table 3.2.

Format	Exponent bias	# bits for S	# bits for E	# bits for F	# total bits
Half	15	1	5	10	16
Single	127	1	8	23	32
Double	1023	1	11	52	64
Quad	16383	1	15	112	128

TABLE 3.2 The IEEE 754 Standard for Floating-Point Number Representation

We can explain the floating-point representation by taking the decimal number -255.25 as an example. Let's take the single floating-point format. The itemized conversion is as follows:

- Decide on the format: Let's pick the "Single" format for this example.
- Represent the integer and fractional parts of the decimal number in binary form: Our number becomes `11111111.01`.
- Decide on the sign bit S: Since the number is negative, $(-1)^1 = -1$, S=1.
- Normalize the number such that the integer part will have one digit: Our number becomes 1.111111101×2^7.
- Find the exponent value: For the single format, the exponent bias is 127. Therefore, the exponent will become $E=127+7=134$ with bias. Or, in binary form $E=10000110$.
- Find the fractional part: Our initial fractional part (after normalization) was `111111101`. Since 23 bits should be used to represent the fractional part of the number in single format, $F=11111110100000000000000$. Remember, since this is the fractional part, we add extra zeros to its right. Hence, the value of the number will not be affected.
- Construct $X = SEF$: Finally, $X=$0xC37F4000 in hexadecimal form.

C language has `float` and `double` data types to represent floating-point numbers. The `float` data type corresponds to the single floating-point format with 32 bits. The `double` data type corresponds to the double floating-point format with 64 bits. Therefore, the little endian representation is used for it.

3.3 ASCII Code

We do not only process numbers in digital systems. For some applications, we may need to handle characters and symbols as well. We know that everything in a digital system is kept in binary form. Therefore, characters and symbols should also be represented as such. One way of representing characters and symbols in binary form is using the ASCII code. ASCII stands for the American Standard Code for Information Interchange. The ASCII code for characters and symbols are given in Table 3.3. In this table, LSP stands for least significant part and MSP stands for most significant part of the code.

To represent a specific character (or symbol), its corresponding code should be given. Let's assume that we would like to represent the @ symbol. Using Table 3.3, the corresponding ASCII code will be 0x40. Since the ASCII code is eight bits, we can use the C data type `char` to represent it.

MSP432 Programming with C

	LSP															
MSP		0	1	2	3	4	5	6	7	8	9	A	B	C	D	E	F
	0	NUL	SOH	STX	ETX	EOT	ENQ	ACK	BEL	BS	HT	LF	VT	FF	CR	SO	SI
	1	DLE	DC1	DC2	DC3	DC4	NAK	SYN	ETB	CAN	EM	SUB	ESC	FS	GS	RS	US
	2		!	"	#	$	%	&	'	(	)	*	+	`	-	.	/
	3	0	1	2	3	4	5	6	7	8	9	:	;	<	=	>	?
	4	@	A	B	C	D	E	F	G	H	I	J	K	L	M	N	O
	5	P	Q	R	S	T	U	V	W	X	Y	Z	[	\	]	^	_
	6	`	a	b	c	d	e	f	g	h	i	j	k	l	m	n	o
	7	p	q	r	s	t	u	v	w	x	y	z	{			}	~	DEL

TABLE 3.3 ASCII Code Table

3.4 Examples on Data Type Usage

We provide examples for the mentioned C data types in Listing 3.3. Here, we define 15 global variables with different types having positive or negative values. To note here, we redefine the character a1 within the code for the compiler to run properly. Besides, there is no other purpose.

We can observe how the defined 15 global variables are represented from the *Expressions* window as shown in Fig. 3.4. This figure provides valuable information as follows. The symbol @ (assigned to variable a2 by single quotes) is saved in memory by its ASCII code 0x40. We can represent a hexadecimal number in C language by adding a prefix 0x as in the variable a3. Positive integer numbers are represented as they are in variables b1, c1, d1, and e1. Negative values for these numbers are represented in two's complement form as can be seen in b2, c2, d2, and e2. The float variable f1 holding −255.25 is represented in memory as 0xC37F4000, which is the single floating-point format as explained in the previous section. The float variable f2 holding 255.25 differs from this representation only by a change in the sign bit. This is also the case for double variables g1 and g2.

To observe the little endian representation, we should look at the memory map. We provide the related memory map for long long variables e1, e2 and double variables g1, g2 as shown in Fig. 3.5. As can be seen in this figure, the little endian representation is used to save these 64-bit numbers.

We assigned valid numbers to variables in Listing 3.3. To observe what happens when an invalid number is assigned to a variable, we provide the C code in Listing 3.4. Here, a number with fractional part is assigned to variable a with type char. A negative number is assigned to variable b with type (unsigned) char. Two numbers that can be represented by at least nine bits are assigned to variables c1 and c2 that can hold at most eight bits. Finally, two numbers with fractional part are assigned to variables d and e of type float and double, respectively.

We provide the *Expressions* window when the C code in Listing 3.4 is run, as shown in Fig. 3.6. As can be seen in this figure, the fractional part of the number 1.2 could not be saved in the variable a since it is of type char. As we try to assign a negative number to a variable with unsigned representation, the number is taken as if it is positive. Hence, the unsigned variable b holds the number 255 instead of −1. If we want to assign a

Listing 3.3 C Program for Data Types, First Example

```c
#include "msp.h"

char a1 = 1;
char a2 = '@';
char a3 = 0xFA;

signed char b1 = 2;
signed char b2 = -2;

unsigned short c1 = 3;
short c2 = -3;

unsigned int d1 = 4;
int d2 = -4;

unsigned long long e1 = 4;
long long e2 = -4;

float f1 = -255.25;
float f2 = 255.25;

double g1 = -255.25;
double g2 = 255.25;

void main(void) {

WDT_A->CTL = WDT_A_CTL_PW | WDT_A_CTL_HOLD;

a1 = 1;

while(1);
}
```

nine-bit number to an eight-bit character variable, the most significant bit will be lost. Therefore, 257 is kept as 1 in variable c1. In a similar manner, -257 is kept as 255 in the variable c2. As for floating-point numbers, we know that number of bits assigned to the fractional part of the number is limited. If the value cannot be represented by these bits, then we will only have an approximation of the number in memory. We can see such examples when 427.37 and 430.57 are saved in variables d and e of type float and double, respectively.

3.5 Arithmetic and Logic Operations

There are addition, multiplication, subtraction, division, mode, and remainder arithmetic operations in C language. These operations are straightforward to use. However, there are some important issues the reader should pay attention to. These can be summarized as overflow, data type of the result, and division by zero. We provide examples to explain these concepts in Listing 3.5.

MSP432 Programming with C   35

Expression	Type	Value	Address
a1	unsigned char	0x01 '\x01' (Hex)	0x20000000
a2	unsigned char	0x40 '@' (Hex)	0x20000001
a3	unsigned char	0xFA '\xfa' (Hex)	0x20000002
b1	char	00000010b '\x02' (Binary)	0x20000003
b2	char	11111110b '\xfe' (Binary)	0x20000004
c1	unsigned short	0000000000000011b (Binary)	0x20000006
c2	short	1111111111111101b (Binary)	0x20000008
d1	unsigned int	0x00000004 (Hex)	0x2000000C
d2	int	0xFFFFFFFC (Hex)	0x20000010
e1	unsigned long long	0x0000000000000004 (Hex)	0x20000018
e2	long long	0xFFFFFFFFFFFFFFFC (Hex)	0x20000020
f1	float	0xC37F4000 (Hex)	0x2000002C
f2	float	0x437F4000 (Hex)	0x20000030
g1	double	0xC06FE80000000000 (Hex)	0x20000038
g2	double	0x406FE80000000000 (Hex)	0x20000040

FIGURE 3.4 The Expressions window when the C program for data types first example is run.

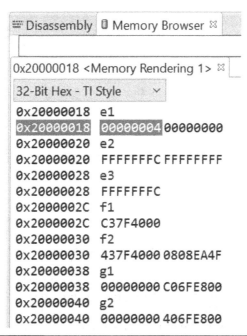

FIGURE 3.5 Part of the memory map when the C program for data types first example is run.

After the C code in Listing 3.5 is run, the variables can be observed in the *Expressions* window as shown in Fig. 3.7. As can be seen in this figure, adding 1 to the integer variable a (which was initially 4294967295) caused an overflow. Therefore, we see 0 for a instead of 4294967296 in the *Expressions* window. For the integer variable b, we assigned

Listing 3.4 C Program for Data Types, Second Example

```c
#include "msp.h"

char a = 1.2;

char b = -1;

char c1 = 257;
char c2 = -257;

float d = 427.37;

double e = 430.57;

void main(void){

WDT_A->CTL = WDT_A_CTL_PW | WDT_A_CTL_HOLD;

a = 1.2;

while(1);
}
```

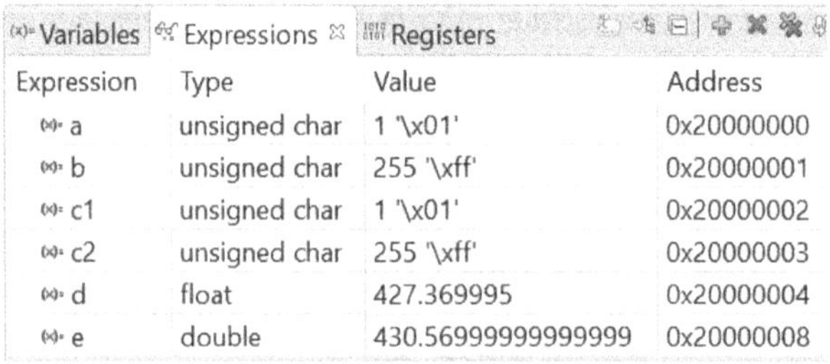

Expression	Type	Value	Address
a	unsigned char	1 '\x01'	0x20000000
b	unsigned char	255 '\xff'	0x20000001
c1	unsigned char	1 '\x01'	0x20000002
c2	unsigned char	255 '\xff'	0x20000003
d	float	427.369995	0x20000004
e	double	430.56999999999999	0x20000008

FIGURE 3.6 The Expressions window when the C program for data types (second example) is run.

a float by a division operation. Here, only the integer part is saved as can be seen in the *Expressions* window. Variables c and d keep hexadecimal values. The first one is defined as unsigned integer. Hence, it can keep four bytes (one word length) of data. The second one is defined as an unsigned character. It can keep one byte of data. To observe these values in the *Expressions* window, we should adjust the format of the number by right-clicking on the variable and selecting the *number format* option. The overflow can also be observed in variables c and d. Variables e, f, and g indicate what happens when a division by zero occurs. The result is zero for the integer variable e. Therefore, the reader cannot detect division by zero. However, the result becomes +Inf for the positive float variable f and -Inf for the negative float variable g. After a 0/0 division, the integer variable h becomes 0. The result becomes NaN for the float variable i. Hence, detecting a division by zero or 0/0 is easier for float variables.

Bitwise logic operations (AND, OR, XOR, NOT) can be done in C language. We provide examples on the usage of these operations in Listing 3.6.

Listing 3.5 Arithmetic Operations

```c
#include "msp.h"

void main(void){

WDT_A->CTL = WDT_A_CTL_PW | WDT_A_CTL_HOLD;

int a = 4294967295;
int b;
unsigned int c = 0xFFFFFFFF;
unsigned char d = 0x00;
int e = 10;
float f = +10.1;
float g = -10.1;
int h = 0;
float i = 0.0;

a += 1;
b = 17/2;
c += 0x0001;
d -= 0x01;
e /= 0;
f /= 0;
g /= 0;
h /= h;
i /= i;

while(1);
}
```

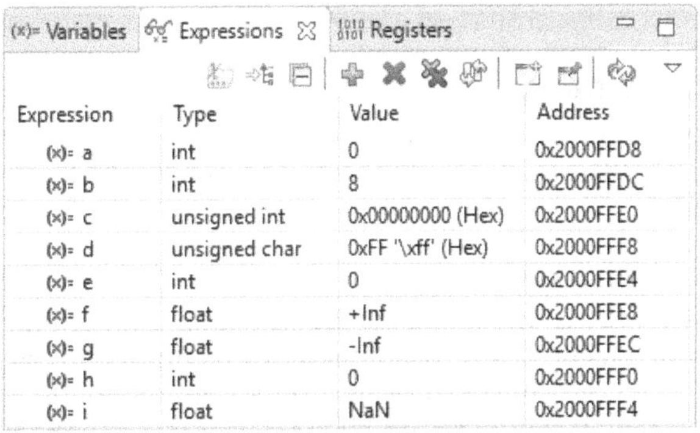

Expression	Type	Value	Address
(x)= a	int	0	0x2000FFD8
(x)= b	int	8	0x2000FFDC
(x)= c	unsigned int	0x00000000 (Hex)	0x2000FFE0
(x)= d	unsigned char	0xFF '\xff' (Hex)	0x2000FFF8
(x)= e	int	0	0x2000FFE4
(x)= f	float	+Inf	0x2000FFE8
(x)= g	float	-Inf	0x2000FFEC
(x)= h	int	0	0x2000FFF0
(x)= i	float	NaN	0x2000FFF4

FIGURE 3.7 The Expressions window after arithmetic operations.

After the C code in Listing 3.6 is run, variables can be observed in the *Expressions* window as shown in Fig. 3.8. As can be seen in this figure, all logic operations are done on a bit basis. It is also possible to observe a variable in binary form in the *Expressions* window by adjusting its number format. This may help the reader to observe the result of the logic operation in a more descriptive manner.

Listing 3.6 Logic Operations

```c
#include "msp.h"

void main(void){

WDT_A->CTL = WDT_A_CTL_PW | WDT_A_CTL_HOLD;

char a = 0x02;
char b = 0xAF;
char c,d,e,f;

c = a|b;
d = a&b;
e = a^b;
f = ~a;

while(1);
}
```

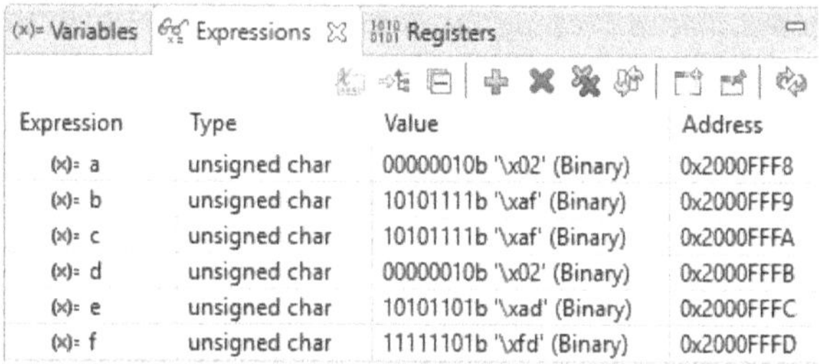

Expression	Type	Value	Address
(x)= a	unsigned char	00000010b '\x02' (Binary)	0x2000FFF8
(x)= b	unsigned char	10101111b '\xaf' (Binary)	0x2000FFF9
(x)= c	unsigned char	10101111b '\xaf' (Binary)	0x2000FFFA
(x)= d	unsigned char	00000010b '\x02' (Binary)	0x2000FFFB
(x)= e	unsigned char	10101101b '\xad' (Binary)	0x2000FFFC
(x)= f	unsigned char	11111101b '\xfd' (Binary)	0x2000FFFD

FIGURE 3.8 The Expressions window after logic operations.

3.6 Control Structures

There are condition check and loop operations under control structures. However, we will not explore these in detail. As mentioned previously, the reader should consult C books for them.

3.6.1 Condition Check

There are two options to check a condition within a program. The first one is the `if` or `if else` statement. The usage of these is straightforward as one of the two options is executed based on a binary decision. The second option to check a condition is the `switch` statement. It should be used if more than two options are available.

Condition checks will be inevitable in our applications. Either we will check the status of a button or we will observe the analog voltage level in a pin. We will perform appropriate actions based on the obtained values. Therefore, we will need condition-check statements for these and similar cases.

3.6.2 Loops

Loop operations can be used if a code block is to be executed more than once. There are three options `for`, `while`, and `do while` as loop operations in C language. They differ

Listing 3.7 Pointer Usage Example

```
#include "msp.h"

int a = 3;
int *a_pointer;

void main(void){

WDT_A->CTL = WDT_A_CTL_PW | WDT_A_CTL_HOLD;

a_pointer = &a;
*a_pointer = 5;

while(1);
}
```

in terms of their starting and stopping conditions. We suggest the reader to check them from a C book.

We will use loop statements in various applications. However, there is one usage that may seem odd. There will be an infinite loop at the end of most of our codes. Or, the main program block will be kept in an infinite loop. This can be performed by a code line `while(1)` or `for(;;)`. The main reason in using such an infinite loop is as follows. The code should run indefinitely for almost all microcontroller applications. In other words, the program should not end after the first run. To perform this, we will let the microcontroller stay in an infinite loop without exiting the program.

3.7 Arrays and Pointers

Arrays and pointers deserve specific consideration in C language. When an array is defined in C language, it is treated as a pointer. Therefore, we consider arrays and pointers together in this section.

Pointers and pointer arithmetic are one of the most confusing topics in C language. Fortunately, we can observe the memory map of the MSP432 directly under CCS. This will help us understand the usage of pointers and pointer arithmetic.

Let's start with the pointer definition. We define a global integer variable, a, and initially assign 3 to it in Listing 3.7. Next, we define a global pointer, a_pointer, with the integer type. We provide the *Expressions* window (after the code is run) in Fig. 3.9. As can be seen in this figure, the variable, a, with value 3 is stored in the memory address 0x20000014. The address of the variable, a, is stored in the pointer, a_pointer, with the code line a_pointer=&a. Therefore, the pointer keeps the memory address. Here, the pointer is kept in another memory address 0x20000020. In the following line, we change the entry of this memory address as *a_pointer=5. As can be seen in the *Expressions* window, this changed the value of the variable a.

Let's consider an array with five integer values and observe operations on it. We provide such a code in Listing 3.8. We first define a global integer array, a, with entries {1, 2, 3, 4, 5}. We provide the *Expressions* window (after the code is run) in Fig. 3.10. As can be seen in this figure, the array, a, is in fact saved by its starting address 0x20000000. We assign this address to a pointer, a_pointer, in the code line a_pointer=a. Then, we reach the fourth element of the array and change it to zero by incrementing the pointer

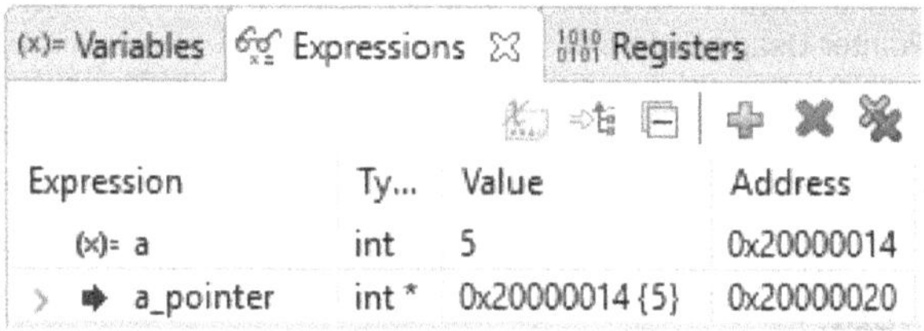

Chapter Three

Expression	Ty...	Value	Address
(x)= a	int	5	0x20000014
➡ a_pointer	int *	0x20000014 {5}	0x20000020

FIGURE 3.9 The Expressions window for the pointer example.

Listing 3.8 Array Usage Example

```
#include "msp.h"

int a[5] = {1,2,3,4,5};
int *a_pointer;

void main(void) {

WDT_A->CTL = WDT_A_CTL_PW | WDT_A_CTL_HOLD;

a_pointer = a;
a_pointer += 3;
*a_pointer = 0;

while(1);
}
```

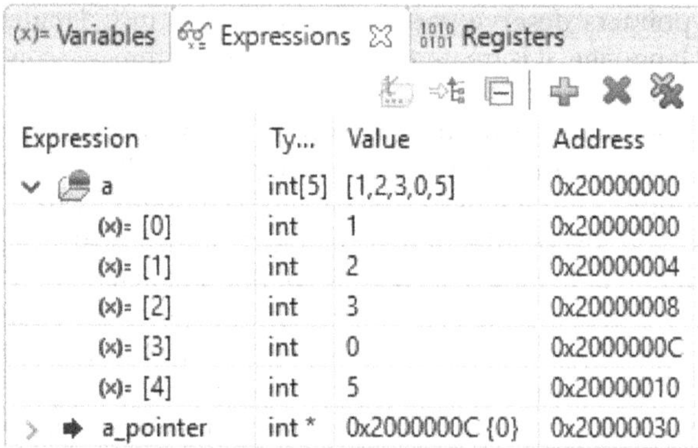

Expression	Ty...	Value	Address
⌄ a	int[5]	[1,2,3,0,5]	0x20000000
(x)= [0]	int	1	0x20000000
(x)= [1]	int	2	0x20000004
(x)= [2]	int	3	0x20000008
(x)= [3]	int	0	0x2000000C
(x)= [4]	int	5	0x20000010
➡ a_pointer	int *	0x2000000C {0}	0x20000030

FIGURE 3.10 The Expressions window for the array and pointer example.

value by 3. There are two important issues here. First, the array can be processed as if it is a pointer as mentioned above. Second, increments and decrements are done relative to the pointer type in pointer arithmetic. Therefore, the code line `a_pointer+=3` increments the pointer value by 12 (3 × 4) bytes, since each integer value occupies four bytes.

Listing 3.9 The Sample Code for the Usage of Structure

```c
#include "msp.h"

void main(void){

WDT_A->CTL = WDT_A_CTL_PW | WDT_A_CTL_HOLD;

struct port_str{
char direction;
char output;
char input;
};

struct port_str Port1;

Port1.direction = 0x02;
Port1.output = 0x04;
Port1.input = 0x08;

while(1);
}
```

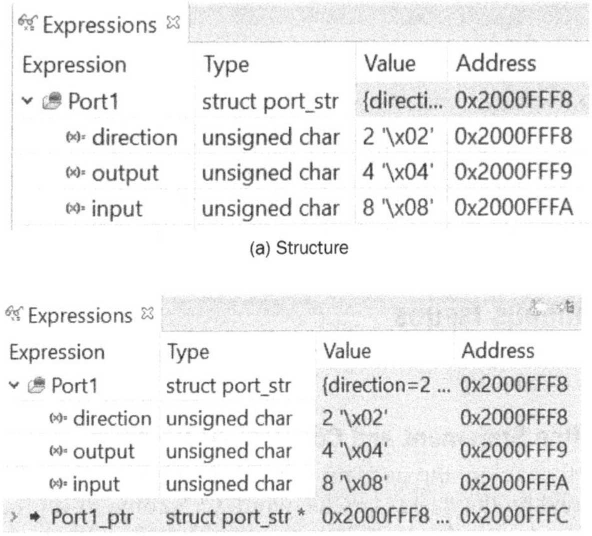

Expression	Type	Value	Address
∨ Port1	struct port_str	{directi...	0x2000FFF8
direction	unsigned char	2 '\x02'	0x2000FFF8
output	unsigned char	4 '\x04'	0x2000FFF9
input	unsigned char	8 '\x08'	0x2000FFFA

(a) Structure

Expression	Type	Value	Address
∨ Port1	struct port_str	{direction=2 ...	0x2000FFF8
direction	unsigned char	2 '\x02'	0x2000FFF8
output	unsigned char	4 '\x04'	0x2000FFF9
input	unsigned char	8 '\x08'	0x2000FFFA
> Port1_ptr	struct port_str *	0x2000FFF8 ...	0x2000FFFC

(b) Pointer to structure

FIGURE 3.11 The Expressions window for the structure examples.

3.8 Structures

We will be using structures in the following chapters while modifying peripheral device registers. Therefore, we provide a sample code on the structure usage in Listing 3.9. In this code, we declare a structure, `port_str`, with three fields. Then, we create a port structure, `Port1`, and assign values to its elements. We provide the *Expressions* window in Fig. 3.11*a* as we run the code.

Listing 3.10 The Sample Code for the Usage of Pointer to Structure

```c
#include "msp.h"

void main(void) {

  WDT_A->CTL = WDT_A_CTL_PW | WDT_A_CTL_HOLD;

  struct port_str{
  char direction;
  char output;
  char input;
  };

  struct port_str Port1;
  struct port_str* Port1_ptr;

  Port1_ptr = &Port1;

  Port1_ptr -> direction = 0x02;
  Port1_ptr -> output = 0x04;
  Port1_ptr -> input = 0x08;

  while(1);
}
```

In fact, we will be using predefined structures while modifying peripheral device registers. To use them, we should know how pointer to structure can be used. We modify the C code in Listing 3.9 accordingly. We provide the new sample code in Listing 3.10. We provide the *Expressions* window as we run the program in Fig. 3.11b.

3.9 Miscellaneous Issues

There are some important miscellaneous issues related to C language. We will cover them in this section.

3.9.1 Define Statement and Constant Declaration

We first briefly explore the define statement and the const declaration. When a constant is defined by the # define keyword, CCS converts all the affected code lines to the final value during the compiling process. Therefore, no more calculations are done in execution. In a similar manner, if a global variable is defined as constant by the const keyword, data it contains are saved in the code memory block.

We provide the sample C code in Listing 3.11 to give examples on the topics considered in this section. To observe how define statements are handled during the compiling process, we provide the *Disassembly* window in Fig. 3.12. As can be seen in this figure, the assembly codes are directly given for the C code lines a=2*CONST and b=4*CONST. In other words, values 2*CONST and 4*CONST are calculated during the compiling process. The end result is assigned directly. We next provide the *Expressions* window in Fig. 3.13 to show that the constant integer variable c is in fact saved in the code memory block.

Listing 3.11 The Sample Code for the Define Statement and Constant Declaration

```c
#include "msp.h"

#define CONST 4

int a = 0, b = 0;
const int c[] = {1,2,3,4};
int d = 32767;

void main(void) {

WDT_A->CTL = WDT_A_CTL_PW | WDT_A_CTL_HOLD;

a = 2*CONST;
b = 4*CONST;
d = d+c[1];

while(1);
}
```

```
            00000676:    0000              movs      r0, r0
         11                WDTCTL = WDTPW|WDTHOLD;
                          main():
            00000678:    4908              ldr       r1, [pc, #0x20]
            0000067a:    F44F40B5          mov.w     r0, #0x5a80
            0000067e:    8008              strh      r0, [r1]
         13                a = 2*CONST;
            00000680:    4907              ldr       r1, [pc, #0x1c]
            00000682:    2008              movs      r0, #8
            00000684:    6008              str       r0, [r1]
         14                b = 4*CONST;
            00000686:    4907              ldr       r1, [pc, #0x1c]
            00000688:    2010              movs      r0, #0x10
            0000068a:    6008              str       r0, [r1]
         15                d = d+c[1];
            0000068c:    4906              ldr       r1, [pc, #0x18]
            0000068e:    4807              ldr       r0, [pc, #0x1c]
            00000690:    680A              ldr       r2, [r1]
            00000692:    6800              ldr       r0, [r0]
            00000694:    4904              ldr       r1, [pc, #0x10]
            00000696:    1880              adds      r0, r0, r2
            00000698:    6008              str       r0, [r1]
         17                while(1);
```

FIGURE 3.12 The Disassembly window for the define statement.

3.9.2 Casting

While performing an arithmetic operation, the operands and the result may not be of the same data type. Under such a circumstance, the result is obtained in terms of the type of the operands. We provide such an example in Listing 3.12. Here, two integer numbers are divided. The result is saved in a float variable. When the division operation is done

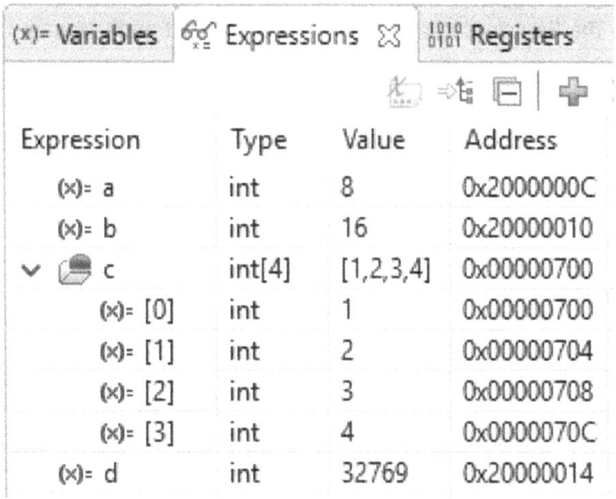
(x)= Variables
Expressions
Registers
Expression
Type
Value
Address
(x)= a int 8 0x2000000C
(x)= b int 16 0x20000010
c int[4] [1,2,3,4] 0x00000700
(x)= [0] int 1 0x00000700
(x)= [1] int 2 0x00000704
(x)= [2] int 3 0x00000708
(x)= [3] int 4 0x0000070C
(x)= d int 32769 0x20000014

FIGURE 3.13 The Expressions window for the constant declaration.

Listing 3.12 The Sample Code for the Usage of Casting in Arithmetic Operations

```
#include "msp.h"

 int a = 17;
 int b = 2;
 float c, d, e;

void main(void){

 WDT_A->CTL = WDT_A_CTL_PW | WDT_A_CTL_HOLD;

 c = a/b;
 d = (float) a / (float) b;
 e =  a / (float) b;

 while(1);
}
```

directly, the result will be an integer converted to float. Hence, it will be false. To correct this miscalculation, we can convert the operand(s) to float by applying casting by adding the keyword `(float)` in front of the integer variable. Hence, the variable is taken as if it is a float variable during the operation. Then, the result becomes as expected. Then, the result becomes as expected as shown in Fig. 3.14.

3.9.3 Reaching a Specific Memory Address

We can reach a specific memory address in MSP432 using pointer arithmetic and casting operations. We provide such an example in Listing 3.13. Here, we modify the memory addresses `0x20000020` and `0x20000024` this way. We provide the memory map after this operation in Fig. 3.15. As can be seen in this figure, the memory content has been updated as expected.

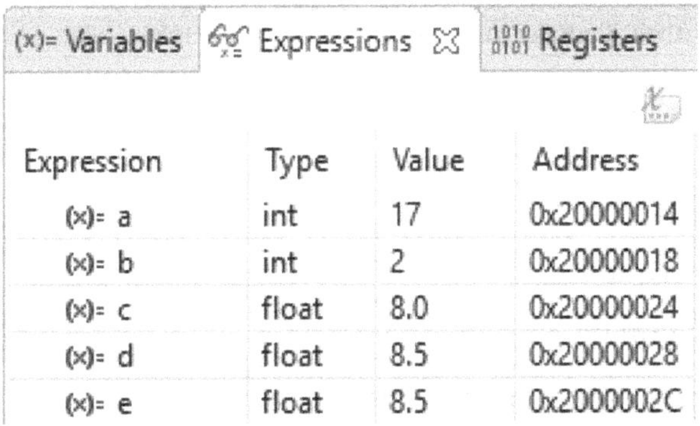

MSP432 Programming with C

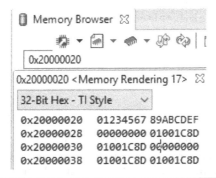

FIGURE 3.14 The Expressions window for the usage of casting in arithmetic operations.

FIGURE 3.15 The memory map after executing the C code in Listing 3.13.

Listing 3.13 The Sample Code to Reach a Specific Memory Address in MSP432

```
#include "msp.h"

int *mem_loc;

void main(void){

WDT_A->CTL = WDT_A_CTL_PW | WDT_A_CTL_HOLD;

mem_loc = (int *) 0x20000020;

*(mem_loc)   = 0x01234567;
*(mem_loc+1) = 0x89ABCDEF;

while(1);
}
```

3.9.4 Advanced Mathematical Operations

The header file math.h can be included to the project for advanced mathematical operations. In Listing 3.14, we use the sin() function, defined under this header file, to fill the array sine_arr with one period of the sine wave. The *Expressions* window can be used to observe the sine_arr entries after the program is executed.

Listing 3.14 The C Code for the Usage of the `math.h` Header File

```c
#include "msp.h"
#include "math.h"

#define M 20
#define PI 3.1415

float sine_arr[M];

void main(void){

  WDT_A->CTL = WDT_A_CTL_PW | WDT_A_CTL_HOLD;

  int count;

  for(count=0;count<M;count++)
  sine_arr[count]=sin(2*PI*count/M);

  while(1);
}
```

3.10 Summary

MSP432 can be programmed by both assembly and C languages. In this chapter, we considered the latter approach. Although we focused on C programming of MSP432, we assumed that the reader has a basic C knowledge. Here, we extended this knowledge on MSP432. Therefore, we first explored the memory management issue. Since CCS allows us to observe the memory map of MSP432, we were able to see how local and global variables are handled. Then, we looked at data types. Afterwards, we reviewed basic C topics such as arithmetic and logic operations, control structures, arrays, pointers, and structures. Then, we summarized miscellaneous issues. All these topics can be mastered on MSP432 since the reader can observe the memory map of the microcontroller directly. Therefore, this chapter provides a good option to strengthen the reader's C knowledge.

3.11 Exercises

3.1 As the C code in Listing 3.15 is run, what will be the value of `num3`, `num4`, and `num5`?

3.2 Write a program in C language such that
   a. it contains a function that calculates twice the square of a given integer.
   b. your program should calculate twice the square of numbers between 0 and 511 using your function. The results should be saved in an array.

3.3 Find the cosine of integer degrees between 0 and 360. The results should be saved in an array. Hint: The `cos()` function in `math.h` header file returns the cosine of a radian angle.

3.4 Write a program in C language such that
   a. it contains a function that calculates the factorial of a given integer number.

Listing 3.15 The C Code for Problem 3.1

```c
#include "msp.h"

char num1=0x0A;
char num2=0x0B;
char num3,num4,num5;

void main(void){

WDT_A->CTL = WDT_A_CTL_PW | WDT_A_CTL_HOLD;

num3=num1+num2;
num4=num1-num2;
num5=num1^num2;

while(1);
}
```

Listing 3.16 The C Code for Problem 3.5

```c
#include "msp.h"

void main (void){

WDT_A->CTL = WDT_A_CTL_PW | WDT_A_CTL_HOLD;

int a = 0xAAAA;
int i = 0;
int mask = 0x0303AF11;
int y = 0xFFFFA5A5;

for(i=0;i<10;i++){
a = !a;
y = i*(y^mask)&a;
}
while(1);
}
```

     b. if an overflow occurs during the calculation, the factorial function should return −1.

     c. Your program should calculate factorial of numbers between 1 and 14 using your function. The results should be saved in an array.

3.5 Find 10 values of the variable y as the code in Listing 3.16 is run.

3.6 What will be the entries of the array arr as the code in Listing 3.17 is run?

3.7 You have to search for the second maximum of a given integer array with 50 elements. The array is saved as a global variable in the program. Write the result to the memory location just after the array ends.

3.8 Memory locations between 0x20000000 and 0x20000200 keep double floating-point numbers in the MSP432 microcontroller memory. They are kept in the standard

Listing 3.17 The C Code for Problem 3.6

```
#include "msp.h"

void main(){

  WDT_A->CTL = WDT_A_CTL_PW | WDT_A_CTL_HOLD;

  float arr[] = {4.24,1.37,9.46,0.0,0.0};
  float *ptr = arr, sum = 0;
  short i;

  for(i=0; i<3; i++)
    sum += *ptr++;

  *ptr++ = sum;
  *ptr = 3*sum+2.33/4-5.55;

  while(1);
}
```

endian format of the microcontroller. Write a C program to change the endiannes of the data kept between the mentioned memory locations.

3.9 Write a C program to scan memory locations between 0x2001000A and 0x200103CC using pointer arithmetic. In scanning, first check for the half-word pattern 0xACCA. If this pattern is observed, increase the value of the variable a by one. Second, check for the word pattern 0xACCAADDA. If this pattern is observed, then increase the value of the variable b by one. First assign zero to variables a and b.

CHAPTER 4
MSP432 Programming with Assembly

Assembly language offers another option to program the MSP432 by reaching the lowest level hardware properties of the microcontroller. Therefore, using assembly language provides a good insight in understanding the inner working principles of the microcontroller. Besides, assembly language allows the programmer to benefit from all possible microcontroller resources for the tiniest and fastest program for a given target. However, performing this is harder to achieve compared to C programming. Besides, recent C compilers perform as good as assembly programming. Hence, this will be the only chapter in which we will be using assembly language for programming purposes. To do so, we will start with the MSP432 core registers. Then, we will analyze the anatomy of an assembly program. Afterwards, we will introduce the instruction set of MSP432. We will also provide practical examples on the usage of these instructions. Finally, we will show how an assembly command can be executed in a C program. To note here, the coverage of assembly language concepts in this chapter is not comprehensive. We will direct the reader to related references for more information on this topic.

4.1 MSP432 Core Registers

Every action within the MSP432 CPU is controlled by registers from a programmer's perspective. Besides, operations can be done on core registers in assembly language. Therefore, we should know what these registers are and how they can be used. We will focus on the MSP432 core registers in this section.

The MSP432 CPU has 21 core registers in which 13 of them (R0 to R12) can be used for general purpose. Remaining eight registers are reserved for specific operations by the CPU. All core registers are listed below:

- **R0–R12**: These are general purpose registers which can be used for data operations. Registers R0–R7 are called low registers. These can be accessed by all instructions. Registers R8–R12 are called high registers. These can be accessed by 32-bit instructions but not 16-bit instructions explained in Sec. 4.3.
- **SP (R13)**: This is the stack pointer (SP) register responsible for holding the stack address. The stack is a last-in first-out list used to store values temporarily. In fact,

there are two stack pointers in MSP432. These are the main stack pointer (MSP) and process stack pointer (PSP). We will extensively use the MSP in this chapter.
- **LR (R14)**: This is the link register (LR) used for holding return address during branching and exception operations.
- **PC (R15)**: This is the program counter (PC) register responsible for holding the address of the instruction to be executed.
- **PSR**: This is the program status register (PSR). It contains three registers as application program status register (APSR), interrupt program status register (IPSR), and execution program status register (EPSR). The APSR contains the current state of condition flags from previous instruction executions. The IPSR holds the exception type number of the current interrupt routine. The EPSR holds the thumb state bit and execution state bits.
- **PRIMASK**: The PRIMASK register enables or disables the exceptions with configurable priority.
- **FAULTMASK**: The FAULTMASK register enables or disables all exceptions except for non-maskable interrupt (NMI).
- **BASEPRI**: The BASEPRI register defines the minimum priority for exception processing.
- **CONTROL**: The control register controls the stack used and the privilege levels. It also holds the floating-point unit (FPU) active status bit.

We should take a closer look at the condition flags available in the APSR register. There are five flags as negative (N), zero (Z), overflow (V), carry (C), and saturation (Q). When an instruction is executed, the status of the result can be observed via these flags. More information on their usage is provided in Sec. 4.3.

4.2 Anatomy of an Assembly Program

The programmer does not pay attention to low-level operations, such as memory management, while writing a C program. These are handled by the C compiler. Unfortunately, it is the duty of the programmer to manage such low-level operations while writing an assembly program. This is done by the help of assembler directives. These directives can be used in tasks such as initializing memory, defining global variables, and allocating the code and data into specific memory area. TI provides a reference for the usage of all directives [9]. We provide only the list of basic directives and their description to be used in this chapter next.

- **.text** assembles the following code into text (executable code) section.
- **.align** defines the length of instructions to be processed in terms of bytes.
- **.field** places a single value into a specified number of bits in the current word.
- **.global** defines one or more global symbols.
- **.end** identifies end of the program.

We provide a template code for assembly programming in Listing 4.1. As can be seen in this code, the first directive `.text` indicates that the following lines correspond

Listing 4.1 Template Code for Assembly Program

```
 .text
 .align  2
 .global main

main:
 ; User code

 .end
```

to assembly program. Hence, they should be placed in the code section of MSP432 memory. The second directive .align defines the length of instructions to be processed in terms of bytes. If the instructions to be executed are of length 16 bits, then we should use .align 2 as in our case. The third directive .global defines a global variable main to be used throughout the program. In fact, this variable will indicate the starting address of the assembly program. Finally, the directive .end indicates the end of the assembly program.

4.3 The Instruction Set

Commands used in assembly language programming of a CPU form its instruction set. Since the MSP432 microcontroller is based on the ARM® Cortex™-M4F processor (more specifically the ARMv7-M architecture), we will focus on its instruction set. The total number of supported instructions in the ARM® Cortex™-M4F processor is 207. These can be divided into six subsets as general data moving and processing, memory access, arithmetic and logic, branch and control, floating-point, and other instructions. We will start with the anatomy of an instruction. Then, we will explore the six instruction groups separately. While doing this, we will only focus on few instructions in each group to show how they work. We will provide reference to the rest of instructions.

4.3.1 Anatomy of an Instruction

ARM assembly syntax is as follows. {label} mnemonic{cc}{S} {Rd}, op1, {op2}; {Comments}. Here, parts with in curly brackets are optional. The label is used as a reference to a memory address. mnemonic is the name of the instruction. cc is the condition code to process the instruction. Condition codes for the instructions are as given in Table 4.1. An instruction may or may not affect the condition flags (N, Z, V, and C) in the APSR register. If the programmer wants the instruction to affect this register, then the letter S should be added to the mnemonic. Rd is the destination register that the result of the operation is written to. If it is omitted, then op1 is used as the destination register. op2 is named as flexible second operand of instruction that is mostly used in data processing instructions. The second operand can be a constant, register or a register with optional shift. Comments in assembly program are written after semicolon (;) which do not affect the execution of the program. There is one important point to be mentioned here. A mnemonic should be written to the source code file by a preceding white space. Otherwise, the compiler raises an error. A label should be written without a white space.

Condition Code	Flags	Brief Description
EQ	Z = 1	Equal
NE	Z = 0	Not equal
CS or HS	C = 1	Higher or same, unsigned
CC or LO	C = 0	Lower, unsigned
MI	N = 1	Negative
PL	N = 0	Positive or zero
VS	V = 1	Overflow
VC	V = 0	No overflow
HI	C = 1 and Z = 0	Higher, unsigned
LS	C = 0 or Z = 1	Lower or same, unsigned
GE	N = V	Greater than or equal, signed
LT	N != V	Less than, signed
GT	Z = 0 and N = V	Greater than, signed
LE	Z = 1 and N != V	Less than or equal, signed
AL	Can have any value	Always

TABLE 4.1 Condition Codes to be Added to a Mnemonic

Mnemonic and Operands	Flags	Brief Description
CLZ Rd, Rm	–	Count leading zeros
CMN{S} Rn, op2	N, Z, C, V	Compare negative
CMP{S} Rn, op2	N, Z, C, V	Compare
MOV{S} Rd, op2	N, Z, C	Move
MOVT Rd, op2	–	Move top
MOVW Rd, op2	N, Z, C	Move 16-bit constant
MVN{S} Rd, op2	N, Z, C	Move NOT
REV Rd, Rn	–	Reverse bytes in word
REV16 Rd, Rn	–	Reverse bytes in each halfword
REVSH Rd, Rn	–	Reverse byte order in bottom halfword and sign extend
RBIT Rd, Rn	–	Reverse bits
ROR{S} Rd, Rm, n	N, Z, C	Rotate right n times. n can be a constant or register
RRX{S} Rd, Rm	N, Z, C	Rotate right with extend

TABLE 4.2 General Data Moving and Processing Instructions

4.3.2 General Data Moving and Processing Instructions

There are a total of 13 instructions in this category. These work on data registers extensively. We provide all data moving and processing instructions with their brief description in Table 4.2. The usage of these instructions alter the APSR flags only if S

Mnemonic and Operands	Mnemonic and Operands	Mnemonic and Operands
ADR Rd, label	LDREX Rt, [Rn {, offset}]	STRHT Rt, [Rn {, offset}]
LDR Rt, [Rn {, offset}]	LDREXB Rt, [Rn]	STRSHT Rt, [Rn {, offset}]
LDRB Rt, [Rn {, offset}]	LDREXH Rt, [Rn]	STREX Rt, [Rn {, offset}]
LDRSB Rt, [Rn {, offset}]	STR Rt, [Rn {, offset}]	STREXB Rt, [Rn]
LDRH Rt, [Rn {, offset}]	STRB Rt, [Rn {, offset}]	STREXH Rt, [Rn]
LDRSH Rt, [Rn {, offset}]	STRSB Rt, [Rn {, offset}]	LDM Rn!, reglist
LDRD Rt, Rt2, [Rn {, offset}]	STRH Rt, [Rn {, offset}]	LDMDB Rn!, reglist
LDRT Rt, [Rn {, offset}]	STRSH Rt, [Rn {, offset}]	STM Rn!, reglist
LDRBT Rt, [Rn {, offset}]	STRD Rt, Rt2, [Rn {, offset}]	STMDB Rn!, reglist
LDRSBT Rt, [Rn {, offset}]	STRT Rt, [Rn {, offset}]	POP {reglist}
LDRHT Rt, [Rn {, offset}]	STRBT Rt, [Rn {, offset}]	PUSH {reglist}
LDRSHT Rt, [Rn {, offset}]	STRSBT Rt, [Rn {, offset}]	CLREX -

TABLE 4.3 Memory Access Instructions

is added to the mnemonic as a suffix. Also, all condition codes can be applied to these instructions.

Let's give two basic examples on the usage of data moving instructions. We can assign a constant to a target register directly. One such example is MOV R0, #10h. Here, the hexadecimal value 10 is assigned to register R0 immediately. The # sign is used before the number to indicate that it is a constant value. In assigning numbers, CCS allows binary, octal, decimal, and hexadecimal values. We can also work between registers. One such example is MOVS R0, R1. Here, the content of register R1 is copied to register R0. This instruction also updates all APSR flags.

4.3.3 Memory Access Instructions

Registers can be loaded or stored from/to memory locations using memory access instructions. There are 36 such instructions as tabulated in Table 4.3. Here, the offset parameter can be a constant or register. reglist represents the register list. Usage of memory access instructions won't alter the APSR flags. However, all condition codes can be applied to these instructions.

We will not consider all instructions in Table 4.3. Instead, we pick four basic memory access instructions as in Table 4.4. We will only deal with these throughout the chapter. We believe usage of these selected instructions will be helpful in understanding other memory access instructions as well. For the description of all memory access instructions, please see [10].

We can reach memory contents in four different ways. The first one is reaching a memory location pointed by a register. An example for this case is LDR R0, [R1]. Here, contents of the memory (with address saved in register R1) is copied to register R0. The second way is reaching a memory location pointed by a register and offset. An example for this case is LDR R0, [R1, #10]. Here, contents of the memory (with

Mnemonic and Operands	Brief Description
LDR Rt, [Rn {, offset}]	Load register with offset, using word
STR Rt, [Rn {, offset}]	Store register with offset, using word
POP {reglist}	Pop registers from stack
PUSH {reglist}	Push registers onto stack

TABLE 4.4 Selected Four Memory Access Instructions

address saved in register R1 plus 10 offset bytes) is copied to register R0. The third way is reaching a memory location pointed by a register and an offset with pre-indexing. An example for this case is LDR R0, [R1, #10]!. Here, the value at register R1 is incremented by 10 bytes. Then, contents of the memory (with address saved in register R1) is copied to register R0. The fourth way is reaching a memory location pointed by a register and offset with post-indexing. An example for this case is LDR R0, [R1], #10. Here, contents of the memory (with address saved in register R1) is copied to register R0. Then, the value at register R1 is incremented by 10 bytes.

Bit-banding is the special operation mode for load/store operations allowing to access a single bit in a given memory address. This method works by reading data from memory to buffer or writing bitwise changed data from buffer to memory using a single instruction. On the other hand, the corresponding normal operation requires read-modify-write sequence to change one bit of memory address that requires three instructions.

For the bit-banding operation, 32-bit data in memory is mapped into a bit-band alias address. There is a predefined bit-band alias region in the memory map for the RAM content. To be more specific, 32 MB memory space between addresses 0x22000000 and 0x23FFFFFF is used as bit-band alias region for the 1-MB RAM region between addresses 0x20000000 and 0x200FFFFF. The mapping formula for the bit-banding operation is as follows:

$$bit\_word\_offset = (byte\_offset \times 32) + (bit\_number \times 4) \quad (4.1)$$
$$bit\_word\_addr = bit\_band\_base + bit\_word\_offset \quad (4.2)$$

where bit_word_offset is the position of the target bit in the bit-band memory region. bit_word_addr is the address of the word in the alias memory region that maps to the targeted bit. bit_band_base is the starting address of the alias region. $byte\_offset$ is the number of the byte in the bit-band region that contains the targeted bit. bit_number is the bit position, [0:7], of the targeted bit. For example, bit [4] of the memory address 0x20000040 is aliased to the bit-band address 0x22000000 + (0x00040 × 32) + 4 × 4 which is equal to 0x22000810.

4.3.4 Instructions for Arithmetic and Logic Operations

The third group of instructions are for arithmetic and logic operations. Within this group, we first consider instructions for arithmetic operations. Then, we evaluate instructions for logic operations. Finally, we consider instructions for saturating arithmetic.

4.3.4.1 Instructions for Arithmetic Operations

There are a total of 75 instructions in this category. These are provided in Table 4.5. The usage of these instructions alter the APSR flags only if S is specified after mnemonic. Also, all condition codes can be applied to these instructions.

MSP432 Programming with Assembly

Mnemonic and Operands	Mnemonic and Operands	Mnemonic and Operands
ADC{S} {Rd,} Rn, op2	SMLADX Rd, Rn, Rm, Ra	SMULWT RdLo, RdHi, Rn, Rm
ADD{S} {Rd,} Rn, op2	SMLAL RdLo, RdHi, Rn, Rm	SMUSD RdLo, RdHi, Rn, Rm
ADDW{S} {Rd,} Rn, op2	SMLALBB RdLo, RdHi, Rn, Rm	SMUSDX RdLo, RdHi, Rn, Rm
ASR{S} Rd Rm, op2	SMLALBT RdLo, RdHi, Rn, Rm	SSUB16 {Rd,} Rn, Rm
MLA Rd, Rn, Rm, Ra	SMLALTB RdLo, RdHi, Rn, Rm	SSUB8 {Rd,} Rn, Rm
MLS Rd, Rn, Rm, Ra	SMLALTT RdLo, RdHi, Rn, Rm	SUB{S} {Rd,} Rn, op2
MUL{S} {Rd,} Rn, Rm	SMLALD RdLo, RdHi, Rn, Rm	SUBW{S} {Rd,} Rn, op2
RSB{S} {Rd,} Rn, op2	SMLALDX RdLo, RdHi, Rn, Rm	UADD16 {Rd,} Rn, Rm
SADD16 {Rd,} Rn, Rm	SMLAWB Rd, Rn, Rm, Ra	UADD8 {Rd,} Rn, Rm
SADD8 {Rd,} Rn, Rm	SMLAWBT Rd, Rn, Rm, Ra	UASX {Rd}, Rn, Rm
SASX {Rd}, Rm, Rn	SMLSD Rd, Rn, Rm, Ra	USAX {Rd}, Rn, Rm
SSAX {Rd}, Rm, Rn	SMLSLD Rd, Rn, Rm, Ra	UDIV {Rd,} Rn, Rm
SBC{S} {Rd,} Rn, op2	SMMLA Rd, Rn, Rm, Ra	UHADD16 {Rd,} Rn, Rm
SDIV {Rd,} Rn, Rm	SMMLS Rd, Rn, Rm, Ra	UHADD8 {Rd,} Rn, Rm
SHADD16 {Rd,} Rn, Rm	SMMLSR Rd, Rn, Rm, Ra	UHASX {Rd}, Rn, Rm
SHADD8 {Rd,} Rn, Rm	SMUAD Rd, Rn, Rm	UHSAX {Rd}, Rn, Rm
SHASX {Rd}, Rn, Rm	SMUADX Rd, Rn, Rm	UHSUB16 {Rd,} Rn, Rm
SHSAX {Rd}, Rn, Rm	SMULBB Rd, Rn, Rm	UHSUB8 {Rd,} Rn, Rm
SHSUB16 {Rd,} Rn, Rm	SMULBT Rd, Rn, Rm	UMAAL RdLo, RdHi, Rn, Rm
SHSUB8 {Rd,} Rn, Rm	SMULTB Rd, Rn, Rm	UMLAL RdLo, RdHi, Rn, Rm
SMLABB Rd, Rn, Rm, Ra	SMULTT Rd, Rn, Rm	UMULL RdLo, RdHi, Rn, Rm
SMLABT Rd, Rn, Rm, Ra	SMMUL Rd, Rn, Rm	USAD8 {Rd,} Rn, Rm
SMLATB Rd, Rn, Rm, Ra	SMMULR Rd, Rn, Rm	USADA8 {Rd,} Rn, Rm, Ra
SMLATT Rd, Rn, Rm, Ra	SMULL RdLo, RdHi, Rn, Rm	USUB16 {Rd,} Rn, Rm
SMLAD Rd, Rn, Rm, Ra	SMULWB RdLo, RdHi, Rn, Rm	USUB8 {Rd,} Rn, Rm

TABLE 4.5 Instructions for Arithmetic Operations

As in the previous section, we do not consider all 75 arithmetic instructions in Table 4.5. Instead, we use only a subset of these instructions as presented in Table 4.6. For the description of all arithmetic operation instructions, please see [10].

In Table 4.6, the reader should observe that all arithmetic operations are performed at least on one register. In other words, it is not possible to perform an arithmetic operation directly on a memory content. The ARM® Cortex™-M4F processor also supports hardware multiply and divide operations that give superior performance in digital signal processing applications. The last five instructions in Table 4.6 correspond to these operations. Here Ra corresponds to the accumulation register used in multiply and accumulate or multiply and subtract operations.

Let's consider the usage of arithmetic instructions. We can add the content of registers R1, R2 and save the result to register R0. The corresponding instruction for this operation is ADD R0, R1, R2. This operation will not alter the flags in APSR. We can perform a similar operation as ADDS R0, R1, R2. This operation will alter the flags in APSR. We can add a condition for addition such as ADDNE R0, R1, R2. There is a NE in the mnemonic. Hence, the instruction will only be executed if the Z flag is equal to 0.

Mnemonic and Operands	Flags	Brief Description
ADC{S} {Rd,} Rn, op2	N, Z, C, V	Add with carry
ADD{S} {Rd,} Rn, op2	N, Z, C, V	Add
ASR{S} Rd, Rm, n	N, Z, C	Arithmetic shift right n times
RSB{S} {Rd,} Rn, op2	N, Z, C, V	Reverse subtract
SBC{S} {Rd,} Rn, op2	N, Z, C, V	Subtract with carry
SUB{S} {Rd,} Rn, op2	N, Z, C, V	Subtract
MLA Rd, Rn, Rm, Ra	-	Multiply with accumulate, 32-bit result
MLS Rd, Rn, Rm, Ra	-	Multiply and subtract, 32-bit result
MUL{S} {Rd,} Rn, Rm	N, Z	Multiply, 32-bit result
SDIV {Rd,} Rn, Rm	-	Signed divide
UDIV {Rd,} Rn, Rm	-	Unsigned divide

TABLE 4.6 Arithmetic Instructions

Mnemonic and Operands	Flags	Brief Description
AND{S} {Rd,} Rn, op2	N, Z, C	Logical AND
BIC{S} {Rd,} Rn, op2	N, Z, C	Bit clear
EOR{S} {Rd,} Rn, op2	N, Z, C	Exclusive OR
LSL{S} Rd, Rm, n	N, Z, C	Logical shift left n times
LSR{S} Rd, Rm, n	N, Z, C	Logical shift right n times
ORN{S} {Rd,} Rn, op2	N, Z, C	Logical OR NOT
ORR{S} {Rd,} Rn, op2	N, Z, C	Logical OR
TEQ Rn, op2	N, Z, C	Test equivalence
TST Rn, op2	N, Z, C	Test

TABLE 4.7 Instructions for Logic Operations

Otherwise, it will be ignored. Here, flags in APRS register will not be altered. Finally, we can use the instruction ADDNES R0, R1, R2. The addition operation will be done if the Z flag is equal to 0. The APSR flags will be set accordingly. If Z is equal to 1, the addition operation will be skipped and the APSR flags remain unaltered.

4.3.4.2 Instructions for Logic Operations
There are a total of nine instructions in this category. These are provided in Table 4.7. The usage of these instructions alter the APSR flags only if S is specified after the mnemonic. Also, all condition codes can be applied to these instructions. We will see the usage of these instructions in Sec. 4.4.

4.3.4.3 Saturating Arithmetic Instructions
Arithmetic operations may result in an overflow. In such a case, the result will be totally different than the one expected. There are 20 specific instructions for this purpose that saturate the result instead of an overflow. We provide all saturation arithmetic instructions in Table 4.8. The n parameter in this table can be a constant that specifies the bit

Mnemonic and Operands	Flags	Brief Description
`SSAT Rd, #n, Rm {, shift #s}`	Q	Signed saturate
`SSAT16 Rd, #n, Rm`	Q	Signed saturate half word
`USAT Rd, #n, Rm {, shift #s}`	Q	Unsigned saturate
`USAT16 Rd, #n, Rm`	Q	Unsigned saturate half word
`QADD {Rd}, Rn, Rm`	Q	Saturating add
`QADD8 {Rd}, Rn, Rm`	-	Saturating add 8
`QADD16 {Rd}, Rn, Rm`	-	Saturating add 16
`QSUB {Rd}, Rn, Rm`	Q	Saturating subtract
`QSUB8 {Rd}, Rn, Rm`	-	Saturating subtract 8
`QSUB16 {Rd}, Rn, Rm`	-	Saturating subtract 16
`QASX {Rd}, Rm, Rn`	-	Saturating add and subtract with exchange
`QSAX {Rd}, Rm, Rn`	-	Saturating subtract and add with exchange
`QDADD {Rd}, Rm, Rn`	Q	Saturating double and add
`QDSUB {Rd}, Rm, Rn`	Q	Saturating double and subtract
`UQADD16 {Rd}, Rn, Rm`	-	Unsigned saturating add 16
`UQADD8 {Rd}, Rn, Rm`	-	Unsigned saturating add 8
`UQASX {Rd}, Rm, Rn`	-	Unsigned saturating add and subtract with exchange
`UQSAX {Rd}, Rm, Rn`	-	Unsigned saturating subtract and add with exchange
`UQSUB16 {Rd}, Rn, Rm`	-	Unsigned saturating subtract 16
`UQSUB8 {Rd}, Rn, Rm`	-	Unsigned saturating subtract 8

TABLE 4.8 Saturating Arithmetic Instructions

position to saturate. The usage of saturating arithmetic instructions alter the APSR Q flag. Also, all condition codes can be applied to these instructions. We will see the usage of these instructions in Sec. 4.4.

4.3.5 Branch and Control Instructions

Branch and control instructions can be used to redirect the program execution flow based on condition codes. In other words, these instructions alter the PC value. We provide the nine branch and control instructions in Table 4.9. Usage of these instructions won't alter the APSR flags. We will see the usage of these instructions in Sec. 4.4.

4.3.6 Floating-Point Instructions

The ARM® Cortex™ M4F processor has an FPU. Hence, floating-point operations can be done in hardware. There are 43 such instructions, given in Table 4.10, to perform these operations. In this table, `Sd` is the destination register for a single word operation; `Sn` is the single-precision floating-point register; `Sm` is the operand floating-point value; `Dd` is the destination register for a double word operation; `list` is a list of extension registers to be loaded or stored. The FPU should be enabled to use these instructions.

Mnemonic and Operands	Brief Description
B label	Branch
BL label	Branch with link
BLX Rm	Branch indirect with link
BX Rm	Branch indirect
CBNZ Rn, label	Compare and branch if nonzero
CBZ Rn, label	Compare and branch if zero
IT	If-then
TBB [Rn, Rm]	Table branch byte
TBH [Rn, Rm, LSL #1]	Table branch half word

TABLE 4.9 Branch and Control Instructions

Mnemonic and Operands	Mnemonic and Operands	Mnemonic and Operands
VABS Sd, Sm	VFMS {Sd,} Sn, Sm	VNMLA {Sd,} Sn, Sm
VADD {Sd,} Sn, Sm	VFNMS {Sd,} Sn, Sm	VNMLS {Sd,} Sn, Sm
VCMP Sd, Sm	VLDM Rn{!}, list	VNMUL {Sd,} Sn, Sm
VCMPE Sd, Sm	VLDR Dd| Sd, [Rn]	VPOP list
VCVT Sd, Sm	VLMA {Sd,} Sn, Sm	VPUSH list
VCVTR Sd, Sm	VLMS {Sd,} Sn, Sm	VSQRT Sd, Sm
VCVTB Sd, Sm	VMOV Dd| Sd, op2	VSTM Rn{!}, list
VCVTT Sd, Sm	VMRS Rt, FPSCR	VSTR Sd, [Rn]
VDIV {Sd,} Sn, Sm	VMSR FPSCR, Rt	VSUB {Sd,} Sn, Sm
VFMA {Sd,} Sn, Sm	VMUL {Sd,} Sn, Sm	
VFNMA {Sd,} Sn, Sm	VNEG Sd, Sm	

TABLE 4.10 All Floating-Point Instructions

Mnemonic and Operands	Brief Description
BFC	Bit field clear
BFI	Bit field insert
SBFX	Signed bit field extract
UBFX	Unsigned bit field extract

TABLE 4.11 Bit Field Instructions

4.3.7 Other Instructions

There are other instructions to be listed besides the ones introduced in previous sections. The first set of these instructions consists of four-bit field operations as in Table 4.11. These instructions are used to extract or clear adjacent bits in registers or bit fields. They do not change condition flags. However, all condition codes can be applied to these instructions.

Mnemonic and Operands	Brief Description
PKH {Rd}, Rn, Rm	Pack half word
SXTAB {Rd,} Rn, Rm	Extend 8 bits to 32 and add
SXTAB16 {Rd,} Rn, Rm	Dual extend 8 bits to 16 and add
SXTAH {Rd,} Rn, Rm	Extend 16 bits to 32 and add
SXTB {Rd,} Rm	Sign extend a byte
SXTB16 {Rd,} Rm	Dual extend 8 bits to 16 and add
SXTH {Rd,} Rm	Sign extend a half word
UXTAB {Rd,} Rn, Rm	Extend 8 bits to 32 and add
UXTAB16 {Rd,} Rn, Rm	Dual extend 8 bits to 16 and add
UXTAH {Rd,} Rn, Rm	Extend 16 bits to 32 and add
UXTB {Rd,} Rm	Zero extend a byte
UXTB16 {Rd,} Rm	Dual zero extend 8 bits to 16 and add
UXTH {Rd,} Rm	Zero extend a half word

TABLE 4.12 Packing and Unpacking Instructions

Mnemonic and Operands	Brief Description
BKPT #imm	Breakpoint
CPSID iflags	Change processor state, disable interrupts
CPSIE iflags	Change processor state, enable interrupts
DMB	Data memory barrier
DSB	Data synchronization barrier
ISB	Instruction synchronization barrier
MRS Rd, spec_reg	Move from special register to register
MSR spec_reg, Rn	Move from register to special register
NOP	No operation
SEV	Send event
SVC #imm	Supervisor call
WFE	Wait for event
WFI	Wait for interrupt
SEL	Select bytes

TABLE 4.13 Miscellaneous Instructions

The next set of instructions are for packing and unpacking sign and extend bits in registers or bit fields. There are 13 such instructions as listed in Table 4.12. These instructions do not change condition flags. However, all condition codes can be applied to these instructions.

We can group the remaining 14 instructions under the miscellaneous category as presented in Table 4.13. These instructions can be used in advanced operations. We will not deal with them in this chapter. To note here, the NOP instruction can be used to add extra delay in some parts of an assembly program. This may be very valuable when adding a breakpoint to a line in the assembly code.

4.4 Sample Programs on Instruction Set Usage

We next provide sample assembly programs to explain how to use the instructions introduced in the previous section. The reader should follow the steps in Sec. 2.1.6 to form an assembly project for each program. Moreover, we suggest executing each program step by step. Hence, the result of each operation can be observed in CCS. Therefore, the Assembly Step Into option in the Debug view should be used.

4.4.1 Data Processing and Memory Access Operations

We first demonstrate the usage of data processing and memory access instructions in Listing 4.2. Here, we first assign three integers to registers R0, R1, and R2. Then, we apply memory access operations including PUSH and POP instructions. Finally, we apply the bit-banding operation to access a single bit of a memory location. This way, we change single bit of the memory content at address 0x20000040. The reader should observe the effect of memory operations by using the Memory Browser window.

Listing 4.2 Usage of Data Processing and Memory Access Instructions

```
 .text
 .align   2

MemAddr .field 0x20000040; Memory address
BitBandAddr .field 0x22000810; Bit-band address

 .global main

main:

;Data moving instructions
 MOV R0, #09h; R0 = 0x09
 MOV R1, #11d; R1 = 0x0B
 MOV R2, #00001010b; R1 = 0x0A

;Memory access instructions
 LDR R3, MemAddr;
 STR R2, [R3];
 STR R1, [R3, #4]!;
 LDR R0, [R3];

 PUSH {R3}
 POP {R4}

;Bit-band operation
 LDR R5, BitBandAddr;
 LDR R6, [R5]
 MOV R6, #1
 STR R6, [R5]

;Forever loop
 B $

 .end
```

4.4.2 Arithmetic and Logic Operations

We apply arithmetic operations to the content of registers R0, R1, and R2 in Listing 4.3. Therefore, we first assign three integers to the mentioned registers. Then, we apply arithmetic operations on them and save the results back to registers R3, R4, R5, and R6. Similarly, we apply logic operations to the content of registers R0, R1, and R2 in Listing 4.3. Thus we reassign three integers to these registers. Then, we apply logic operations on them and save the results back to registers R3, R4, R5, R6, and R7. Finally, we apply saturating arithmetic operations to the content of registers R0 and R1. To demonstrate the

Listing 4.3 Usage of Arithmetic and Logic Instructions

```
.text
.align   2
.global main

main:

;Arithmetic operation instructions
MOV R0, #09h; R0 = 0x09
MOV R1, #11d; R1 = 0x0B
MOV R2, #00001010b; R1 = 0x0A

ADD R3, R0, R1; R3 = R0 + R1 = 0x14
SUB R4, R2, #0002h; R4 = R2 - 0x02 = 0x08
MOV R5, R2; R5 = R2
MUL R5, R0; R5 = R5 * R0 = 0x5A
SDIV R6, R5, R1; R6 = R5 / R1 = 0x08

;Logic operation instructions
MOV R0, #09h; R0 = 0x09
MOV R1, #11d; R1 = 0x0B
MOV R2, #00001010b; R1 = 0x0A

AND R3, R0, R1; R3 = R0 + R1 = 0x14
ORR R4, R1, #0002h; R4 = R2 - 0x02 = 0x08
EOR R5, R0, R2; R5 = R2
LSL R6, R0, #2; R5 = R5 * R0 = 0x5A
LSR R7, R0, #2; R5 = R5 * R0 = 0x5A

;Saturating arithmetic instructions
MOV  R0, #0FFFFh;  R0 = 0x0000FFFF
MOVT R0, #7FFFh;   R0 = 0x7FFFFFFF
MOV  R1, #0FFFFh;  R1 = 0x0000FFFF
MOVT R1, #7FFFh;   R1 = 0x7FFFFFFF

ADDS R2, R1, R0;  R2 = 0xFFFFFFFE, overflow
QADD R3, R1, R0;  R3 = 0x7FFFFFFF, saturation

;Forever loop
B $

.end
```

difference between saturating and overflow operations, we both use saturating and non-saturating instructions. We load registers with 0x7FFFFFFF which is the largest positive two's complement integer that can be represented in 32-bit arithmetic. Therefore, using non-saturating ADD instruction triggers the overflow flag V, but not the unsigned overflow (or carry) flag. The result, 0xFFFFFFFE, is correct if interpreted as an unsigned quantity, but represents a negative value (-2) if interpreted as a signed quantity. On the other hand, QADD triggers the saturation flag and the result saturates to the largest positive two's complement integer, 0x7FFFFFFF.

4.4.3 Branch and Control Structures

Control structures are not explicitly defined in assembly language. Therefore, we provided several examples on forming C like control structures using assembly instructions in Listing 4.4. In all these examples, branch and control operations and condition codes are mandatory. To note here, these are not the only control structures in assembly language. The reader can form his or her structure using different branch instructions.

Listing 4.4 Branch and Control Structures in Assembly Language

```
 .text
 .align  2
 .global main

main:

;if else
 MOV R4, #05h; R4 = 0x05
 CMPS R4, #06h;
 BHS Greater; Branch if higher or same
 ADD R4, #-1; R4 = R4 - 1
 B Done1
Greater:
 ADD R4, #1; R4 = R4 + 1
Done1:

;for
 MOV R4, #0Ah; R4 = 0x0A
Loop1:
 ADDS R4, #-1; R4 = R4 - 1
 BPL Loop1

;while
 MOV R4, #0006h; R4 = 0x06
 MOV R5, #0002h; R5 = 0x02
Loop2:
 ADDS R4, #-1; R4 = R4 - 1
 CMPS R4, R5;
 BNE Loop2

;forever loop
 B $

 .end
```

We construct an if-else loop in the first part of code in Listing 4.4. This is done by first loading register R4 and comparing it with an immediate value. To alter the APSR bits, the compare instruction CMP with S option is used. By using the B instruction with HS condition code, the PC is changed only if the APSR C bit is one and program jumps to label Greater. Otherwise, PC stays unchanged. The second part of the code provides a for loop like structure. Here, register R4 is used as loop counter. It is summed with an immediate value, -1, using the ADDS instruction. Since the S option is used, APSR bits are changed according to the result of the operation. If the negative flag N is zero, then the B instruction with PL condition code moves the PC to Loop1 label. Otherwise, it is ignored. The third part of the code in Listing 4.4 forms a structure similar to a while loop. Here, the register R4 is used as loop variable and register R5 is used as compare value. At the end of the loop, R4 is compared with R5 using the CMPS instruction. Since the S option is used, the APSR bits are modified according to comparison. If the equal flag is zero, then B instruction with NE condition code moves PC to label Loop2. Otherwise, it is ignored. The last part is a forever loop similar to while(1) loop used in C language. Here, the branch unconditionally instruction (B) is used. $ represents the generic label of the current memory address. Hence, the branch operation loops on itself.

We provide an example on the usage of subroutine (function) in Listing 4.5. Here, the content of registers R5, R6, and R7 are swapped. This operation is done in the user-defined assembly subroutine.

4.4.4 Usage of Other Instructions

We finally consider the usage of instructions discussed in Sec. 4.3.7 in Listing 4.6. Here, we first demonstrate bit field instructions on the content of register R6. Bits from 8 to 20 are cleared by using the BFC instruction. Then, the bits from 4 to 8 of register R6 are extracted and saved to register R7. Afterwards, we apply the packing and unpacking operations to the content of registers R8 and R9. Therefore, we first assign two integers

---

Listing 4.5 Usage of Subroutine in Assembly Language

```
.text
.align   2
.global main

main:
MOV R5, #05h; R5 = 0x05
MOV R6, #06h; R6 = 0x06
MOV R7, #07h; R7 = 0x07
BL Replace;

B $;

Replace:
MOV R4, R7
MOV R7, R6
MOV R6, R5
MOV R5, R4
BX   LR

.end
```

Listing 4.6 Usage of Other Instructions in Assembly Language

```
    .text
    .align  2
    .global main

main:

;Bitfield instructions
    MOV R6, #0A12Fh; R6 = 0xA12F
    BFC R6, #8, #12; R6 = 0x2F
    SBFX R7, R6, #4, #8; R7 = 0x2

;Packing and unpacking instructions
    MOV R8, #0AABBh; R8 = 0xAABB
    MOV R9, #0CCDDh; R9 = 0xCCDD

    PKHBT R10, R8, R9, LSL #16; R9 = 0xCCDDAABB
    SXTH R11, R10, ROR #16; R10 = 0xFFFFCCDD

;Misc instructions
    NOP
    DMB
    DSB
    MRS R12, APSR
    NOP

    .end
```

Listing 4.7 Usage of the Function __asm() to Change the Value of the Register R3

```c
#include "msp.h"

void main(void) {

    WDT_A->CTL = WDT_A_CTL_PW | WDT_A_CTL_HOLD;

    __asm(" MOV R3, #0ABCh");

    while(1);
}
```

to these registers. Then, we pack half words of these registers and save the result to register R10. Before the packing operation, register R9 is left shifted by 16 bits. Afterwards, we apply the sign extend operation to register R10. Finally, we demonstrate no operation, data memory barrier, data synchronization barrier, and reading of a special register using NOP, DMB, DSB, and MRS instructions, respectively.

4.5 Inline Assembly

CCS allows adding assembly instructions to a C program. This can be done by the intrinsic function __asm(). Then, the assembly code line can be executed. This operation is called inline assembly since the assembly instruction is executed within the C code.

MSP432 Programming with Assembly

Listing 4.8 Usage of the Function __asm() to Calculate the Sum of Two Values in Different Registers

```
#include "msp.h"

void main(void){

WDT_A->CTL = WDT_A_CTL_PW | WDT_A_CTL_HOLD;

__asm(" MOV R0, #0003h \n"
      " MOV R1, #0005h \n"
      " ADD R2, R0, R1");

while(1);
}
```

Listing 4.9 Usage of the Function __asm() to Calculate the Sum of Two Values in Different Memory Addresses

```
#include "msp.h"

unsigned int a;
int *mem_loc;

void main(void){

WDT_A->CTL = WDT_A_CTL_PW | WDT_A_CTL_HOLD;

mem_loc = (int *) 0x20000040;

*(mem_loc)   = 0x0001;
*(mem_loc+1) = 0x0002;

__asm(" MOV R7, #0040h \n"
      " MOVT R7, #2000h \n"
      " LDR R0, [R7], #4 \n"
      " LDR R1, [R7], #4 \n"
      " ADD R2, R1, R0 \n"
      " STR R2, [R7], #4 \n");

a=*(mem_loc+2);

while(1);
}
```

We provide a sample program on the usage of this operation to alter the value of register R3 within the C code in Listing 4.7.

If more than one assembly code line is needed, CCS asks for a slightly different structure. Here, each assembly instruction should end with the new line symbol \n. We provide such an example in Listing 4.8. Here, assembly instructions are used to calculate the sum of two values in different registers.

One major problem remains. How can we use C variables in assembly instructions? Unfortunately, the solution of this problem is not straightforward. Besides, it depends on the compiler used. Our simple solution to this problem is turning back to the roots of variable assignment. As mentioned in Chap. 3, each variable in fact represents a memory address. Therefore, we can benefit from the pointer usage in Sec. 3.9 to operate on C variables within the assembly code indirectly. We provide such an example in Listing 4.9. Here, assembly instructions are used to calculate the sum of two values in different memory addresses. The result is written to another memory address that is associated with the C variable a.

4.6 Summary

Although the C language can be used in most operations in MSP432, assembly programming gives insight on the basics of the microcontroller. Therefore, we considered the instruction set and basic assembly language programming in this chapter. To do so, we started with the core registers available in MSP432. Then, we briefly analyzed the anatomy of an assembly program. Afterwards, we listed the instruction set of MSP432. Here, we only focused on the subset of these, and direct the reader to references for the rest of instructions. This will be the only chapter on assembly language programming since C language is more appropriate for almost all operations to be done through the book. Therefore, we kindly ask the reader to consult related references for advanced assembly language usage

4.7 Exercises

4.1 We would like to add the odd and even entries (each having number of bits as word length) separately in 10 successive memory locations starting from 0x20000000. Write an assembly program to perform this operation.

4.2 50 hexadecimal 32-bit numbers are written to the RAM of the MSP432 microcontroller. The starting address is 0x20000000. Write an assembly program for the following operations.
    a. Your code should scan the memory location of all 50 numbers. If a memory entry has a value between 0x00000080 and 0x00001FFF, then save it starting from the memory location 0x20000100.
    b. Scan the saved numbers. If the value of a memory entry is odd, do nothing. If the value is even, divide it by two iteratively until the result becomes odd or zero. Save the result to the same memory address.
    c. Save the number of divisions (for each memory entry) starting from the memory address 0x20000200. If the initial memory entry was odd, the number of divisions will be zero by default.

4.3 Write an assembly program such that it scans memory locations between 0x20000000 and 0x20000100. If the entry of a memory location is a number in ASCII format, then it will be replaced by the ASCII character *.

4.4 Write an assembly program to calculate the sum of integer numbers 113, -15, 117, -9 using loop operations. Write the sum to an appropriate register at your will. Add a subroutine to your assembly program to calculate the average of the given four numbers.

4.5 Write an assembly program with below specifications:
   a. Fill memory locations between 0x20000000 and 0x200000FE with numbers 20, 20, 0, 0 in a periodic manner.
   b. Take the average of the numbers in successive memory locations within the given memory range and write the result to the first address. As an example, take the average of numbers at 0x20000000 and 0x20000004. Write the result to the memory address 0x20000000.
   c. For three successive memory locations (within the given memory range) multiply the value in the first memory address with 1, the second with 2, and the third with 1. Sum these values, divide the result by four and write it to the first address.

   This operation is similar to filtering in digital signal processing. What do you observe when you check the output of the program?

4.6 Write a complete assembly program such that:
   a. you will write your birth year to an appropriate memory location.
   b. You will multiply this number by two in a loop structure until an overflow occurs.
   c. The number of multiplications until an overflow occurs will be stored in another appropriate memory location. Check whether an overflow occurs from the status register.

4.7 Write a complete assembly program such that:
   a. it will contain an array (starting from memory address 0x20000000) that will be filled with the ASCII representation of the characters in the string micro.
   b. Write this array in reverse order (orcim) to the memory addresses starting from 0x20000020 by using the "offset with pre-indexing" and a loop structure.
   c. The first array should not change during this operation.

4.8 Write a program in assembly language such that:
   a. two binary numbers are saved in two separate memory locations.
   b. as the numbers are added, no overflow occurs.
   c. as the numbers are added, overflow occurs.
   d. as the numbers are added, saturation occurs. Check whether an overflow or saturation occurs from the status register.

4.9 Repeat Problem 4.8 by using the subtraction operation.

4.10 Write a complete assembly program such that:
   a. you will write your birth year to an appropriate memory location.
   b. subtract 1024 from your birth year using only XOR operations.
   c. subtract 1024 from your birth year using bit-banding.

4.11 Write an assembly program with the following specifications.
   a. It contains a subroutine that calculates the square of the number in register R6 and saves result to R7.
   b. Calculate the square of numbers in five successive memory locations (starting from memory address 0x20000000) using your subroutine.
   c. Save the results to five successive memory locations (starting from memory address 0x20000020).

Listing 4.10 The Assembly Code for Problem 4.13

```
.text
.align   2
.global main

main:
 MOV R4, #0B1h;
 MOV R5, R4;
 ADD R5, R4, #16;
 EOR R6, R5, #27h;
 LSL R7, R4, #4;
 MUL R6, R7, R6;
 UDIV R6, R6, R4;
 B $

.end
```

4.12 Write a program in assembly language to calculate the first 20 elements of the Fibonacci series.
   a. The user only provides the first two entries of the series.
   b. The rest will be calculated by the program.
   c. Use appropriate memory locations.
   d. The numbers can be represented in hexadecimal form.

4.13 What will the register values R4, R5, R6, and R7 be as the program in Listing 4.10 is run?

4.14 Write a C program with the following specifications:
   a. It contains a function that calculates the square of the given number. This function will be written in in-line assembly.
   b. Calculate the square of numbers from 1 to 10. Save the results in an array.

CHAPTER 5

Digital Input and Output

A microcontroller interacts with the outside world through its input and output ports. This interaction can be in either analog or digital form. The latter is easy to implement and understand. Therefore, we will focus on the digital input and output (digital I/O) characteristics of the MSP432 microcontroller in this chapter. To do so, we will first introduce methods to reach ports and pins of the microcontroller. Here, we will evaluate register-level, DriverLib, and Energia-based methods. Then, we will use these in practical applications. Finally, we will provide a comprehensive example on digital I/O concepts.

5.1 Pin Layout for Digital I/O

Digital input and output is the simplest form of communication between the microcontroller and outside world. The input or output is either logic level 0 or 1 in this form. In other words, the input or output is either ground (0 V) or V_{CC} (3.3 V).

The MSP432 microcontroller I/O pins are arranged into 11 ports called P1 to P10 and PJ. These are generally called Px. Ports P1 to P9 have eight pins. These are called P1.0–P1.7 to P9.0–P9.7. Ports P10 and PJ have six pins. These are called P10.0–P10.5 and PJ.0–PJ.5. All of these pins can be used for digital input and output.

We cannot reach all pins of the MSP432 microcontroller on the MSP432 LaunchPad. We provide available pins in Table 5.1. The MSP432 LaunchPad also has wired pins as follows. Pin P1.0 is used for the onboard red LED (labeled as LED1 on the board). Pins P2.0, P2.1, and P2.2 are used for onboard RGB LED (commonly labeled as LED2 on the board). These four pins can be used for other purposes by removing their jumper. Pins P1.1 and P1.4 are used for onboard push buttons (labeled S1 and S2 on the board). These pins cannot be used for other purposes. Besides, pins P1.2 and P1.3 are reserved for the UART communication between the MSP432 LaunchPad and host PC. The reader can observe all the mentioned pins in Fig. 1.3.

We diagram the basic hardware for a generic pin Px.y in Fig. 5.1. This diagram serves two purposes. First, it gives insight on the circuit layout available on the pin. Second and more importantly, it associates digital I/O registers with the hardware. Next, we will focus on these registers.

Port Name	Available Pins
Port P1	P1.5–P1.7
Port P2	P2.3–P2.7
Port P3	P3.0, P3.2–P3.3, P3.5–P3.7
Port P4	P4.0–P4.7
Port P5	P5.0–P5.7
Port P6	P6.0–P6.7
Port P7	P7.0–P7.7
Port P8	P8.0, P8.2–P8.7
Port P9	P9.0–P9.7
Port P10	P10.0–P10.5

TABLE 5.1 Pins Available on the MSP432 LaunchPad

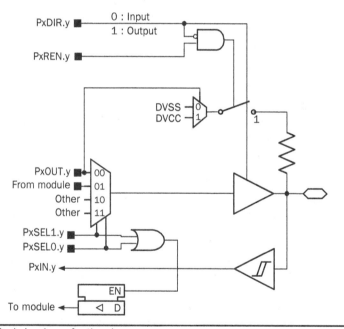

FIGURE 5.1 Basic hardware for the pins.

5.2 Digital I/O Registers

There are eight 8-bit registers available to control the basic digital I/O operations of each port. These are named PxDIR, PxIN, PxOUT, PxDS, PxREN, PxSEL1, PxSEL2, and PxSELC for a generic port Px. We will not use PxDS and PxSELC registers in this book. For further information on these, please see [4].

Each port register will be taken as a pointer for structure as explained in Sec. 3.8. Structure elements related to the registers are DIR, IN, OUT, REN, SEL1, and SEL2. We will use these in modifying and reaching a specific register for a given port.

5.2.1 Direction Register

Each pin in a port can be set either as input or output. This is done by the register PxDIR. To set a specific pin as input, the corresponding bit in this register should be reset (to 0). In a similar manner, to set a specific pin as output, the corresponding bit in the PxDIR register should be set (to 1). This can be done in two ways. Either we can apply bit-banding (as explained in Sec. 4.3.3) to reach a specific bit or we can take all bits in the direction register in the operation. In fact, this is applicable for all port registers. Throughout the book, we will apply the second method extensively. As an example, let's say that we want to assign the first pin of the first port (P1.0) as output. We would also like to assign the second pin of the same port (P1.1) as input. To do so, we should assign the hexadecimal number 0x01 to P1DIR. Therefore, the code line P1->DIR = 0x01 does the job. Now, the input can be connected to pin P1.1 and output can be connected to pin P1.0.

5.2.2 Input Register

The pin P1.1 is set as input in the previous section. The P1IN register should be checked to read values from this pin. In general, we will call this register PxIN. The digital input fed to the microcontroller is directly observed from this register. We have two options to observe a specific pin in this register. Either we can apply bit-banding or we can take an indirect approach by applying a binary mask to the register to extract the desired input value. We only need the value of the pin P1.1 for our example. In binary mask-based operation, we should apply an AND operation between the P1IN register and the hexadecimal mask 0x02. This will be done by the code line P1->IN & 0x02. While processing digital input values, please also take into account the active high and low settings (to be discussed in Sec. 5.5).

5.2.3 Output Register

To feed output value to pin P1.0, the P1OUT register is used. In general, we will call this register as PxOUT. If we want to feed 0 V to output from a specific pin, the corresponding bit value in PxOUT should be reset (to 0). We should set the pin (to 1) to feed V_{CC} to output. Again, we can apply bit-banding or the byte-based operation. As an example to the byte-based operation, we should assign 0x01 to P1OUT to feed V_{CC} to output from pin P1.0. The corresponding code line will be P1->OUT = 0x01.

5.2.4 Pull-Up/Down Resistor Register

Each pin in the ports of the MSP432 microcontroller has a pull-up/down resistor. These are controlled by the PxREN register. These resistors are disabled by default. To enable a resistor connected to a specific pin, the corresponding bit in PxREN should be set (to 1). As an example, the code line P1->REN = 0x02 should be used to enable the pull-up/down resistor of the pin P1.1. After the pull-up/down resistor is enabled, the selection between the pull-up and pull-down option is done with the PxOUT register. If the related bit of the PxOUT register is set, the internal resistor will be used as pull-up. Otherwise, it will be used as pull-down.

Application	PxSEL	PxSEL2
Digital I/O	0	0
Primary module function	1	0
Secondary module function	0	1
Tertiary module function	1	1

TABLE 5.2 Application Type–Based PxSEL and PxSEL2 Settings

5.2.5 Select Register

MSP432 pins can be used for more than one purpose. The registers PxSEL and PxSEL2 are for selecting the usage area of the ports. If a specific pin will be used for digital I/O, the corresponding bits in PxSEL and PxSEL2 should be reset. If the pin will be used for a specific purpose other than digital I/O, then PxSEL and PxSEL2 should be set accordingly. We provide these settings for different application types, which will be explored in the following chapters, in Table 5.2.

5.3 Digital I/O via DriverLib Functions

MSP432 pins can also be controlled via DriverLib functions. The complete function list and related macro definitions can be found in [11]. We will only use a subset of these functions given below. Here, `selectedPort` represents the selected port in the form of `GPIO_PORT_Px`. `selectedPin` corresponds to the selected pin in the selected port in the form of `GPIO_PINx`. More than one pin can be used in a function by combining them using the logical OR operation.

```
void GPIO_setAsOutputPin (uint_fast8_t selectedPort, uint_fast16_t
    selectedPins)
/*
Configures the selected pin(s) direction as output by modifying PxDIR
    and PxSEL registers.
*/

void GPIO_setAsInputPin (uint_fast8_t selectedPort, uint_fast16_t
    selectedPins)
/*
Configures the selected pin(s) direction as input by modifying PxDIR,
    PxREN, and PxSEL registers.
*/

void GPIO_setAsInputPinWithPullDownResistor (uint_fast8_t selectedPort,
    uint_fast16_t selectedPins)
/*
Configures the selected pin(s) direction as input and enables the
    internal pull-down resistor by modifying PxDIR, PxREN, and PxSEL
    registers.
*/
```

```
void GPIO_setAsInputPinWithPullUpResistor (uint_fast8_t selectedPort,
    uint_fast16_t selectedPins)
/*
Configures the selected pin(s) direction as input and enables the
    internal pull-up resistor by modifying PxDIR, PxREN, and PxSEL
    registers.
*/

void GPIO_setOutputHighOnPin (uint_fast8_t selectedPort, uint_fast16_t
    selectedPins)
/*
Configures the selected output pin(s) to give logic level 1 by
    modifying the PxOUT register.
*/

void GPIO_setOutputLowOnPin (uint_fast8_t selectedPort, uint_fast16_t
    selectedPins)
/*
Configures the selected output pin(s) to give logic level 0 by
    modifying the PxOUT register.
*/

void GPIO_toggleOutputOnPin (uint_fast8_t selectedPort, uint_fast16_t
    selectedPins)
/*
Configures the selected output pin(s) to give logical NOT of the
    present value by modifying the PxOUT register.
*/

uint8_t GPIO_getInputPinValue (uint_fast8_t selectedPort, uint_fast16_t
    selectedPins)
/*
Reads the selected input pin(s) value through the PxIN register.
*/
```

5.4 Digital I/O via Energia Functions

Energia functions can also be used to control MSP432 pins. We provide related functions below. Here, pin represents the selected pin. More information on Energia functions can be found in [12].

```
pinMode(pin, mode)
/* Configures the selected pin direction as input or output.
mode: is the selected mode. It can be INPUT,  OUTPUT, INPUT_PULLUP, or
    INPUT_PULLDOWN.
*/

digitalWrite(pin, value)
/*
Configures the selected output pin(s) to give logic level 0 or 1.
value: is the logic level value. It can be LOW (0) or HIGH (1).
*/
```

74 Chapter Five

```
digitalRead(pin)
/*
Reads the selected input pin logic level value. It returns either LOW
    or HIGH.
*/
```

5.5 Digital I/O Hardware Issues

There are two major hardware issues to be dealt with when using digital I/O. The first is the definition and setup of active high/low input. The second is switch bouncing. We will explain them next.

5.5.1 Active High/Low Input

There are two setup options to use a push button in a digital circuit. In the first setup, the microcontroller gets V_{CC} (logic level 1) on its pin when the button is pressed. This is called active high input. In the second setup, the microcontroller gets 0 V (logic level 0) on its pin when the button is pressed. This is called active low input. To note here, active high or low inputs are not related to the microcontroller. They are based on the connection type to the digital I/O pin. The circuit diagrams for the active high/low input setup are as in Fig. 5.2.

The preferred setup is active low on the MSP432 LaunchPad. Therefore, when a button is pressed on the LaunchPad, it will generate logic level 0. When the button is released, it will give logic level 1. This should be taken into account while reading the value from the onboard button-connected pins.

5.5.2 Switch Bouncing

The second hardware issue on digital I/O is switch (button) bouncing. This problem may occur due to the following reason. If the button is pressed once, it may generate output more than once depending on its physical characteristics. Therefore, the microcontroller may see one input and its successive shadow versions. Either a software- or hardware-based solution can be used to eliminate this effect. In the software-based solution, a delay should be added to the button input reading part of the code. The input will not be

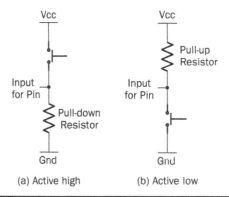

FIGURE 5.2 Active-high and low-input circuit diagrams for the push button.

observed and possible shadow inputs will be eliminated during this delay. Although the software-based solution is easier to implement, actual inputs will also be eliminated during delay. Therefore, it should be used with precaution. We provide sample codes on this issue in the following section.

There are also hardware-based solutions for switch bouncing. The most feasible circuitry is a low-pass RC filter (composed of a resistor and capacitor) followed by a Schmitt trigger for MSP432. This setup is given in Fig. 5.3. Each digital I/O pin of MSP432 has a Schmitt trigger [4]. The internal pull-up resistor available at the pin hardware can be used in forming the RC filter. The user should only connect an external capacitor to solve the switch-bouncing problem by hardware.

In Fig. 5.3, the low-pass RC filter is used to eliminate the high-frequency shadow inputs coming from the button. The remaining glitches are eliminated by the Schmitt trigger. In this circuitry, the time constant of the filter ($\tau = R \times C$) must be larger than the switch-bouncing time to eliminate all shadow inputs. The internal pull-up resistor values may vary between 20 and 40 kΩ. We will assume that a 2- to 4-millisecond time constant is enough to eliminate switch bouncing throughout the book. Therefore, a 100-ηF external capacitor should be used. The time constant of the filter can be adjusted by this capacitance value depending on other constraints.

5.6 Coding Practices for Digital I/O

The best way to understand digital I/O usage is through programming. Therefore, we provide sample codes (using registers directly, via DriverLib and Energia functions) related to digital I/O in this section. While doing this, we will take digital input values from the buttons S1 and S2 connected to pins P1.1 and P1.4 on the MSP432 LaunchPad, respectively. We will feed digital output to the LED1 and red, green colors of LED2 connected to pins P1.0, P2.0, and P2.1 on the MSP432 LaunchPad, respectively. Therefore, the MSP432 LaunchPad will be sufficient as a hardware platform in executing the sample codes.

5.6.1 Register-Level Examples

The first digital I/O example turns on LED1 when the button S1 is pressed. When the button S1 is released, LED1 turns off. We provide the register-level C code for this application in Listing 5.1.

In Listing 5.1, the pin directions for the LED1 and button S1 are set first. Initially, the LED1 is turned off. Also the pull-up resistor is enabled for P1.1 to the active-low

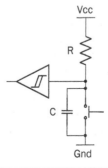

FIGURE 5.3 Hardware solution of the switch-bouncing problem.

Listing 5.1 Turning on LED1 When the Button S1 Is Pressed, Using Registers

```c
#include "msp.h"

#define LED1 BIT0
#define S1 BIT1

void main(void) {

  WDT_A->CTL = WDT_A_CTL_PW | WDT_A_CTL_HOLD;

  P1->DIR = LED1; // Set P1.0 as output and P1.1 as input
  P1->REN = S1; // Enable pull-up/down resistor for P1.1
  P1->OUT = S1; // Pull-up/down resistor is set as pull-up

  while(1){
   if((P1->IN & S1) == 0x00) // Active low input
      P1->OUT |= LED1; // Turn on the LED1
    else
      P1->OUT &= ~LED1; // Turn off the LED1
}}
```

configuration. Then, the input from the button S1 is checked in an infinite loop. Switch-bouncing problem is not taken into account since the LED1 turns on only during the button press. Finally, bitwise operations are used for turning on/off the LED1 so as not to change the initial setup of the pull-up resistor.

The next digital I/O example is more complex. Here, the red color of LED2 is toggled every time the button S1 is pressed. The green color of LED2 is toggled every time the button S2 is pressed. We provide the register-level C code for this application in Listing 5.2.

In Listing 5.2, the pin directions for the LED2, S1, and S2 are set first. Initially, LED2 is turned off. Then, inputs from buttons S1 and S2 are checked in an infinite loop. When one of the buttons is pressed, the code enters a for loop to eliminate the switch-bouncing problem. The code checks for the button condition after this delay. If the button is still pressed, then the related color of LED2 is toggled. Afterwards, the code waits in a while loop for the release of the button. This step ensures that LED2 cannot be toggled again unless the button is released.

5.6.2 DriverLib Examples

We can redo the application in Listing 5.1 using DriverLib functions. We provide the corresponding C code in Listing 5.3. As can be seen in this example, the DriverLib function usage simplified the code written for the same application compared to the register-level implementation. The DriverLib function `WDT_A_holdTimer()` in Listing 5.3 stops the watchdog timer, which we will discuss in Chap. 7. Till that chapter, please take this function as granted.

We next consider the application in Listing 5.2 using DriverLib functions. We provide the corresponding C code in Listing 5.4. Again, this version of the implementation is more compact compared to its register-level implementation.

Listing 5.2 Toggling the Red and Green Colors of LED2 with Buttons S1 and S2, Using Registers

```c
#include "msp.h"

#define LED2_RED BIT0
#define LED2_GREEN BIT1
#define S1 BIT1
#define S2 BIT4
#define DELAY 500
int i;

void main(void){

  WDT_A->CTL = WDT_A_CTL_PW | WDT_A_CTL_HOLD;

  P1->DIR &= ~(S1 | S2); // Set P1.1 and P1.4 as input
  P1->REN = S1 | S2; // Enable pull-up/down resistor for P1.1 and P1.4
  P1->OUT = S1 | S2; // Pull-up/down resistor is set as pull-up
  P2->DIR = LED2_RED | LED2_GREEN; // Set P2.0 and P2.1 as output
  P2->OUT = 0x00; // LED2 is turned off

  while(1){
   if((P1->IN & S1) == 0x00){ // Active low input
     for(i=0;i<DELAY;i++){} // Delay for debounce
     if((P1->IN & S1) == 0x00){
       P2->OUT ^= LED2_RED; // Toggle P2.0
       while((P1->IN & S1) == 0x00);}} // Wait until S1 is released
   if((P1->IN & S2) == 0x00){ // Active low input
    for(i=0;i<DELAY;i++){} // Delay for debounce
    if((P1->IN & S2) == 0x00){
      P2->OUT ^= LED2_GREEN; // Toggle P2.1
      while((P1->IN & S2) == 0x00);}} // Wait until S2 is released
}}
```

Listing 5.3 Turning on LED1 When the Button S1 Is Pressed, Using DriverLib Functions

```c
#include <ti/devices/msp432p4xx/driverlib/driverlib.h>

void main(void){

 WDT_A_holdTimer();

 GPIO_setAsOutputPin(GPIO_PORT_P1, GPIO_PIN0);
 GPIO_setAsInputPinWithPullUpResistor(GPIO_PORT_P1, GPIO_PIN1);

 while (1){
  if(GPIO_getInputPinValue(GPIO_PORT_P1, GPIO_PIN1) ==
      GPIO_INPUT_PIN_LOW)
   GPIO_setOutputHighOnPin(GPIO_PORT_P1, GPIO_PIN0);
  else
   GPIO_setOutputLowOnPin(GPIO_PORT_P1, GPIO_PIN0);
}}
```

Listing 5.4 Toggling the Red and Green Colors of LED2 with Buttons S1 and S2, Using DriverLib Functions

```c
#include <ti/devices/msp432p4xx/driverlib/driverlib.h>

#define DELAY 500
int i;

void main(void){

 WDT_A_holdTimer();

 GPIO_setAsOutputPin(GPIO_PORT_P2,GPIO_PIN0|GPIO_PIN1);
 GPIO_setOutputLowOnPin(GPIO_PORT_P2,GPIO_PIN0|GPIO_PIN1);
 GPIO_setAsInputPinWithPullUpResistor(GPIO_PORT_P1,GPIO_PIN1|
    GPIO_PIN4);

 while (1){
  if(GPIO_getInputPinValue(GPIO_PORT_P1,GPIO_PIN1)==GPIO_INPUT_PIN_LOW)
    {
   for(i=0;i<DELAY;i++);
    if(GPIO_getInputPinValue(GPIO_PORT_P1,GPIO_PIN1)==
      GPIO_INPUT_PIN_LOW){
    GPIO_toggleOutputOnPin(GPIO_PORT_P2, GPIO_PIN0);
    while(GPIO_getInputPinValue(GPIO_PORT_P1, GPIO_PIN1)==
      GPIO_INPUT_PIN_LOW);}}
  if(GPIO_getInputPinValue(GPIO_PORT_P1,GPIO_PIN4)==GPIO_INPUT_PIN_LOW)
    {
   for(i=0;i<DELAY;i++);
    if(GPIO_getInputPinValue(GPIO_PORT_P1,GPIO_PIN4)==
      GPIO_INPUT_PIN_LOW){
    GPIO_toggleOutputOnPin(GPIO_PORT_P2,GPIO_PIN1);
    while(GPIO_getInputPinValue(GPIO_PORT_P1,GPIO_PIN4)==
      GPIO_INPUT_PIN_LOW);}}
 }}
```

5.6.3 Energia Examples

We finally consider the applications in previous sections using Energia. Therefore, we first repeat the application in Listing 5.1 using Energia functions. We provide the corresponding code in Listing 5.5.

As can be seen in Listing 5.1, pin P1.0 is set as output by the code line `pinMode(YELLOW_LED, OUTPUT)`. The pin P1.1 is set as input with enabled internal pull-up resistor by the code line `pinMode(PUSH1, INPUT_PULLUP)`. Then, the button is checked in an infinite loop with the `digitalRead(PUSH1)` function. The LED1 output is set with `digitalWrite(YELLOW_LED, HIGH)` and `digitalWrite(YELLOW_LED, LOW)` functions. To note here, LED1 is called `YELLOW_LED` under Energia. As a follow up, the red, green, and blue colors of LED2 are called `RED_LED`, `GREEN_LED`, and `BLUE_LED`, respectively under Energia.

We can repeat the application in Listing 5.2 using Energia functions. We provide the corresponding code in Listing 5.6. As can be seen here, the representation via Energia functions is even simpler than the DriverLib function–based implementation.

Listing 5.5 Turning on LED1 When the Button S1 Is Pressed, Using Energia Functions

```
void setup(){
 pinMode(YELLOW_LED, OUTPUT);
 pinMode(PUSH1, INPUT_PULLUP);
}

void loop(){
 if (digitalRead(PUSH1) == LOW)
  digitalWrite(YELLOW_LED, HIGH);
 else
  digitalWrite(YELLOW_LED, LOW);
}
```

Listing 5.6 Toggling the Red and Green Colors of LED2 with Buttons S1 and S2, Using Energia Functions

```
boolean RED_LED_state = LOW;
boolean GREEN_LED_state = LOW;

void setup(){
 pinMode(RED_LED, OUTPUT);
 pinMode(GREEN_LED, OUTPUT);
 pinMode(PUSH1, INPUT_PULLUP);
 pinMode(PUSH2, INPUT_PULLUP);
}

void loop(){
 if (digitalRead(PUSH1) == LOW){
  delay(100);
  if (digitalRead(PUSH1) == LOW){
   RED_LED_state = !RED_LED_state;
   digitalWrite(RED_LED, RED_LED_state);
   while(digitalRead(PUSH1) == LOW);}}
 if (digitalRead(PUSH2) == LOW){
  delay(100);
  if (digitalRead(PUSH2) == LOW){
   GREEN_LED_state = !GREEN_LED_state;
   digitalWrite(GREEN_LED, GREEN_LED_state);
   while(digitalRead(PUSH2) == LOW);}}
}
```

5.7 Application: Home Alarm System

The aim in this application is learning how to use the digital I/O pins of the MSP432 microcontroller. As a real-life application, we will design a home alarm system. Our system will check the three windows and door of a house. In this section, we provide the equipment list, layout of the circuit, and system design specifications.

5.7.1 Equipment List

The equipment to be used in this application, besides the LEDs and button on the MSP432 LaunchPad, is a magnetic door switch. In fact, we will use four switches (for three windows and the door to be monitored) in the application.

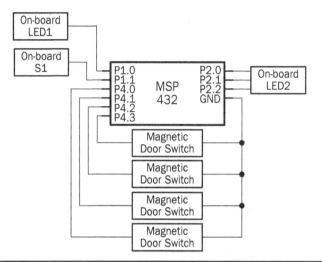

FIGURE 5.4 Layout of the home alarm system.

We can use magnetic door switches to implement the home alarm system. When one of the terminals of the magnetic door switch is connected to ground and a pull-up resistor is used at the other terminal, it provides output logic level 1 when the window or door is opened. Otherwise, output of the magnetic door switch will be at logic level 0. In our application, we should connect the reed relay and the magnet to each window and its frame. The same setting should also be done to the door and its frame. To be more precise, the reed part of the magnetic door switch should be connected to the frame. The magnet part of the magnetic door switch should be connected to the window or door. This setup can be used to detect whether a window or door is opened or closed.

5.7.2 Layout

The layout of the home alarm system is given in Fig. 5.4.

5.7.3 System Design Specifications

The design of the home alarm system will have three main blocks. These are listed below:

- **Block 1**: One magnetic door switch will be attached to home door. When the door is opened, the LED1 connected to pin P1.0 on the MSP432 LaunchPad will turn on. When the door is closed, the LED1 will turn off. Once the door open-close cycle is complete, a counter will be increased by one.
- **Block 2**: Three magnetic door switches will be attached to three windows. When one of the three windows are opened; the red, green, and blue colors of LED2 connected to pins P2.0, P2.1, and P2.2 on the MSP432 LaunchPad will turn on, respectively. When a window is closed afterwards, the LEDs will remain turned on.
- **Block 3**: When the button connected to pin P1.1 on the MSP432 LaunchPad is pressed, the alarm system will be reset. All LEDs will turn off, and the counter will be cleared.

We could not add the C code of the project here due to page limitations. However, we provided the complete CCS project for this application in the companion website for the reader. We strongly suggest implementing it.

5.8 Summary

A microcontroller interacts with the outside world through its ports. In this chapter, we focused on the digital input and output in MSP432. We first reviewed specific registers to setup digital I/O properties. Then, we explored digital I/O usage via predefined Driver-Lib and Energia functions. We also considered two important hardware issues related to push buttons as active high/low input and switch bouncing. Finally, we designed a real-life application (home alarm) using digital I/O. We provided all the hardware and software design specifications related to it. We hope that this will encourage the reader to develop new projects using digital I/O.

5.9 Exercises

5.1 Write a C program to find the R, G, and B letters in an array. If the letters are detected, the red, green, and blue colors of LED2 connected to pins P2.0, P2.1, and P2.2 on the MSP432 LaunchPad will turn on, respectively. Your program should count the number of occurrences of each letter in the array.

5.2 Write a C program to detect a specified pattern with three entries in an array. The pattern is saved in another array. When the pattern is detected, the green color of LED2 connected to pin P2.1 on the MSP432 LaunchPad will turn on. If the pattern is not detected, the red color of LED2 connected to pin P2.0 on the MSP432 LaunchPad will turn on at the end of the program. Your program should count the number of occurrences of the pattern.

5.3 We would like to add the odd and even integer entries separately in an array ARR with 10 values. Write a C program using array operations. If the sum of odd entries is greater than the sum of even entries, then the red color of LED2 connected to pin P2.0 on the MSP432 LaunchPad will turn on. Otherwise, the green color of LED2 connected to pin P2.1 on the MSP432 LaunchPad will turn on. You do not have to worry about overflow based on the values in the array.

5.4 Write a C program to calculate the root mean square of numbers in an array. If the result is less than the first predefined value, then the red color of LED2 connected to pin P2.0 on the MSP432 LaunchPad will turn on. If the result is between the first and second predefined values, then the green color of LED2 connected to pin P2.1 on the MSP432 LaunchPad will turn on. If the result is larger than the second predefined value, then the blue LED connected to pin P2.2 on the MSP432 LaunchPad will turn on.

5.5 We would like to turn on and off the red, green, and blue colors of LED2 in a sequence. Therefore, write a C program with the following specifications.
   a. When the button S1 connected to pin P1.1 on the MSP432 LaunchPad is pressed, turn on the red, green, and blue colors of LED2 connected to pins P2.0, P2.1, and P2.2 on the MSP432 LaunchPad with the following order, R-RG-RGB-GB-B-RB.
   b. This procedure is repeated until the button is pressed again. Hint: Use loop operations to generate delay times.

Chapter Five

5.6 We would like to form a color palette selector using the red, green, and blue colors of LED2 connected to pins P2.0, P2.1, and P2.2 on the MSP432 LaunchPad. Therefore, write a C program with the following specifications.
  a. When the button S1 connected to pin P1.1 on the MSP432 LaunchPad is pressed, a color will be picked from the red, green, and blue color palette.
  b. When the button S2 connected to pin P1.4 on the MSP432 LaunchPad is pressed, then the red, green, or blue colors of LED2 will be toggled based on the selected color.

5.7 We would like to develop a simple calculator with the following specifications.
  a. There will be three arrays each with 10 elements. The first two arrays will have integer numbers. The third array will be empty. We will apply arithmetic operations on the first two arrays.
  b. When the button S1 connected to pin P1.1 on the MSP432 LaunchPad is pressed once, a counter will be incremented once. This will correspond to the addition operation. The red color of LED2 connected to pin P2.0 on the MSP432 LaunchPad will turn on.
  c. When the button S1 is pressed twice, the counter will be incremented twice. This will correspond to the subtraction operation. The green color of LED2 connected to pin P2.1 on the MSP432 LaunchPad will turn on.
  d. When the button S1 is pressed three times, the counter will be incremented thrice. This will correspond to the multiplication operation. The blue color of LED2 connected to pin P2.2 on the MSP432 LaunchPad will turn on.
  e. When the button S2 connected to pin P1.4 on the MSP432 LaunchPad is pressed, the selected operation will be applied on the two elements of the first and second arrays. The result will be saved in the third array and all LEDs will turn off. The operation will be applied to all remaining array elements. When all the results are saved, the LED1 connected to P1.0 on the MSP432 LaunchPad will turn on.

CHAPTER 6

Exceptions and Interrupts

The usage of the MSP432 microcontroller was based on sequential code execution up to now. However, the microcontroller may need to respond to an external or internal signal by executing a code block. This can only be done by the help of exceptions. In other words, if the user wants to write a program to react to predefined actions, the only solution is using exceptions. This chapter will focus on interrupts as a subset of exceptions. Therefore, we will first provide the common and distinct features between an interrupt and exception. Then, we will briefly evaluate what happens when an interrupt occurs in the microcontroller. Afterwards, we will introduce the nested vectored interrupt controller (NVIC) module that is responsible for exception and interrupt handling in MSP432. Then, we will focus on port interrupts and sample codes related to them. Finally, we will provide a comprehensive example on interrupt concepts.

6.1 Interrupt as an Exception

When the MSP432 microcontroller works in normal conditions, we call it to be in thread mode. This was the case up to now. We had a program to be executed. The flow of the program was known and it was sequential. If another source changes the flow of the program, then we call that the microcontroller to be in handler mode. The general name for such a source to change the program flow is called exception in the ARM literature. The program block (or "function" more specifically) executed during this time is called the exception handler.

The source changing the program flow can be external or internal with respect to the CPU. The external source can be a signal originating from a port, ADC, digital communication, or another peripheral. This exception type is specifically called interrupt. The exception handler executed during this time is called interrupt service routine (ISR). Therefore, interrupt is in fact a subset of exception in the ARM literature. The exception is specifically used for program flow changes originating from the CPU. In other words, exception originates from internal sources such as faulty operations or RTOS-based implementation (to be explored in Chap. 13. Throughout the book, we will only focus on interrupt and RTOS-based exceptions. We direct the reader to Yiu's [13] excellent book on ARM Cortex-M3 and Cortex-M4 microcontrollers for exceptions originating from faulty operations.

6.2 Interrupts in Our Daily Lives

We experience interrupts in our daily lives. If we are working on a problem and another source distracts us from doing it, then we had an interrupt. For some cases, it is up to us to respond to this distraction. For others, we must respond to the interrupt.

We can explain the interrupt concept on an actual example. Assume that an instructor starts the class with a planned slide show to be presented to the students. The instructor may inform the students that they can ask questions during the class hour. Therefore, he or she allows a possible break from the usual course flow. If a student raises a hand and asks a question, the instructor stops the sequential flow of the course, answers the question, and resumes the flow afterwards.

In terms of interrupts, this scenario corresponds to the following operations. The instructor starts the class in thread mode allowing an interrupt to occur. The student interrupts the flow of the course by raising a hand (or let's call it flag). In response to this, the instructor goes to the handler mode to answer the question (ISR). The instructor asks the student to lower his or her hand (or flag). After the ISR is executed, as the student gets his or her answer, the flow of the course continues where it is left from.

The instructor may not allow any interrupts from students. This is called masking the interrupt. However, fire alarm may be activated (for a fire drill) during the class hour. In this case, the instructor must halt what he or she is doing and everyone should leave the class. Therefore, this interrupt should be served under any circumstances. This is called a non-maskable interrupt. After the fire drill is done, the course resumes where it was left from.

Let's further assume that more than one question is asked during the class hour. Then, the instructor should answer the question with higher priority. In other words, the instructor selects which interrupt to serve. Related to this scenario, while one question is being answered, or while in handler mode (ISR), another question may emerge. If the second question has higher priority than the one being handled, the instructor should stop answering the present question and turn his or her attention to the more important one. After responding to this question, the one with lower priority can be answered. This is called nested operation.

The interrupt service pattern is the same for the microcontroller. Let's summarize what happens when an interrupt occurs in the CPU. First of all, the interrupt request must come from an external source (such as port). The user should enable the interrupt option for the desired source to process the request. As the interrupt comes, the CPU stops what it is doing. The CPU stores necessary data to stack without executing the next instruction. It then jumps to the ISR, which is in fact a function to be processed in response to the interrupt. The user is responsible for the code block to be written in the ISR. The CPU recalls the stored data from the stack as the ISR is executed. Hence, it turns back to the main program and continues executing the next instruction there.

Let's analyze a simple example to clarify what happens when an interrupt occurs in the MSP432. Assume that we want to turn on LED1 (connected to pin P1.0 on the MSP432 LaunchPad) when the button S1 (connected to pin P1.1 on the MSP432 LaunchPad) is pressed. Since the aim is using the interrupt, it will originate from the button S1. Here, we assume that the port settings are done as in Chap. 5. Besides, the pin P1.1 should be able to generate an interrupt. Or, its interrupt generation property should be enabled, which is explained in Sec. 6.4 in detail. If we only want to turn on LED1 when the button S1 is pressed, the main program will just have an infinite loop. In other words, the CPU will wait in an infinite loop doing nothing. When the user pushes the button S1, an interrupt will be generated. The CPU will stop the infinite loop and will reach the associated ISR. As the programmer, it is our responsibility to write the code block in the ISR. Since our task is just to turn on LED1 when an interrupt occurs, the code in the ISR will just set the pin P1.0 to logic level 1. As this code block is executed, the CPU will turn back from the ISR and continue waiting in the infinite loop. We will explore all these steps in detail in the following sections.

6.3 The Nested Vectored Interrupt Controller

Exceptions and interrupts are handled by the nested vectored interrupt controller (NVIC) module in ARM Cortex-M4 processors. This module can be used to enable or disable a specific interrupt, assign a priority level to an interrupt, and allow nesting successive interrupts. The NVIC supports 64 interrupt inputs, with up to eight programmable priority levels in the MSP432 microcontroller. We can analyze the working principles of the NVIC under two scenarios. The first scenario is when a single interrupt request comes. The second scenario is when more than one interrupt request comes.

6.3.1 Serving a Single Interrupt Request

The first scenario for the NVIC is when a single interrupt request comes. The Cortex-M4 processor supports two types of interrupt requests as level-sensitive and pulse. A level-sensitive interrupt is asserted by the peripheral and held asserted until the peripheral de-asserts the interrupt signal. A pulse interrupt signal on the other hand is sampled synchronously on the rising edge of the processor clock. To ensure detecting the pulse interrupt, the signal width must be at least one clock cycle. Hence, the NVIC detects the pulse and latches the interrupt.

When an interrupt request is detected (an interrupt flag is set), the CPU first checks whether that interrupt type and in general all interrupts are enabled. If this is the case, then CPU performs two operations. First, necessary key register content is saved to the stack that will be recalled after serving the ISR. Second, the CPU checks for the address of the ISR to be executed in response to the corresponding interrupt request. For each interrupt type, the address of the ISR to be executed is saved in the vector table that associates the interrupt request type and the ISR to be executed. This information is kept in the `startup_msp432p401r_ccs.c` file which is automatically generated by each CCS project. We provide subpart of this file in Listing 6.1. As can be seen here, the first entry is the address of the stack pointer in the interrupt vector table. Then comes 10 exceptions related to system operations starting with the reset handler. Afterwards, the interrupts are listed. In the original file, there are a total of 42 interrupt request definitions. We only provided a few related to port interrupts to be evaluated in this chapter.

The time needed to reach the ISR is called interrupt latency. More specifically, the interrupt latency is defined as the number of clock cycles needed by the processor to stop the running main code (thread) and jump into the appropriate ISR. The interrupt latency on the Cortex-M4 processors is fixed to 12 clock cycles. This value is independent of the code executed in main loop and will be executed in ISR. Therefore, deterministic operations can be performed while using interrupts.

The ISR is in fact a C function. Associating it with the corresponding interrupt request can be done by using the name in the interrupt vector. As an example, if an ISR needs to be constructed for an interrupt originating from port P1, its ISR name in the main

Listing 6.1 Subpart of the Automatically Generated MSP432 Startup Code

```
/* Interrupt vector table.*/

void (* const interruptVectors[])(void) =
{

(void (*)(void))((uint32_t)&__STACK_END),
/* The initial stack pointer */
Reset_Handler,          /* The reset handler        */
NMI_Handler,            /* The NMI handler          */
HardFault_Handler,      /* The hard fault handler   */
MemManage_Handler,      /* The MPU fault handler    */
BusFault_Handler,       /* The bus fault handler    */
UsageFault_Handler,     /* The usage fault handler  */
0,                      /* Reserved                 */
0,                      /* Reserved                 */
0,                      /* Reserved                 */
0,                      /* Reserved                 */
SVC_Handler,            /* SVCall handler           */
DebugMon_Handler,       /* Debug monitor handler    */
0,                      /* Reserved                 */
PendSV_Handler,         /* The PendSV handler       */
SysTick_Handler,        /* The SysTick handler      */
PSS_IRQHandler,         /* PSS ISR                  */
    ...
    ...
    ...
PORT1_IRQHandler,       /* PORT1 ISR                */
PORT2_IRQHandler,       /* PORT2 ISR                */
PORT3_IRQHandler,       /* PORT3 ISR                */
PORT4_IRQHandler,       /* PORT4 ISR                */
PORT5_IRQHandler,       /* PORT5 ISR                */
PORT6_IRQHandler        /* PORT6 ISR                */
};
```

program is PORT1_IRQHandler as given in Listing 6.1. It is the programmer's responsibility to define what will be done in the ISR. After the ISR is executed, the corresponding interrupt flag should be reset so that the CPU will not evaluate the same interrupt request twice.

6.3.2 Serving Multiple Interrupt Requests

The second scenario for the NVIC is when multiple interrupt requests come. This scenario is more complex compared to the single interrupt case. The reason for this complexity comes from the interrupt priority definition and nested interrupt operations. Let's analyze these next.

When more than one interrupt request is present, the NVIC should decide which one to serve. This is done by the priority level of the interrupts (or exceptions in general). We provide system exceptions and interrupts with their priority level in Table 6.1. As can be seen in this table, the first three exceptions (Reset, Non-Maskable Interrupt, and Hard Fault) have the highest priority with negative values. The remaining exceptions and interrupts have programmable priority levels. Therefore, the user can set their priority level based on his or her needs. Here, the lower the priority level, the higher is

Exceptions and Interrupts

Exception Type	Vector Number	Priority
–	0	–
Reset	1	−3
Non-Maskable Interrupt (NMI)	2	−2
Hard Fault	3	−1
Memory Management	4	Programmable
Bus Fault	5	Programmable
Usage Fault	6	Programmable
–	7–10	Programmable
SVCall	11	Programmable
Debug Monitor	12	Programmable
–	13	Programmable
PendSV	14	Programmable
SysTick	15	Programmable
Interrupts	16 and above	Programmable

TABLE 6.1 Exceptions with Priority Levels

the corresponding interrupt priority. However, these priority levels cannot have a negative value. In other words, the priority level of a programmable exception or an interrupt cannot be higher than Reset, Non-Maskable Interrupt, or Hard Fault.

NVIC, as its name implies, supports nesting of interrupts. This means that while serving a low-priority interrupt, a high-priority interrupt request can pre-empt. Then, the execution of the low-priority interrupt (ISR) is suspended and the high-priority interrupt (ISR) is served. Low-priority ISR resumes when execution of the high-priority ISR ends.

When a present ISR has completed and if there is a second ISR waiting to be served, the CPU will switch to it immediately by skipping some of the unstacking and stacking operations that are normally needed. This is called tail chaining that lowers the interrupt latency to six cycles. This also makes the CPU much more energy efficient by avoiding unnecessary memory access operations.

If a high-priority interrupt request arrives during the stacking stage of a lower priority interrupt, the high-priority interrupt will always be serviced first. This is called late arrival that ensures high-priority interrupts are serviced quickly. It also avoids another level of stacking operation during the nested interrupt handling process. Additionally, this will save energy (due to less access to memory) and less stack space too.

If an interrupt request arrives just as another ISR is exiting and the unstacking process is underway, the unstacking sequence is stopped and the ISR for the new interrupt is entered as soon as possible. This is called pop pre-emption. Again, this avoids unnecessary unstacking and stacking that reduces power consumption and latency.

6.3.3 NVIC Registers

There are 27 registers available in the MSP432 microcontroller to control the NVIC module. These registers are ISERx, ICERx, ISPRx, ICPRx, IABRx, IPRy, and STIR. In these, x can get values 0 and 1; y can get values 0 to 15. A structure named NVIC is used to control all registers. We will only consider ISERx, ICERx, and IPRy registers in this book. They are named ISER[x], ICER[x], and IP[y] in the NVIC structure. In these, x can get values 0 and 1; y can get values from 0 to 63. For further information on the remaining registers, please see [4].

ISER[x] and ICER[x] registers are used to enable or disable an interrupt in the NVIC, respectively. Interrupt requests from interrupt vector table entries 0 to 31 are controlled by ISER[0] and ICER[0] registers. Similarly, interrupt requests from interrupt vector table entries 32 to 63 are controlled by ISER[1] and ICER[1] registers. Interrupt request (IRQ) numbers and their source for the NVIC module of the MSP432 microcontroller are as in Table 6.2. In the same table, we also provide related ISER[x] and ICER[x] registers. Constant values for these are also tabulated there. The user can use predefined macros instead of these constant values. However, they will need bit shifting operations. Therefore, we will not use macro-based register adjustments in this book. To note here, IRQ numbers from 41 to 63 are not used in the MSP432 microcontroller.

The IP[y] register is used to set the programmable priority level for the desired interrupt. Here, y value is set according to the IRQ number in Table 6.2. Then, the priority level can be set from 0 (highest priority) to 255 (lowest priority). Also, the lowest bit of the PRIMASK register (part of the core register) is used to control interrupts with programmed priorities. If this bit is reset, interrupts with programmed priorities can be used. Otherwise, interrupts with programmed priorities are not serviced.

Apart from the NVIC module, there are two predefined intrinsic functions to control all interrupts in the MSP432 microcontroller. These are _enable_interrupts() and _disable_interrupts(). These functions are used to enable or disable all interrupts, respectively. Therefore, the function _enable_interrupts() must be called in the C code before using any interrupts listed in Table 6.2.

6.3.4 Interrupts via DriverLib Functions

The NVIC and MSP432 core interrupt properties can be also controlled via DriverLib functions. The complete function list and related macro definitions can be found in [11]. We will only use a subset of these functions given below.

```
bool Interrupt_enableMaster (void)
/*
Enables the NVIC to pass interrupts to the CPU.
It is the DriverLib version of the intrinsic function
    _enable_interrupts().
*/

bool Interrupt_disableMaster (void)
/*
Disables the NVIC to pass interrupts to the CPU.
It is the DriverLib version of the intrinsic function
    _disable_interrupts().
*/

void Interrupt_enableInterrupt (uint32_t interruptNumber)
/*
Enables the interrupt in the NVIC modifying ISER register.
interruptNumber: the interrupt to be enabled.
*/
```

IRQ Number	Source	ISER-ICER Registers	Register Inputs	Macro
0	PSS	ISER[0]-ICER[0]	0x00000001	PSS_IRQn
1	CS	ISER[0]-ICER[0]	0x00000002	CS_IRQn
2	PCM	ISER[0]-ICER[0]	0x00000004	PCM_IRQn
3	WDT_A	ISER[0]-ICER[0]	0x00000008	WDT_A_IRQn
4	FPU_INT	ISER[0]-ICER[0]	0x00000010	FPU_IRQn
5	FLCTL	ISER[0]-ICER[0]	0x00000020	FLCTL_IRQn
6	COMP_E0	ISER[0]-ICER[0]	0x00000040	COMP_E0_IRQn
7	COMP_E1	ISER[0]-ICER[0]	0x00000080	COMP_E1_IRQn
8	Timer_A0	ISER[0]-ICER[0]	0x00000100	TA0_0_IRQn
9	Timer_A0	ISER[0]-ICER[0]	0x00000200	TA0_N_IRQn
10	Timer_A1	ISER[0]-ICER[0]	0x00000400	TA1_0_IRQn
11	Timer_A1	ISER[0]-ICER[0]	0x00000800	TA1_N_IRQn
12	Timer_A2	ISER[0]-ICER[0]	0x00001000	TA2_0_IRQn
13	Timer_A2	ISER[0]-ICER[0]	0x00002000	TA2_N_IRQn
14	Timer_A3	ISER[0]-ICER[0]	0x00004000	TA3_0_IRQn
15	Timer_A3	ISER[0]-ICER[0]	0x00008000	TA3_N_IRQn
16	eUSCI_A0	ISER[0]-ICER[0]	0x00010000	EUSCIA0_IRQn
17	eUSCI_A1	ISER[0]-ICER[0]	0x00020000	EUSCIA1_IRQn
18	eUSCI_A2	ISER[0]-ICER[0]	0x00040000	EUSCIA2_IRQn
19	eUSCI_A3	ISER[0]-ICER[0]	0x00080000	EUSCIA3_IRQn
20	eUSCI_B0	ISER[0]-ICER[0]	0x00100000	EUSCIB0_IRQn
21	eUSCI_B1	ISER[0]-ICER[0]	0x00200000	EUSCIB1_IRQn
22	eUSCI_B2	ISER[0]-ICER[0]	0x00400000	EUSCIB2_IRQn
23	eUSCI_B3	ISER[0]-ICER[0]	0x00800000	EUSCIB3_IRQn
24	ADC14	ISER[0]-ICER[0]	0x01000000	ADC14_IRQn
25	Timer32_INT1	ISER[0]-ICER[0]	0x02000000	T32_INT1_IRQn
26	Timer32_INT2	ISER[0]-ICER[0]	0x04000000	T32_INT2_IRQn
27	Timer32_INTC	ISER[0]-ICER[0]	0x08000000	T32_INTC_IRQn
28	AES256	ISER[0]-ICER[0]	0x10000000	AES256_IRQn
29	RTC_C	ISER[0]-ICER[0]	0x20000000	RTC_C_IRQn
30	DMA_ERR	ISER[0]-ICER[0]	0x40000000	DMA_ERR_IRQn
31	DMA_INT3	ISER[0]-ICER[0]	0x80000000	DMA_INT3_IRQn
32	DMA_INT2	ISER[1]-ICER[1]	0x00000001	DMA_INT2_IRQn
33	DMA_INT1	ISER[1]-ICER[1]	0x00000002	DMA_INT1_IRQn
34	DMA_INT0	ISER[1]-ICER[1]	0x00000004	DMA_INT0_IRQn
35	I/O Port P1	ISER[1]-ICER[1]	0x00000008	PORT1_IRQn
36	I/O Port P2	ISER[1]-ICER[1]	0x00000010	PORT2_IRQn
37	I/O Port P3	ISER[1]-ICER[1]	0x00000020	PORT3_IRQn
38	I/O Port P4	ISER[1]-ICER[1]	0x00000040	PORT4_IRQn
39	I/O Port P5	ISER[1]-ICER[1]	0x00000080	PORT5_IRQn
40	I/O Port P6	ISER[1]-ICER[1]	0x00000100	PORT6_IRQn

TABLE 6.2 Interrupt Requests for the NCIV Module of the MSP432 Microcontroller

```
void Interrupt_disableInterrupt (uint32_t interruptNumber)
/*
Disables the interrupt in NVIC modifying ICER register.
interruptNumber: the interrupt to be disabled.
*/

void Interrupt_setPriority (uint32_t interruptNumber, uint8_t priority)
/*
Sets the priority of the interrupt.
interruptNumber: the interrupt to be modified.
priority: the priority level.
*/
```

6.4 Port Interrupts

The interrupt may be generated from several sources as mentioned in the previous section. Since digital input and output is considered in the previous chapter, here we focus on the interrupts from ports. We will focus on interrupts from other sources in the following chapters.

Only ports P1, P2, P3, P4, P5, and P6 can be used as an interrupt source for MSP432. Each pin of these ports can be used for a different interrupt. Unfortunately, these share the same interrupt vector, hence the same ISR. Interrupt flags can be used to discriminate the actual source. Using them, the task of each pin interrupt can be controlled separately by the condition of the related interrupt flag. We will see such an application in Sec. 6.5.

6.4.1 Port Interrupt Registers

There are three 8-bit registers defined to control digital I/O interrupt operations of each port. They are called PxIE, PxIES, and PxIFG. Here, x can get values from 1 to 6. A structure named Px is used to control port registers. The registers of interest are named IE, IES, and IFG in this structure. Again, x can get values from 1 to 6.

The PxIE register is used to enable the interrupt for the associated pin. To enable the port interrupt from a specific pin, the corresponding bit in PxIE should be set. To disable the interrupt from the same pin, the corresponding bit should be reset.

The PxIES register is used to select the signal edge in which the interrupt occurs (on a specific pin). The interrupt occurs when the input goes from low to high, if the bit corresponding to the pin is reset. The interrupt occurs when the input goes from high to low, if the bit is set.

The PxIFG register is used to check the interrupt condition. When an interrupt occurs from a pin, the related interrupt flag is set. PxIFG must be cleared in the ISR to allow a new interrupt. In the same way, all interrupt flags should be reset at the beginning of the program to avoid any confusion.

6.4.2 Port Interrupts via DriverLib Functions

Port interrupts can also be controlled via DriverLib functions. The complete function list and related macro definitions can be found in [11]. We will only use a subset of these functions given below. Here, selectedPort represents the selected port in the form of GPIO_PORT_Px. selectedPin corresponds to the selected pin in the selected port in the form of GPIO_PINx. More than one pin can be used in a function by combining them by the logical OR operation.

```
void GPIO_enableInterrupt (uint_fast8_t selectedPort, uint_fast16_t
    selectedPins)
/*
Enables the port interrupt for selected pins by modifying the PxIE
    register.
 */

void GPIO_disableInterrupt (uint_fast8_t selectedPort, uint_fast16_t
    selectedPins)
/*
Disables the port interrupt for selected pins by modifying the PxIE
    register.
 */

uint_fast16_t GPIO_getInterruptStatus (uint_fast8_t selectedPort,
    uint_fast16_t selectedPins)
/*
Reads the port interrupt status for selected pins by modifying the
    PxIFG register.
 */

void GPIO_clearInterruptFlag (uint_fast8_t selectedPort, uint_fast16_t
    selectedPins)
/*
Clears the port interrupt flag for selected pins by modifying the PxIFG
    register.
 */

void GPIO_interruptEdgeSelect (uint_fast8_t selectedPort, uint_fast16_t
    selectedPins, uint_fast8_t edgeSelect)
/*
Configures the transition edge to set the port interrupt flag by
    modifying the PxIES register.
edgeSelect: can be one of GPIO_LOW_TO_HIGH_TRANSITION or
    GPIO_HIGH_TO_LOW_TRANSITION values.
*/
```

6.4.3 Port Interrupts via Energia Functions

Port interrupts can also be controlled via Energia. We provide related functions below. More information on Energia functions can be found in [12].

```
attachInterrupt(interrupt, function, mode)
/*
Attaches an ISR function to interrupt by combining the functionality of
    multiple Energia functions.
interrupt: the interrupt source pin.
function: ISR function to be called when interrupt occurs.
mode: the transition edge or level to trigger interrupt event.
*/
```

```
detachInterrupt(interrupt)
/*
Detaches the interrupt function attached to interrupt source by
    combining the functionality of multiple Energia functions.
interrupt: the interrupt source pin.
*/
```

6.5 Coding Practices for Port Interrupts

We provide sample codes for port interrupts using registers directly, via DriverLib and Energia functions in this section. In all these examples, we will use LED1 connected to pin P1.0, red color of LED2 connected to pin P2.0, and green color of LED2 connected to pin P2.1 on the MSP432 LaunchPad. We will also use the buttons S1 and S2 connected to pins P1.1 and P1.4 on the MSP432 LaunchPad, respectively.

6.5.1 Register-Level Examples

The first C code on port interrupts, in Listing 6.2, toggles LED1 when the button S1 (connected with the pull-up resistor) is released since the interrupt edge is selected as low to high. Here, it is important to note that all interrupts are enabled by the intrinsic

Listing 6.2 Toggle LED1 by an Interrupt, Using Registers

```c
#include "msp.h"

#define LED1 BIT0
#define S1 BIT1
#define DELAY 500
int i;

void main(void){

  WDT_A->CTL = WDT_A_CTL_PW | WDT_A_CTL_HOLD;
  P1->DIR = LED1;
  P1->REN = S1;
  P1->OUT = S1;
  P1->IE = S1; // Port interrupt is enabled for P1.1
  P1->IES = 0x00; // Interrupt edge select from low to high
  P1->IFG = 0x00; // Interrupt flag is cleared

  NVIC->ISER[1] = 0x00000008; // Port P1 interrupt is enabled in NVIC
  _enable_interrupts(); // All interrupts are enabled

  while(1);
}

void PORT1_IRQHandler(void){
  if(P1->IFG & S1) P1->OUT ^= LED1; // LED1 is toggled
  for(i=0;i<DELAY;i++){} // Delay for debounce
  P1->IFG &= ~S1; // Interrupt flag is cleared for P1.1
}
```

function `_enable_interrupts()`. Also, Port1 interrupt in NVIC is enabled with the line `_NVIC->ISER[1] = 0x00000008`.

In the second C code, in Listing 6.3, the number of button S1 presses are counted within the ISR. Interrupt edge is selected as low to high, which means the `count` variable is increased when S1 is released. Here, the variable `count` is specifically defined as global. Therefore, it can be kept between successive interrupts.

In the third C code, in Listing 6.4, LED1 and green color of LED2 are turned based on the total number of button S1 presses. The interrupt edge is selected as low to high, which means `count` variable is increased when S1 is released. Again, the variable `count` is specifically defined as global for the same reason as in Listing 6.3.

In the fourth C code, in Listing 6.5, red color of LED2 is toggled when the button S1 is released and the green color of LED2 is toggled when the button S2 is released (since interrupt edges are selected as low to high). Here, we used two different interrupts (from pins P1.1 and P1.4) from the same interrupt source (Port P1). Therefore, we had to check interrupt flags in the ISR and perform the related operation.

In the fifth C code, in Listing 6.6, the red color of LED2 toggles when the button S1 is released and the green color of LED2 when an external push button connected to pin P3.6 is released. Here, we set the priority of Port1 interrupt to the highest level and priority of the Port3 interrupt to 64 by using the IP[y] registers of NVIC module. We deliberately added a long delay function to see the nested interrupt operation. If the button S1 is released while the CPU is in handler mode in Port3 ISR, then Port3 ISR is held and CPU

Listing 6.3 Count the Number of Button Presses by Interrupts, Using Registers

```c
#include "msp.h"

#define S1 BIT1
#define DELAY 500
int i;
int count = 0;

void main(void){

 WDT_A->CTL = WDT_A_CTL_PW | WDT_A_CTL_HOLD;
 P1->DIR = 0x00;
 P1->REN = S1;
 P1->OUT = S1;
 P1->IE = S1; // Port interrupt is enabled for P1.1
 P1->IES = 0x00; // Interrupt edge select from low to high
 P1->IFG = 0x00; // Interrupt flag is cleared

 NVIC->ISER[1] = 0x00000008; // Port P1 interrupt is enabled in NVIC
 _enable_interrupts(); // All interrupts are enabled

 while(1);
}
void PORT1_IRQHandler(void){
 if(P1->IFG & S1) count++;
 for(i=0;i<DELAY;i++){} // Delay for debounce
 P1->IFG &= ~S1; // Interrupt flag is cleared for P1.1
}
```

Listing 6.4 Turn On and Off LEDs by the Total Number of Interrupt Requests, Using Registers

```c
#include "msp.h"

#define LED1 BIT0
#define LED2_GREEN BIT1
#define S1 BIT1
#define DELAY 500
int i;
int count = 0;

void main(void){

 WDT_A->CTL = WDT_A_CTL_PW | WDT_A_CTL_HOLD;
 P1->DIR = LED1;
 P2->DIR = LED2_GREEN;
 P1->REN = S1;
 P1->OUT = S1;
 P2->OUT = 0x00;
 P1->IE = S1; // Port interrupt is enabled for P1.1
 P1->IES = 0x00; // Interrupt edge select from low to high
 P1->IFG = 0x00; // Interrupt flag is cleared

 NVIC->ISER[1] = 0x00000008; // Port P1 interrupt is enabled in NVIC
 _enable_interrupts(); // All interrupts are enabled

 while(1);
}
void PORT1_IRQHandler(void){
 if(P1->IFG & S1){
  count++;
  if(count == 4){
   P1->OUT |= LED1; // Turn on LED1
   P2->OUT = 0x00;} // Turn off LED2_GREEN
  if(count == 6){
   P1->OUT &= ~LED1; // Turn off LED1
   P2->OUT = LED2_GREEN; // Turn on LED2_GREEN
   count = 0;}}

  for(i=0;i<DELAY;i++){} // Delay for debounce
   P1->IFG &= ~S1; // Interrupt flag is cleared for P1.1
}
```

handles the Port1 ISR. During this operation, both the red and green colors of LED2 are turned on at the same time. When the handling process is completed, the red color of LED2 is turned off and CPU handles the Port3 ISR where it is left from.

6.5.2 DriverLib Examples

We repeat the example in Listing 6.2 using DriverLib functions. We provide the corresponding C code in Listing 6.7. Here, we first set LED1 as output and button S1 as input with the pull-up resistor. Then, we clear and enable interrupt for S1 using GPIO_clearInterruptFlag and GPIO_enableInterrupt functions. Finally, we

Listing 6.5 Toggle LEDs by Using Two Different Buttons Having the Same Interrupt Vector, Using Registers

```c
#include "msp.h"

#define LED2_RED BIT0
#define LED2_GREEN BIT1
#define S1 BIT1
#define S2 BIT4
#define DELAY 500
int i;

void main(void){

  WDT_A->CTL = WDT_A_CTL_PW | WDT_A_CTL_HOLD;
  P1->DIR = 0x00;
  P2->DIR = LED2_RED|LED2_GREEN;
  P1->REN = S1|S2;
  P1->OUT = S1|S2;
  P2->OUT = 0x00;
  P1->IE = S1|S2; // Port interrupt is enabled for P1.1 and P1.4
  P1->IES = 0x00; // Interrupt edge selects from low to high
  P1->IFG = 0x00; // Interrupt flags are cleared

  NVIC->ISER[1] = 0x00000008; // Port P1 interrupt is enabled in NVIC
  _enable_interrupts(); // All interrupts are enabled

  while(1);
}
void PORT1_IRQHandler(void){
 if(P1->IFG & S1){
  P2->OUT ^= LED2_RED; // Toggle LED2_RED
  for(i=0;i<DELAY;i++){} // Delay for debounce
   P1->IFG &= ~S1;} // Interrupt flag is cleared for P1.1
 if(P1->IFG & S2){
  P2->OUT ^= LED2_GREEN; // Toggle LED2_GREEN
  for(i=0;i<DELAY;i++){} // Delay for debounce
   P1->IFG &= ~S2;} // Interrupt flag is cleared for P1.4
}
```

enable the port interrupt on the NVIC and let it pass interrupts to CPU by the functions Interrupt_enableInterrupt and Interrupt_enableMaster. In the ISR, we first read the interrupt status and clear flags by using the functions GPIO_get EnabledInterruptStatus and GPIO_clearInterruptFlag. Then, we check whether the interrupt is coming from S1. If this is the case, we toggle LED1 accordingly.

We next repeat the example in Listing 6.3 using DriverLib functions. We provide the corresponding C code in Listing 6.8. Similar to the previous example in Listing 6.7, we set LED1 as output and S1 as input with the pull-up resistor and enable the port interrupt. In the ISR, we increase the counter according to the interrupt status.

We repeat the example in Listing 6.4 using DriverLib functions. We provide the corresponding C code in Listing 6.9. Different from the previous example in Listing 6.8,

Listing 6.6 Toggle LEDs by Nested Interrupt Operations, Using Registers

```c
#include "msp.h"

#define LED2_RED BIT0
#define LED2_GREEN BIT1
#define S1 BIT1
#define S_ext BIT6

int i;

void delay_ms(uint16_t ms);

void main(void){

  WDT_A->CTL = WDT_A_CTL_PW | WDT_A_CTL_HOLD;

  P2->DIR = LED2_RED | LED2_GREEN;
  P2->OUT = 0x00;
  P1->REN = S1;
  P1->OUT = S1;
  P3->REN = S_ext;
  P3->OUT = S_ext;
  P1->IE = S1; // Port interrupt is enabled for P1.1
  P1->IES = 0x00; // Interrupt edge select from low to high
  P1->IFG = 0x00; // Interrupt flag is cleared
  P3->IE = S_ext; // Port interrupt is enabled for P3.6
  P3->IES = 0x00; // Interrupt edge select from low to high
  P3->IFG = 0x00; // Interrupt flag is cleared

  NVIC->IP[35] = 0; // Set the interrupt priority to 0 for Port P1 interrupt
  NVIC->IP[37] = 64; // Set the interrupt priority to 64 for Port P3
      interrupt

  NVIC->ISER[1] = 0x00000008; // Port P1 interrupt is enabled in NVIC
  NVIC->ISER[1] = 0x00000020; // Port P3 interrupt is enabled in NVIC
  _enable_interrupts(); // All interrupts are enabled


  while (1);
}
void PORT1_IRQHandler(void){
  P2->OUT ^= LED2_RED; // Toggle LED2_RED
  delay_ms(3000);
  P2->OUT ^= LED2_RED; // Toggle LED2_RED
  P1->IFG &= ~S1; // Interrupt flag is cleared for P1.1
}
void PORT3_IRQHandler(void){
  P2->OUT ^= LED2_GREEN; // Toggle LED2_GREEN
  delay_ms(3000);
  P2->OUT ^= LED2_GREEN; // Toggle LED2_GREEN
  P3->IFG &= ~S_ext; // Interrupt flag is cleared for P3.6
}
void delay_ms(uint16_t ms){
  uint16_t delay;
  volatile uint32_t i;
  for (delay = ms; delay >0 ; delay--)
    for (i=300; i >0;i--);
}
```

Listing 6.7 Toggle LED1 by an Interrupt, with DriverLib Functions

```c
#include <ti/devices/msp432p4xx/driverlib/driverlib.h>

int main(void){

WDT_A_holdTimer();

GPIO_setAsOutputPin(GPIO_PORT_P1, GPIO_PIN0);
GPIO_setAsInputPinWithPullUpResistor(GPIO_PORT_P1, GPIO_PIN1);
GPIO_interruptEdgeSelect(GPIO_PORT_P1, GPIO_PIN1,
    GPIO_LOW_TO_HIGH_TRANSITION);
GPIO_clearInterruptFlag(GPIO_PORT_P1, GPIO_PIN1);
GPIO_enableInterrupt(GPIO_PORT_P1, GPIO_PIN1);
Interrupt_enableInterrupt(INT_PORT1);
Interrupt_enableMaster();

while (1);
}
void PORT1_IRQHandler(void){
uint32_t status;

status = GPIO_getEnabledInterruptStatus(GPIO_PORT_P1);
GPIO_clearInterruptFlag(GPIO_PORT_P1, status);

if(status & GPIO_PIN1)
  GPIO_toggleOutputOnPin(GPIO_PORT_P1, GPIO_PIN0);
}
```

we add conditions to toggle red and green colors of LED2 according to the total number of interrupt requests in the ISR code.

We next repeat the example in Listing 6.5 using DriverLib functions. We provide the corresponding C code in Listing 6.10. Here, we also set S2 as input with the pull-up resistor and enable interrupt for both buttons on the MSP432 LaunchPad. We also set the red and green colors of LED2 as output. In the ISR code, we check whether the interrupt is coming from S1 or S2 by looking at the interrupt flag and toggle LEDs accordingly. Then, we clear the interrupt flag as we leave the ISR.

We finally repeat the example in Listing 6.6 using DriverLib functions. We provide the corresponding C code in Listing 6.11. This code toggles the red color of LED2 when the button S1 is released and the green color of LED2 when an external push button connected to pin P3.6 is released. Here we set the priority of Port1 interrupt to the highest level and priority of the Port3 interrupt to 64 by using the function `Interrupt_setPriority`. We deliberately added a long delay function to see the nested interrupt operation. If the button S1 is released while the CPU is in handler mode in Port3 ISR, then Port3 ISR is held and CPU handles the Port1 ISR. During this operation, both the red and green colors of LED2 are turned on at the same time. When the handling process is completed, the red color of LED2 is turned off and CPU handles the Port3 ISR where it is left from.

Listing 6.8 Count the Number of Button Presses by Interrupts, with DriverLib Functions

```c
#include <ti/devices/msp432p4xx/driverlib/driverlib.h>

int count = 0;

void main(void) {

 WDT_A_holdTimer();

 GPIO_setAsInputPinWithPullUpResistor(GPIO_PORT_P1, GPIO_PIN1);
 GPIO_interruptEdgeSelect(GPIO_PORT_P1, GPIO_PIN1,
     GPIO_LOW_TO_HIGH_TRANSITION);
 GPIO_clearInterruptFlag(GPIO_PORT_P1, GPIO_PIN1);
 GPIO_enableInterrupt(GPIO_PORT_P1, GPIO_PIN1);
 Interrupt_enableInterrupt(INT_PORT1);
 Interrupt_enableMaster();

 while (1);
}

void PORT1_IRQHandler(void) {
 uint32_t status;

 status = GPIO_getEnabledInterruptStatus(GPIO_PORT_P1);
 GPIO_clearInterruptFlag(GPIO_PORT_P1, status);

 if(status & GPIO_PIN1) count++;
}
```

6.5.3 Energia Examples

We repeat the previous example in Listing 6.2 with Energia functions. We provide the corresponding C code in Listing 6.12.

As can be seen in Listing 6.12, after setting LED1 as output and S1 as input with the pull-up resistor, the `toggle` function is attached as interrupt function. It is set to trigger at falling edges with the `attachInterrupt` function. In the `toggle` function, LED1 is toggled according to status of the software flag `LEDstate`.

We next repeat the previous example in Listing 6.3 with Energia functions. We provide the corresponding C code in Listing 6.13. Here, we count the button presses in the attached interrupt function.

We repeat the previous example given in Listing 6.4 with Energia functions. We provide the corresponding C code in Listing 6.14. Here, we count the button presses in the attached interrupt function and toggle red and green colors of LED2 according to the total count value.

We finally repeat the previous example in Listing 6.5 with Energia functions. We provide the corresponding C code in Listing 6.15. Here, we attach two different interrupt functions to S1 and S2 buttons separately. In these functions, the red and green colors of LED2 are toggled.

Listing 6.9 Turn On and Off LEDs by the Total Number of Interrupt Requests, with DriverLib Functions

```c
#include <ti/devices/msp432p4xx/driverlib/driverlib.h>

int count = 0;

void main(void){

WDT_A_holdTimer();

GPIO_setAsOutputPin(GPIO_PORT_P2, GPIO_PIN0 | GPIO_PIN1);
GPIO_setAsInputPinWithPullUpResistor(GPIO_PORT_P1, GPIO_PIN1);
GPIO_interruptEdgeSelect(GPIO_PORT_P1, GPIO_PIN1,
    GPIO_LOW_TO_HIGH_TRANSITION);
GPIO_clearInterruptFlag(GPIO_PORT_P1, GPIO_PIN1);
GPIO_enableInterrupt(GPIO_PORT_P1, GPIO_PIN1);
Interrupt_enableInterrupt(INT_PORT1);
Interrupt_enableMaster();

while (1);
}
void PORT1_IRQHandler(void){
 uint32_t status;

 status = GPIO_getEnabledInterruptStatus(GPIO_PORT_P1);
 GPIO_clearInterruptFlag(GPIO_PORT_P1, status);

 if(status & GPIO_PIN1){
  count++;
  if(count == 4){
   GPIO_setOutputLowOnPin(GPIO_PORT_P2, GPIO_PIN0 | GPIO_PIN1);
   GPIO_setOutputHighOnPin(GPIO_PORT_P2, GPIO_PIN0);}
  if(count == 6){
   count = 0;
   GPIO_setOutputLowOnPin(GPIO_PORT_P2, GPIO_PIN0 | GPIO_PIN1);
   GPIO_setOutputHighOnPin(GPIO_PORT_P2, GPIO_PIN1);}}
}
```

6.6 Application: Home Entrance System

The aim of this application is to learn how to use the port interrupts on the MSP432 microcontroller. As a real-life application, we will design a home entrance system. In this section, we provide the equipment list, layout of the circuit, and system design specifications.

6.6.1 Equipment List

Below, we provide the equipment list to be used in the home entrance system.

- 4×4 matrix keypad
- 74c922 16 key encoder IC

Listing 6.10 Toggle LEDs by Using Two Different Buttons Having the Same Interrupt Vector, with DriverLib Functions

```c
#include <ti/devices/msp432p4xx/driverlib/driverlib.h>

void main(void){

 WDT_A_holdTimer();

 GPIO_setAsOutputPin(GPIO_PORT_P2, GPIO_PIN0 | GPIO_PIN1);
 GPIO_setOutputLowOnPin(GPIO_PORT_P2, GPIO_PIN0 | GPIO_PIN1);
 GPIO_setAsInputPinWithPullUpResistor(GPIO_PORT_P1, GPIO_PIN1 |
    GPIO_PIN4);
 GPIO_interruptEdgeSelect (GPIO_PORT_P1, GPIO_PIN1 | GPIO_PIN4,
    GPIO_HIGH_TO_LOW_TRANSITION);
 GPIO_clearInterruptFlag(GPIO_PORT_P1, GPIO_PIN1 | GPIO_PIN4);
 GPIO_enableInterrupt(GPIO_PORT_P1, GPIO_PIN1 | GPIO_PIN4);
 Interrupt_enableInterrupt(INT_PORT1);
 Interrupt_enableMaster();

 while (1);
}

void PORT1_IRQHandler(void){

 uint32_t status;

 status = GPIO_getEnabledInterruptStatus(GPIO_PORT_P1);
 GPIO_clearInterruptFlag(GPIO_PORT_P1, status);

 if(status & GPIO_PIN1)
  GPIO_toggleOutputOnPin(GPIO_PORT_P2, GPIO_PIN0);
 else if(status & GPIO_PIN4)
  GPIO_toggleOutputOnPin(GPIO_PORT_P2, GPIO_PIN1);
}
```

- One 1-μF capacitor
- One 10-μF capacitor

We will use the 4×4 matrix keypad in our home entrance system. This keypad has buttons as numbers (0-9), symbols (*, #), and letters (A-D). The matrix keypad works as follows. Logic level 1 should be fed to its rows sequentially. Only one of the keypad rows can be at logic level 1 at a time. Other rows should be at logic level 0. When a button on the keypad is pressed, logic level 1 can be read from its corresponding column input while its row output is at logic level 1.

The 74c922 IC encodes the 4×4 matrix keypad button press information into a four-bit data. This four-bit data is fed to output via parallel output pins. The IC additionally gives data-ready output that can be used as interrupt source to read the encoded key data.

6.6.2 Layout

The layout of home entrance system is given in Fig. 6.1.

Listing 6.11 Toggle LEDs by Nested Interrupt Operations, with DriverLib Functions

```c
#include <ti/devices/msp432p4xx/driverlib/driverlib.h>

#define DELAY 15000

int i;

void delay_ms(uint16_t ms);

void main(void){

 WDT_A_holdTimer();

 GPIO_setAsOutputPin(GPIO_PORT_P2, GPIO_PIN0 | GPIO_PIN1);
 GPIO_setOutputLowOnPin(GPIO_PORT_P2, GPIO_PIN0 | GPIO_PIN1);
 GPIO_setAsInputPinWithPullUpResistor(GPIO_PORT_P1, GPIO_PIN1);
 GPIO_setAsInputPinWithPullUpResistor(GPIO_PORT_P3, GPIO_PIN6);
 GPIO_clearInterruptFlag(GPIO_PORT_P1, GPIO_PIN1);
 GPIO_clearInterruptFlag(GPIO_PORT_P3, GPIO_PIN6);
 GPIO_enableInterrupt(GPIO_PORT_P1, GPIO_PIN1);
 GPIO_enableInterrupt(GPIO_PORT_P3, GPIO_PIN6);

 Interrupt_setPriority(INT_PORT1, 0);
 Interrupt_setPriority(INT_PORT3, 64);
 Interrupt_enableInterrupt(INT_PORT1);
 Interrupt_enableInterrupt(INT_PORT3);
 Interrupt_enableMaster();

 while (1);
}
void PORT1_IRQHandler(void){
 uint32_t status = GPIO_getEnabledInterruptStatus(GPIO_PORT_P1);
 GPIO_clearInterruptFlag(GPIO_PORT_P1, status);

 if(status & GPIO_PIN1){
  GPIO_toggleOutputOnPin(GPIO_PORT_P2, GPIO_PIN0);
  delay_ms(3000);
  GPIO_toggleOutputOnPin(GPIO_PORT_P2, GPIO_PIN0);}
}
void PORT3_IRQHandler(void){
 uint32_t status = GPIO_getEnabledInterruptStatus(GPIO_PORT_P3);
 GPIO_clearInterruptFlag(GPIO_PORT_P3, status);

 if(status & GPIO_PIN6){
  GPIO_toggleOutputOnPin(GPIO_PORT_P2, GPIO_PIN1);
  delay_ms(3000);
  GPIO_toggleOutputOnPin(GPIO_PORT_P2, GPIO_PIN1);}
}
void delay_ms(uint16_t ms){
 uint16_t delay;
 volatile uint32_t i;
 for (delay = ms; delay >0 ; delay--)
  for (i=300; i >0;i--);
}
```

Listing 6.12 Toggle LED1 by an Interrupt, with Energia Functions

```
boolean LEDstate = LOW;

void setup(){
 pinMode(YELLOW_LED, OUTPUT);
 pinMode(PUSH1, INPUT_PULLUP);
 attachInterrupt(PUSH1, toggle, FALLING);
}

void loop(){}

void toggle(){
 delay(100);
 LEDstate = !LEDstate;
 digitalWrite(YELLOW_LED, LEDstate);
}
```

Listing 6.13 Count the Number of Button Presses by Interrupts, with Energia Functions

```
int count = 0;

void setup(){
 pinMode(PUSH1, INPUT_PULLUP);
 attachInterrupt(PUSH1, counter, FALLING);
}

void loop(){}

void counter(){
 count++;
}
```

6.6.3 System Design Specifications

The design of the home entrance system will have four main blocks. These are listed below:

- **Block 1**: At startup, the user should press the *A* button on the keypad. Then, he or she can enter a password (having maximum 10 digits) using number buttons. Afterwards, the user can lock the system using the *B* button. The blue color of LED2 connected to pin P2.2 on the MSP432 LaunchPad will turn on. The system will wait for an input.

- **Block 2**: If the user wants to unlock the system, first he or she should press the *C* button and enter the correct password using number buttons. Then, the *D* button should be pressed to unlock the system. If the entered password is correct, the green color of LED2 connected to pin P2.1 on the MSP432 LaunchPad will turn on. Otherwise, the red color of LED2 connected to pin P2.0 on the MSP432 LaunchPad will turn on and the system will wait for the correct password.

Exceptions and Interrupts

Listing 6.14 Turn On and Off LEDs by the Total Number of Interrupts, with Energia Functions

```
int count = 0;

void setup(){
 pinMode(RED_LED, OUTPUT);
 pinMode(GREEN_LED, OUTPUT);
 pinMode(PUSH1, INPUT_PULLUP);
 attachInterrupt(PUSH1, toggle, FALLING);
}

void loop(){}

void toggle(){
 count++;
 if(count == 4){
  digitalWrite(GREEN_LED, LOW);
  digitalWrite(RED_LED, HIGH);}
 if(count == 6){
  count = 0;
  digitalWrite(RED_LED, LOW);
  digitalWrite(GREEN_LED, HIGH);}
}
```

Listing 6.15 Toggle LEDs by Using Two Different Buttons Having the Same Interrupt Vector, with Energia Functions

```
boolean LED1state = LOW;
boolean LED2state = LOW;

void setup(){
 pinMode(RED_LED, OUTPUT);
 pinMode(GREEN_LED, OUTPUT);
 pinMode(PUSH1, INPUT_PULLUP);
 pinMode(PUSH2, INPUT_PULLUP);
 attachInterrupt(PUSH1, toggle1, FALLING);
 attachInterrupt(PUSH2, toggle2, FALLING);
}

void loop(){}

void toggle1(){
 delay(100);
 LED1state = !LED1state;
 digitalWrite(RED_LED, LED1state);
}

void toggle2(){
 delay(100);
 LED2state = !LED2state;
 digitalWrite(GREEN_LED, LED2state);
}
```

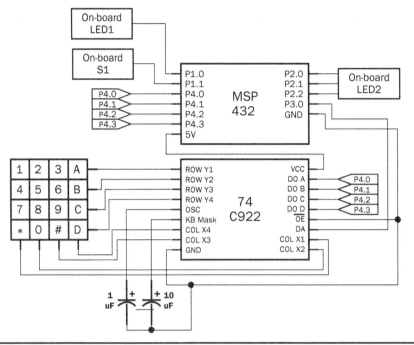

FIGURE 6.1 Layout of the home entrance system.

- **Block 3**: If the wrong password is entered to the system three times consecutively, LED1 connected to pin P1.0 on the MSP432 LaunchPad will turn on to indicate an alarm condition. When the button S1 connected to pin P1.1 of the MSP432 Launchpad is pressed, the alarm condition will be cleared.
- **Block 4**: After the first entry, the password can only be changed if the previous password is entered correctly.

We could not add the C code of the project here due to page limitations. However, we provided the complete CCS project for this application in the companion website for the reader. We strongly suggest implementing it.

6.7 Summary

This chapter is about the interrupt-based programming of MSP432. We first explained the difference between an exception and interrupt. Then, we provided examples related to interrupts from our daily lives. We also explored the NVIC module that is responsible for all interrupt-based operations in the MSP432 microcontroller. Afterwards, we focused on port interrupts. We provided sample C codes using port interrupts. Finally, we provided a real-life application (home entrance system) using port interrupts. We will be using interrupt-based operations in the following chapters. Therefore, we strongly suggest the reader to master the interrupt concept discussed in this chapter.

6.8 Exercises

6.1 Write a C program to count the difference between the number of times the buttons S1 and S2, connected to pins P1.1 and P1.4 on the MSP432 LaunchPad, are pressed.
   a. The button pressing operation should be defined in an ISR.
   b. Observe the count value from the watch window.
   c. If the button S1 is pressed more than the button S2, then turn on the red color of LED2 connected to pin P2.0 on the MSP432 LaunchPad. Otherwise, turn on the green color of LED2 connected to pin P2.1 on the MSP432 LaunchPad. If the button press numbers are equal, then turn on the blue color of LED connected to pin P2.2 on the MSP432 LaunchPad.

6.2 Write a C program to design a parallel to serial converter such that:
   a. When the button S1 connected to pin P1.1 on the MSP432 LaunchPad is pressed, eight-bit data will be read from all pins of P4 (the MSB will be P4.7 and LSB will be P4.0). The parallel to serial conversion operation should be defined in an ISR.
   b. Then LED1 connected to pin P1.0 on the MSP432 LaunchPad will be toggled in sequence according to bits of data. Hint: Use loop operations to generate waiting times.

6.3 Repeat Problem 5.5 using interrupts.

6.4 Repeat Problem 5.6 using interrupts.

6.5 Repeat Problem 5.7 using interrupts.

6.6 Write a C program such that the global integer array x with 10 elements will be filled initially. For this problem, fill it at the beginning of the code. When an interrupt comes from the button S1 connected to pin P1.1 on the MSP432 LaunchPad, the ISR will be called. The ISR will calculate the global integer array y defined as y[n] = 2*x[n] - x[n-1] where n is the index for the array. In fact, this is a simple filtering operation working with interrupts. In the actual application, the interrupt should come from some other source. The array y should also be filled by other module or a peripheral (such as ADC).

CHAPTER 7
Power Management and Timing Operations

Recent microcontroller applications heavily depend on either battery or energy harvesting systems (such as solar panel) as power source. Especially, microcontroller systems used in IoT applications should be working stand-alone. Therefore, their power consumption becomes extremely important. The MSP432 microcontroller has multiple power modes. Hence, the system can work under different conditions such as active and low power modes. The MSP432 microcontroller has a dedicated power control module (PCM) for this purpose. PCM controls the device power modes and allows switching between them using the power supply system (PSS) and clock system (CS) of the device. We will first introduce the PSS in this chapter. Power management is interwoven with the clock system of the microcontroller. Therefore, we will also introduce the clock system and associated timer modules of the MSP432 microcontroller in this chapter. There are five different timer modules in the MSP432 microcontroller. These are the watchdog timer (WDT_A), system timer (SysTick), Timer32, Timer_A, and real-time clock (RTC). All these modules are controlled and configured individually. Using these modules, we will achieve time-based operations in applications.

7.1 Power Supply System

The power supply system (PSS) is responsible for all functions related to power supply in the MSP432 microcontroller. It primarily generates the core voltage (V_{core}) required by the CPU, memory, and other modules. PSS also controls the distribution of the supply voltage (V_{CC}) to digital I/O and analog peripherals. The PSS has supervisor blocks to detect low voltage levels. Hence, when a voltage level falls below a specific threshold, a reset signal is generated.

The PSS is controlled by its dedicated registers. However, we will not consider them here. The PCM module handles all low-level operation and adjusts PSS registers required by the selected power mode. For more information on PSS registers, please see [4].

Clock Source	Frequency Range	Brief Explanation
DCO	1–48 MHz	Internal high-frequency clock source
HFXT	1–48 MHz	High-frequency external crystal clock source
LFXT	32 kHz	Low-frequency external crystal clock source
MODOSC	24 MHz	Internal clock source suitable for ADC or flash
SYSOSC	5 MHz	Internal clock source suitable for peripherals
REFO	32–128 kHz	Low-frequency internal clock source
VLO	10 kHz	Very low frequency internal clock source

TABLE 7.1 Clock Sources Available in the MSP432 Microcontroller

7.2 Clock System

The clock system (CS) is responsible for the clock generation and distribution in the MSP432 microcontroller. It contains flexible clock sources for both low power and high performance applications. Clock sources in the CS can generate wide range of clock signals starting from 10 kHz up to 48 MHz. They can be used to drive different clocks on the MSP432 microcontroller. The CS also contains built-in features to ensure the robust operation such as fail-safe mechanisms and on the fly configuration.

7.2.1 Clock Sources

There are seven clock sources available in the MSP432 microcontroller. They are tabulated in Table 7.1 with their properties. Here, the digitally controlled oscillator (DCO) is an internal clock source that allows generating high-frequency clock. The external counterpart of this clock is the high-frequency external crystal clock source (HFXT). Both DCO and HFXT can operate at 48 MHz. Low-frequency crystal (LFXT) is the low-frequency counterpart of HFXT. It can generate clock signals up to 32 kHz. The module oscillator (MODOSC) is an internal clock source that can operate at 24 MHz. It is suitable for internal analog operations, especially for analog-to-digital conversion (to be introduced in Chap. 8). For peripheral operations, 5 MHz clock source system oscillator (SYSOSC) can be used. The REFO is the internal low-frequency clock source that can generate a 128-kHz clock. This clock can be divided down to 32 kHz. VLO is the very low frequency clock source used for ultra low power and low-cost applications. It generates a 10-kHz clock.

7.2.2 Clocks

Five clocks are available in the MSP432 microcontroller. They are tabulated in Table 7.2 with their properties. The first one is the master clock (MCLK), which is mainly used by the CPU and direct memory access (DMA) module (to be introduced in Chap. 12). MSP432 uses the sub-main clock (SMCLK) and high-speed sub-main clock (HSMCLK) to drive peripheral units. The other two clocks are the low-frequency auxiliary clock (ACLK) and backup domain clock (BCLK). These two clocks are extensively used for low-frequency and low-power operations since they are optimized for low power modes to be introduced in the following sections. Besides the BCLK, all clocks have associated frequency divider modules. Therefore, the clock frequency for a selected clock can be divided by 2, 4, 8, 16, 32, 64, or 128.

Power Management and Timing Operations

Clock	Used by	Sourced from
MCLK	CPU, DMA	LFXT, VLO, REFO, DCO, MODSC, HFXT
SMCLK	Peripherals	LFXT, VLO, REFO, DCO, MODSC, HFXT
HSMCLK	Peripherals	LFXT, VLO, REFO, DCO, MODSC, HFXT
ACLK	Peripherals	LFXTCLK, VLOCLK, or REFOCLK
BCLK	Peripherals	LFXT or REFO

TABLE 7.2 Clocks Available in the MSP432 Microcontroller

Bits	23	22	18-16	9-0
	DCOEN	DCORES	DCORSEL	DCOTUNE

TABLE 7.3 CS Control Register 0 (CSCTL0)

7.2.3 Clock System Faults

External crystal oscillators may cause errors due to their start-up stabilization time or by a failure during operation. When such an oscillator fault occurs, clock(s) using it also malfunction. In such a case, if the LFXT was used, then the associated clock is automatically switched to REFOCLK for its clock source. If the HFXT was used, then the associated clock is automatically switched to SYSOSC for its clock source. These switching operations are also valid for DCO external resistor and external clock signal faults. When such a fault is detected, the corresponding crystal oscillator fault flag is set. It remains set until software clears it. If the application clears the corresponding fault flag, but the fault condition still exists, the fault flag is automatically set again. For more information on this issue, please see [4].

7.2.4 Clock System Registers

Thirteen registers are available in the MSP432 microcontroller to control the CS module. Out of these, we will only focus on CSKEY, CSCTL0, CSCTL1, CSCTL2, CSCLKEN, CSIFG, and CSCLRIFG registers in this book. For further information on the remaining registers, please see [4]. A structure named CS is used to control the mentioned registers.

The CSKEY register is used to allow or prevent access to the CS module for register configuration. When the hexadecimal number 0x695A (or the corresponding constant CS_KEY_VAL) is written to this register, all CS registers will be available for configuration. After the configuration process is done, 0x0000 must be written to this register to prevent faulty access.

The CSCTL0 register is used to configure the DCO. Entries of this register are tabulated in Table 7.3. Here, DCORSEL bits are used to set DCO frequency to 1.5, 3, 6, 12, 24, or 48 MHz. The constant CS_CTL0_DCORSEL_X can be used to set these frequency values where X can get values between 0 (for 1.5 MHz) and 5 (for 48 MHz). The default value for the DCORSEL bits is CS_CTL0_DCORSEL_1. For further information on other bits of this register, please see [4].

The CSCTL1 register is used to select the source and frequency divider values for all five clocks. The entries of this register are tabulated in Table 7.4. Here, DIVS, DIVA, DIVHS, and DIVM bits are used for the frequency division operation of the SMCLK, ACLK, HSMCLK, and MCLK, respectively. Instead of setting these bits separately, constants CS_CTL1_DIVS_X, CS_CTL1_DIVA_X, CS_CTL1_DIVHS_X, and CS_CTL1_

110 Chapter Seven

Bits	30-28	26-24	22-20	18-16	12	10-8	6-4	2-0
	DIVS	DIVA	DIVHS	DIVM	SELB	SELA	SELS	SELM

TABLE 7.4 CS Control Register 1 (CSCTL1)

Bits	Clock	Constant	Source Used
SELB	BCLK	CS_CTL1_SELB	REFO
		~CS_CTL1_SELB	LFXT (Default)
SELA	ACLK	CS_CTL1_SELA_0	LFXT if available, otherwise REFO (Default)
		CS_CTL1_SELA_1	VLO
		CS_CTL1_SELA_2	REFO
SELS		CS_CTL1_SELS_0	LFXT if available, otherwise REFO
		CS_CTL1_SELS_1	VLO
	SMCLK	CS_CTL1_SELS_2	REFO
	HSMCLK	CS_CTL1_SELS_3	DCO (Default)
		CS_CTL1_SELS_4	MODOSC
		CS_CTL1_SELS_5	HFXT if available, otherwise DCO
SELM	MCLK	CS_CTL1_SELM_0	LFXT if available, otherwise REFO
		CS_CTL1_SELM_1	VLO
		CS_CTL1_SELM_2	REFO
		CS_CTL1_SELM_3	DCO (Default)
		CS_CTL1_SELM_4	MODOSC
		CS_CTL1_SELM_5	HFXT if available, otherwise DCO

TABLE 7.5 Clock Source Constants Used in SELB, SELA, SELS, and SELM Bits in CSCTL1

DIVM_X can be used. Here, X can get values between 0 and 7 to divide the clock frequency by 2^X.

In Table 7.4, the SELB, SELA, SELS, and SELM bits are used to select the clock source for the BCLK, ACLK, SMCLK, HSMCLK, and MCLK, respectively. We provide the constants used in setting these bits in Table 7.5. As can be seen in this table, each clock has a default clock source associated with it. Besides, different clock sources can be used for the selected clock.

The CSCTL2 register is used to select and configure external crystal oscillators. The entries of this register are tabulated in Table 7.6. Here, HFXT_EN and LFXT_EN bits are used to enable HFXT and LFXT, respectively. Constants for these bits are CS_CTL2_HFXT_EN and CS_CTL2_LFXT_EN. HFXTFREQ bits are used to select a proper frequency range according to frequency of the external crystal used. Constants for these bits are CS_CTL2_HFXTFREQ_X. Here, X can get values between 0 and 6 to select frequency ranges 1–4, 4–8, 8–16, 16–24, 24–32, 32–40, and 40–48 MHz, respectively. For further information on the remaining bits of this register, please see [4].

Power Management and Timing Operations

Bits	25	24	22-20	16	9	8	1-0
	HFXTBYPASS	HFXT_EN	HFXTFREQ	HFXTDRIVE	LFXTBYPASS	LFXT_EN	LFXTDRIVE

TABLE 7.6 CS Control Register 2 (CSCTL2)

Bits	15	10	9	8	3	2	1	0
	REFOFSEL	MODOSC_EN	REFO_EN	VLO_EN	SMCLK_EN	HSMCLK_EN	MCLK_EN	ACLK_EN

TABLE 7.7 CS Clock Enable Register (CSCLKEN)

Bits	9	8	6	5	1	0
	FCNTHFIFG	FCNTLFIFG	DCOR_OPNIFG	DCOR_SHTIFG	HFXTIFG	LFXTIFG

TABLE 7.8 CS Interrupt Flag Register (CSIFG)

Bits	9	8	6	1	0
	CLR_FCNTHFIFG	CLR_FCNTLFIFG	CLR_DCOR_OPNIFG	CLR_HFXTIFG	CLR_LFXTIFG

TABLE 7.9 CS Clear Interrupt Flag Register (CSCLRIFG)

The CSCLKEN register is used to enable some clocks and select REFO frequency. Entries of this register are tabulated in Table 7.7. Here, REFOFSEL bit is used to select the frequency for REFO. Frequency of the REFO is set to 128 kHz if the constant CS_CLKEN_REFOFSEL is used for the REFOFSEL bit. If the constant is not used, the frequency will be set to 32768 Hz which is the default value. For further information on the remaining bits of this register, please see [4].

The CSIFG register is used to control clock system fault interrupt flags. Entries of this register are tabulated in Table 7.8. Here, HFXTIFG and LFXTIFG bits are the flags for HFXT and LFXT oscillator faults, respectively. These bits are set when a fault condition is detected and they must be cleared by software. They are set again if the fault condition still exists. Constants for these bits are CS_IFG_HFXTIFG and CS_IFG_LFXTIFG. For further information on the remaining bits of this register, please see [4].

The CSCLRIFG register is used to clear the clock system fault interrupt flags. The entries of this register are tabulated in Table 7.9. Here, CLR_HFXTIFG and CLR_LFXTIFG bits are used to clear the flags for HFXT and LFXT oscillator faults, with constants CS_CLRIFG_CLR_HFXTIFG and CS_CLRIFG_CLR_LFXTIFG, respectively. For further information on the remaining bits of this register, please see [4].

7.2.5 Clock System Usage via DriverLib Functions

The CS can also be controlled via DriverLib functions. We provide the necessary functions to be used for this purpose below. The complete function list for the CS can be found in [11].

```
void CS_setExternalClockSourceFrequency (uint32_t lfxt_XT_
   CLK_frequency, uint32_t hfxt_XT_CLK_frequency)
/*
Sets the external crystal frequency values for LFXT and HFXT
   oscillators.
This function must be called if an external crystal LFXT or HFXT is
   used. It does not change any register.
lfxt_XT_CLK_frequency: the LFXT frequency in Hz.
hfxt_XT_CLK_frequency: the HFXT frequency in Hz.
*/

void CS_initClockSignal (uint32_t selectedClockSignal, uint32_t
   clockSource, uint32_t clockSourceDivider)
/*
Initializes each clock from selected clock source with optional clock
   divider by modifying CSKEY, CSCTL0, CSCTL1, CSCTL2, and CSCLKEN
   registers.
selectedClockSignal: the clock to be started. It can be one of CS_ACLK,
   CS_MCLK, CS_HSMCLK, CS_SMCLK, or CS_BCLK.
clockSource: the clock source which can be one of LFXT, HFXT, VLOCLK,
   DCO, REFO or MODSC. Please see DriverLib user guide for macro
   definitions.
clockSourceDivider: the clock divider which can be one of 1, 2, 4, 8,
   16, 32, 64 or 128. Please see DriverLib user guide for macro
   definitions.
*/

bool CS_startHFXT (bool bypassMode)
/*
Starts the HFXT crystal oscillator by modifying CSKEY, CSCTL2, and
   CSIFG registers.
bypassMode: If it is true, the HFXT can be driven with external square
   wave.
*/

bool CS_startLFXT (uint32_t xtDrive)
/*
Starts the LFXT crystal oscillator by modifying CSKEY, CSCTL2, and
   CSIFG registers.
xtDrive: the drive strength to drive the oscillator.
*/

void CS_setReferenceOscillatorFrequency (uint8_t referenceFrequency)
/*
Sets the internal REFO clock source to the given frequency by modifying
   CSKEY and CSCLKEN registers.
referenceFrequency: It can be either CS_REFO_32KHZ or CS_REFO_128KHZ.
*/

void CS_setDCOFrequency (uint32_t dcoFrequency)
/*
Sets the DCO to the given frequency by modifying CSKEY and CSCTL0
   registers. It automatically tunes the frequency if necessary.
dcoFrequency: the frequency in Hz that DCO will be set.
*/
```

```
void FlashCtl_setWaitState (uint32_t bank, uint32_t waitState)
/*
Changes the number of wait states used by the flash controller for read
    operations. When changing clock frequency, this function must be
    used to allow readable flash memory.
bank: Flash bank to set wait state which can be one of FLASH_BANK0 or
    FLASH_BANK1 values.
waitState: The number of wait states to set. It can be 0, 1, 2 or 3.
    Please see DriverLib user guide and datasheet for more detail.
*/
```

7.3 Power Control Module

The power control module (PCM) is the main interface between the PSS and CS to adjust the power settings of the MSP432 microcontroller. Hence, it directly controls the power consumption of the device. Related to this, the PCM is responsible for managing the active and low-power mode operations of MSP432. The PCM performs this operation automatically. Therefore, the user does not have to deal with low-level operations.

7.3.1 Power Modes

The PCM offers several power modes for operation. These can be divided into two main groups as active mode (AM) and low-power mode (LPM). The CPU and peripherals will be active in AM. On the other hand, the CPU will be turned off in LPM. Only selected peripherals will be active in LPM. To note here, direct transition between some power modes is not allowed in MSP432. For more information on this issue, please see [4].

Before explaining the power modes in MSP432, we should first take a closer look at the voltage ranges available in the microcontroller. The MSP432 can be supplied with a wide voltage values from 1.62 to 3.65 V. However, the CPU, memory, and digital peripherals run with lower voltage levels than the supply voltage. Hence, a voltage regulator is required to generate core voltage (V_{core}) from the supply voltage. There are two different voltage regulators in MSP432 for this purpose. The first one is a low-dropout (LDO) linear voltage regulator. The second one is a DC-DC switching voltage regulator. The LDO is integrated in the device and does not need any external components. It acts like a variable resistor to regulate voltage at its output. Its low dropout feature allows generating the required voltage even when the supply voltage is very close to the output voltage. However, LDO has low power efficiency. The DC-DC switching voltage regulator continuously turns on and off a switching element (typically a power transistor) to connect its DC input to output. This way, it generates regulated output voltage. The DC-DC switching voltage regulator requires a large external inductor and capacitor to filter switching noise and gives required current to output load. On the other hand, it has high power efficiency.

The MSP432 supports two user-selectable core voltage levels. The first one is the core voltage level 0 in which $V_{core} = 1.2$ V. The second one is the core voltage level 1 in which $V_{core} = 1.4$ V. When a low-frequency operation is to be done, core voltage level 0 can be selected to reduce the power consumption. When a high-frequency operation is to be done, core voltage level 1 must be selected.

Six active modes (AMs) are available in the MSP432 as tabulated in Table 7.10 with their properties. AMs in this table can be grouped into two main categories by the V_{core}

AM	Supplier	Voltage Level	Max. CPU Frequency
AM_LDO_VCORE0	LDO	0	24 MHz
AM_DCDC_VCORE0	DC-DC	0	24 MHz
AM_LDO_VCORE1	LDO	1	48 MHz
AM_DCDC_VCORE1	DC-DC	1	48 MHz
AM_LF_VCORE0	LDO	0	128 kHz
AM_LF_VCORE1	LDO	1	128 kHz

TABLE 7.10 AMs Available in MSP432

LPM	Group	LPM Label
LPM0_LDO_VCORE0	Sleep	LPM0
LPM0_DCDC_VCORE0	Sleep	LPM0
LPM0_LDO_VCORE1	Sleep	LPM0
LPM0_DCDC_VCORE1	Sleep	LPM0
LPM0_LF_VCORE0	Sleep	LPM0
LPM0_LF_VCORE1	Sleep	LPM0
LPM3_VCORE0	Deep Sleep	LPM3
LPM3_VCORE1	Deep Sleep	LPM3
LPM4_VCORE0	Deep Sleep	LPM4
LPM4_VCORE1	Deep Sleep	LPM4
LPM3.5	Shutdown	LPM3.5
LPM4.5	Shutdown	LPM4.5

TABLE 7.11 LPMs Available in MSP432

voltage level they are fed with. At each voltage level, different maximum frequency of operation is supported. The core voltage can be supplied either by the LDO linear or DC-DC switching voltage regulator. As an example, the AM represented by AM_LDO_VCORE0 in Table 7.10 indicates that the LDO linear voltage regulator is used with core voltage level 0 in operation. The maximum CPU frequency in this mode will be 24 MHz.

There are 12 low-power modes (LPMs) in MSP432. These are tabulated in Table 7.11 with their properties. In this table, LPMs can be grouped into three categories as sleep, deep sleep, and shutdown mode. In LPMs, the core voltage level can be 0 or 1. The core voltage can be supplied either by the LDO linear or DC-DC switching voltage regulator at sleep modes. Deep sleep and shutdown modes allow only the LDO linear voltage regulator to generate the core voltage. As an example, the LPM represented by LPM0_LDO_VCORE0 in Table 7.10 indicates that the LDO regulator is used with core voltage level 0 in sleep mode.

In the first four LPMs in Table 7.11, LDO linear and DC-DC switching voltage regulators can be used to generate core voltage levels 0 and 1. In these modes, all low- and high-frequency clocks are active. The flash is also active in these modes. Maximum peripheral clock frequency can be 12 MHz for the first two LPMs. This value increases to 24 MHz for the following two LPMs. In the fifth and sixth LPMs, only the LDO linear voltage regulator can be used to generate core voltage levels 0 and 1. In both modes, all low-frequency clocks are active. On the other hand, high-frequency clocks are disabled. Flash is active in both modes. Maximum peripheral clock frequency can be 128 kHz in these modes. These six low-power modes (labeled as LPM0) are useful to save power when the CPU operation is not required. Using these, interrupt-driven applications can be run with low power, since all the low- and high-frequency clock sources and all peripherals are active.

In the seventh and eighth LPMs, only the LDO linear voltage regulator can be used to generate core voltage levels 0 and 1. In both modes, all low-frequency clocks are active and high-frequency clocks are disabled. Moreover, flash is inactive and pin states are latched for these modes. These two low-power modes (labeled LPM3) can be used to save power when the CPU operation, peripheral usage, and high-frequency clocks are not required for a long time period. In the seventh and eighth LPMs, only certain low-frequency modules are active. Hence, the CPU can be activated only in certain time instants.

In the ninth and tenth LPMs, only the LDO linear voltage regulator can be used to generate core voltage levels 0 and 1. In both modes, all low and high frequency clocks are disabled. Moreover, flash is inactive and pin states are latched for these modes. These two low-power modes (labeled LPM4) can be used to save power when the CPU operation and peripheral usage is not required for a long time period. Since no modules are active in these modes, only external interrupts can wake up the CPU.

In the eleventh LPM, only the LDO linear voltage regulator can be used to generate core voltage level 0. In this mode, CPU, flash, and all peripherals are inactive except Bank0 of the SRAM (to be introduced in Chap. 11). This low-power mode (labeled LPM3.5) can be used to save power when the CPU, peripherals, and high-frequency clocks are not required for a long time period. In this mode, only certain low-frequency modules are active. Hence, the CPU can be activated only in certain time instants. Different from LPM3, the peripheral register data are not retained. Therefore, when the device wakes up, it is required to reconfigure all peripherals to be used in application. Only Bank0 of SRAM can be used as data backup storage while entering this LPM.

In the final LPM, core voltage is turned off. All low- and high-frequency clocks are inactive. CPU, SRAM, flash, and all peripherals are inactive. This low-power mode (labeled LPM4.5) can be used to save power when the CPU and peripheral operations are not required for a long time period. Since no modules are active in this mode, only external interrupts can wake up the CPU. Different from LPM4, the peripheral register data is not retained. When the device wakes up, it is required to reconfigure all peripherals to be used in application. Also different from the LPM3.5 mode, no SRAM data are retained. Hence, all data should be saved to flash before entering to this LPM.

7.3.2 Power Control Module Registers

Five registers are available to control the PCM in the MSP432 microcontroller. We will only explain PCMCTL0 and PCMCTL1 registers in this book. For further information on the remaining registers, please see [4]. A structure named PCM is used to control the above-mentioned registers.

The PCMCTL0 register is used to change the active power mode or check the current power mode of the system. It can also be used to make some LPM requests. The entries of this register are tabulated in Table 7.12. Here, PCMKEY bits are used to access this register to change the power mode. The desired power mode can be selected when

Bits	31-16	13-8	7-4	3-0
	PCMKEY	CPM	LPMR	AMR

TABLE 7.12 PCM Control Register 0 (PCMCTL0)

Constant	Power Mode
PCM_CTL0_CPM_0	AM_LDO_VCORE0
PCM_CTL0_CPM_1	AM_LDO_VCORE1
PCM_CTL0_CPM_4	AM_DCDC_VCORE0
PCM_CTL0_CPM_5	AM_DCDC_VCORE1
PCM_CTL0_CPM_8	AM_LF_VCORE0
PCM_CTL0_CPM_9	AM_LF_VCORE1
PCM_CTL0_CPM_16	LPM0_LDO_VCORE0
PCM_CTL0_CPM_17	LPM0_LDO_VCORE1
PCM_CTL0_CPM_20	LPM0_DCDC_VCORE0
PCM_CTL0_CPM_21	LPM0_DCDC_VCORE1
PCM_CTL0_CPM_24	LPM0_LF_VCORE0
PCM_CTL0_CPM_25	LPM0_LF_VCORE1

TABLE 7.13 Constants for the CPM

Constant	AM
PCM_CTL0_AMR_0	AM_LDO_VCORE0
PCM_CTL0_AMR_1	AM_LDO_VCORE1
PCM_CTL0_AMR_4	AM_DCDC_VCORE0
PCM_CTL0_AMR_5	AM_DCDC_VCORE1
PCM_CTL0_AMR_8	AM_LF_VCORE0
PCM_CTL0_AMR_9	AM_LF_VCORE1

TABLE 7.14 Constants for AMR

the hexadecimal number 0x695A (or the predefined constant PCM_CTL0_KEY_VAL) is written to these bits. CPM bits are used to obtain the current power mode. Constants for these bits and related power modes are tabulated in Table 7.13. LPMR bits are used to make a LPM request. Constants for these bits are PCM_CTL0_LPMR_0 (for LPM3), PCM_CTL0_LPMR_10 (for LPM3.5), and PCM_CTL0_LPMR_12 (for LPM4.5). AMR bits are used to make an AM request. Constants for these bits and related AMs are tabulated in Table 7.14.

The PCMCTL1 register is used to check the power mode changing process or force the system to enter LPM3 (or any LPM with lower value). The entries of this register are tabulated in Table 7.15. Here, the PCMKEY bits are used to access this register. When the hexadecimal number 0x695A (or the predefined constant PCM_CTL1_KEY_VAL) is written to these bits, the configuration in this register is allowed. The PMR_BUSY bit

Bits	31-16	8	2	1	0
	PCMKEY	PMR_BUSY	FORCE_LPM_ENTRY	LOCKBKUP	LOCKLPM5

TABLE 7.15 PCM Control Register 1 (PCMCTL1)

Bits	4	2	1
	SEVONPEND	SLEEPDEEP	SLEEPONEXIT

TABLE 7.16 System Control Register (SCR)

is the busy flag for power mode request. This bit is set when an active or LPM request is processed. It is reset when the transition is done. Constant for this bit is PCM_CTL1_PMR_BUSY. The FORCE_LPM_ENTRY bit is used to force the CPU to enter the desired deep sleep or shutdown mode even essential active clock criteria is not ensured. Constant for this bit is PCM_CTL1_FORCE_LPM_ENTRY. For further information on the remaining bits in this register, please see [4].

The predefined __sleep() function is used to put the device in a desired LPM. If the user wants to use one of the deep sleep or shutdown modes, he or she must perform some modifications in the system control block (SCB) beforehand. The entries of the system control register (SCR) of SCB are tabulated in Table 7.16. Here, the SLEEPDEEP bit is used to differentiate sleep and deep sleep/shutdown modes. Constant for this bit is SCB_SCR_SLEEPDEEP_Msk. This bit must be set before using the __sleep() function to use deep sleep or shutdown modes. The SLEEPONEXIT bit is used to put the device into sleep mode again when exiting from ISR. If the user wants to perform some actions after returning from ISR, he or she must not set this bit. Constant for this bit is SCB_SCR_SLEEPONEXIT_Msk. For further information on the remaining bits of the SCR, please see [4].

7.3.3 Power Control Module Usage via DriverLib Functions

The PCM can also be controlled via DriverLib functions. We provide functions of interest below. The complete function list and macro definitions can be found in [11].

```
bool PCM_setPowerState(uint_fast8_t powerState)
/*
Adjusts the power mode (including low power and shutdown modes) and
    core voltage levels, automatically switches from one mode to
    another if necessary by modifying PCMCTL0 and PCMCTL1 registers.
    This function combines all functionality of other functions to
    control power modes and voltage levels in DriverLib.
powerState: the power state to be set. It can be one of
    PCM_AM_LDO_VCORE0, PCM_AM_LDO_VCORE1, PCM_AM_DCDC_VCORE0,
    PCM_AM_DCDC_VCORE1, PCM_AM_LF_VCORE0, PCM_AM_LF_VCORE1,
    PCM_LPM0_LDO_VCORE0, PCM_LPM0_LDO_VCORE1, PCM_LPM0_DCDC_VCORE0,
    PCM_LPM0_DCDC_VCORE1, PCM_LPM0_LF_VCORE0, PCM_LPM0_LF_VCORE1,
    PCM_LPM3, PCM_LPM35_VCORE0, PCM_LPM4 or PCM_LPM45
*/

void Interrupt_enableSleepOnIsrExit(void)
/*
```

```
    Enables entering sleep mode when return from an ISR function by
        modifying SLEEPONEXIT bit of the SCR.
*/

void Interrupt_disableSleepOnIsrExit(void)
/*
Disables entering sleep mode when return from an ISR function by
    modifying SLEEPONEXIT bit of the SCR.
*/
```

7.4 Watchdog Timer

The watchdog timer (WDT_A) module performs controlled system restart when a predefined time interval passes. This property can be used to reset the CPU if a software problem occurs. This can be done as follows. The WDT_A module works independent of the CPU. It should be disabled every time before generating the reset interrupt within the program. If the CPU cannot achieve this disabling action in time due to a software problem, then the WDT_A module resets the CPU. The program restarts again from the beginning. This way, the fault may be bypassed during operation. We provide the block diagram of the WDT_A module in Fig. 7.1.

The WDT_A module is active by default. However, it can be disabled if it is not needed in an application. WDT_A can also be configured as an interval timer. Hence, it can generate interrupts at selected time intervals.

7.4.1 Watchdog Timer Registers

One 16-bit register is available in the MSP432 microcontroller to control the WDT_A module. This register is called WDTCTL. A structure named WDT_A is used to control this register.

The WDTCTL register is used to configure and control the WDT_A module. Entries of this register are tabulated in Table 7.17. Here, the WDTPW bits are used for entering the password. To stop the reset signal (power up clear, PUC), the hexadecimal number 0x05A (or the predefined constant WDT_A_CTL_PW) should be written to it. By setting the WDTHOLD bit (or using the predefined constant WDT_A_CTL_HOLD), the WDT_A module can be stopped. The WDTSSEL bits select the WDT_A module clock. Constants for these bits are WDT_A_CTL_SSEL_0 (for SMCLK), WDT_A_CTL_SSEL_1 (for ACLK), WDT_A_CTL_SSEL_2 (for VLOCLK), and WDT_A_CTL_SSEL_3 (for BCLK). The WDTTMSEL bit is used for mode selection. When this bit is set (or the predefined constant WDT_A_CTL_TMSEL is used), the WDT_A module can be used as an interval timer (without any watchdog operation). Otherwise, it is used as watchdog. The WDTCNTCL bit clears the watchdog counter to 0x0000. Constant for this bit is WDT_A_CTL_CNTCL. The WDTIS bits are used for interval select (both for watchdog and timer operations). Constant for these bits is WDT_A_CTL_IS_X where X can get values between 0 and 7 to divide the clock frequency by 2^{31}, 2^{27}, 2^{23}, 2^{19}, 2^{15}, 2^{13}, 2^9, and 2^6, respectively.

Bits	15-8	7	6-5	4	3	2-0
	WDTPW	WDTHOLD	WDTSSEL	WDTTMSEL	WDTCNTCL	WDTIS

TABLE 7.17 WDT Control Register (WDTCTL)

7.4.2 Watchdog Timer Usage via DriverLib Functions

The WDT_A module can also be controlled with predefined DriverLib functions. We provide functions of interest below. The complete function list and macro definitions can be found in [11].

```
void WDT_A_holdTimer (void)
/*
Pauses watchdog timer counting by modifying WDTPW and WDTHOLD bits
    of the WDTCTL register.
*/

void WDT_A_startTimer (void)
/*
Starts watchdog timer counting by modifying WDTPW and WDTHOLD bits
    of the WDTCTL register.
*/

void WDT_A_clearTimer (void)
```

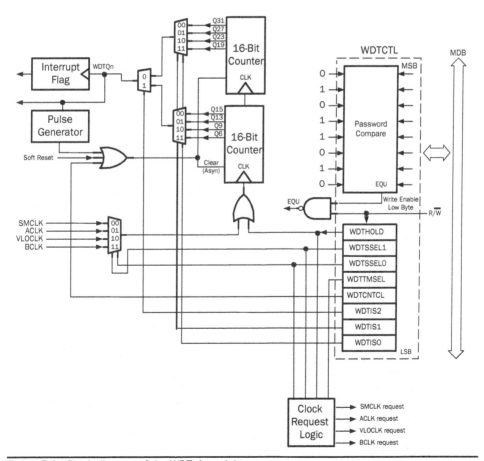

FIGURE 7.1 Block diagram of the WDT_A module.

```
/*
Resets watchdog timer count by modifying WDTCNTCL bit of the WDTCTL
    register.
*/

void WDT_A_initWatchdogTimer (uint_fast8_t clockSelect, uint_fast8_t
    clockIterations)
/*
Initializes the watchdog timer in watchdog mode by modifying WDTSSEL,
    WDTTMSEL, and WDTIS bits of the WDTCTL register.
clockSelect: the clock for the watchdog timer. It can be one of SMCLK,
    ACLK, VLOCLK or BCLK. Please see DriverLib user guide for macro
    definitions.
clockIterations: the timer tick to be counted before resetting the
    device. Possible values are 64, 512, 8192, 32K, 512K, 8192K, 128M,
    and 2G. Please see DriverLib user guide for macro definitions.
*/

void WDT_A_initIntervalTimer (uint_fast8_t clockSelect, uint_fast8_t
    clockDivider)
/*
Initializes the watchdog timer in interval timer mode by modifying
    WDTSSEL, WDTTMSEL, and WDTIS bits of the WDTCTL register.
clockSelect: the clock for the watchdog timer. It can be one of SMCLK,
    ACLK, VLOCLK or BCLK. Please see DriverLib user guide for macro
    definitions.
clockDivider: the timer tick to be counted before overflow. Possible
    values are 64, 512, 8192, 32K, 512K, 8192K, 128M, and 2G. Please
    see DriverLib user guide for macro definitions.
*/
```

7.4.3 Coding Practices for the Watchdog Timer Module

We can explain the working principles of the WDT_A module by using it. Therefore, we provide register-level and DriverLib examples in this section. We use the WDT_A module both in watchdog and timer modes in the provided examples. Here, we use LED1 connected to pin P1.0 on the MSP432 LaunchPad as output. The button S1 connected to pin P1.1 on the MSP432 LaunchPad is set as input.

7.4.3.1 Register-Level Examples

We first provide the C code in which the WDT_A module is used as watchdog in Listing 7.1. In this code, initially the WDT_A module and LED1 are off. Also, the system is in LPM0_LDO_VCORE0 mode waiting for a port interrupt. As we press the button S1, the system goes to the port ISR in which LED1 turns on and WDT_A module starts running. We use the frequency-divided ACLK (supplied by REFO) for the WDT_A module. This gives a 4-second delay with selected interval. Then, the system returns from the ISR and goes back to LPM0_LDO_VCORE0 mode. Meanwhile, the WDT_A module works in background. As the WDT_A module resets the microcontroller, the program counter goes to the beginning of the main function. Then, the WDT_A module and LED1 are turned off again. The system waits for another button press to repeat the procedure.

Listing 7.1 Usage of the WDT_A Module in Watchdog Mode, Using Registers

```c
#include "msp.h"

#define LED1 BIT0
#define S1 BIT1

void main(void) {

  WDT_A->CTL = WDT_A_CTL_PW | WDT_A_CTL_HOLD;

  P1->DIR = LED1;
  P1->REN = S1;
  P1->OUT = S1;
  P1->IE = S1;
  P1->IES = 0x00;
  P1->IFG = 0x00;

  CS->KEY = CS_KEY_VAL;
  CS->CTL1 |= CS_CTL1_SELA_2 | CS_CTL1_DIVA_2;
  CS->KEY = 0;

  NVIC->ISER[1] = 1 << ((PORT1_IRQn) & 31);
  __enable_interrupts();

  SCB->SCR |= SCB_SCR_SLEEPONEXIT_Msk;
  __sleep();
}

void PORT1_IRQHandler(void) {
  __delay_cycles(5000);
  P1IFG = 0x00;
  P1OUT ^= LED1;
  WDT_A->CTL=WDT_A_CTL_PW|WDT_A_CTL_SSEL_1|WDT_A_CTL_CNTCL|
      WDT_A_CTL_IS_4;
}
```

In Listing 7.2, we provide a sample code for the usage of the WDT_A module in timer mode. Here, LED1 toggles every second by using the ACLK with selected interval. All clocks of the system are supplied from REFO and active mode for the system is changed to AM_LF_VCORE0. Between LED toggle operations, the system is put in LPM3 mode since the WDT_A module can be used in LPM3 with timer mode.

7.4.3.2 DriverLib Examples

We repeat the previous example in Listing 7.1 with DriverLib functions. We provide the corresponding C code in Listing 7.3. Here, we first set ACLK (to be supplied from 32-kHz REFO clock source) using `CS_setReferenceOscillatorFrequency` and `CS_initClockSignal` functions. Afterwards, we set LED1 as output and S1 as input with the pull-up resistor. We also enable the port interrupt. We put the system on LPM0_LDO_VCORE0 mode using `PCM_setPowerState` function. In the ISR, we turn on LED1. Then, we enable and start the WDT_A module in watchdog mode using `WDT_A_initWatchdogTimer` and `WDT_A_startTimer` functions.

Listing 7.2 Usage of the WDT_A Module in Timer Mode, Using Registers

```c
#include "msp.h"

#define LED1 BIT0

void main(void) {

  WDT_A->CTL = WDT_A_CTL_PW | WDT_A_CTL_HOLD;

  P1->DIR = LED1;

  CS->KEY = CS_KEY_VAL;
  CS->CTL1 |= CS_CTL1_SELB;
  CS->CTL1 |= CS_CTL1_SELA_2;
  CS->CTL1 &= ~(CS_CTL1_SELS_MASK | CS_CTL1_DIVS_MASK);
  CS->CTL1 |= CS_CTL1_SELS_2;
  CS->CTL1 &= ~(CS_CTL1_SELM_MASK | CS_CTL1_DIVM_MASK);
  CS->CTL1 |= CS_CTL1_SELM_2;
  CS->KEY = 0;

  while (PCM->CTL1 & PCM_CTL1_PMR_BUSY);
    PCM->CTL0 = PCM_CTL0_KEY_VAL | PCM_CTL0_AMR_8;
  while (PCM->CTL1 & PCM_CTL1_PMR_BUSY);
    WDT_A->CTL=WDT_A_CTL_PW|WDT_A_CTL_SSEL_1|WDT_A_CTL_TMSEL|
       WDT_A_CTL_CNTCL|WDT_A_CTL_IS_4;

  NVIC->ISER[0] = 1 << ((WDT_A_IRQn) & 31);
  __enable_irq();

  SCB->SCR |= SCB_SCR_SLEEPONEXIT_Msk | SCB_SCR_SLEEPDEEP_Msk;
  __sleep();
}
void WDT_A_IRQHandler(void) {
  P1->OUT ^= LED1;
}
```

We repeat the example in Listing 7.2 using DriverLib functions. We provide the corresponding C code in Listing 7.4. Different from the previous example in Listing 7.3, we initialize and start the WDT_A module in timer mode using `WDT_A_initIntervalTimer` and `WDT_A_startTimer` functions in the main function. Then, we enable the watchdog interrupt and toggle LED1 in the watchdog ISR.

7.5 System Timer

The ARM Cortex-M4 core (hence the MSP432) has an integrated system timer module called SysTick. This module has a flexible control mechanism. It has a 24-bit counter counting downwards to zero. The main usage area of the SysTick module is providing periodic interrupt signals for an RTOS (to be considered in Chap. 13). However, it can also be used for other simple timing purposes as well.

Listing 7.3 Usage of the WDT_A Module in Watchdog Mode, Using DriverLib Functions

```
#include <ti/devices/msp432p4xx/driverlib/driverlib.h>

void main(void){

 WDT_A_holdTimer();

 CS_setReferenceOscillatorFrequency(CS_REFO_32KHZ);
 CS_initClockSignal(CS_ACLK, CS_REFOCLK_SELECT, CS_CLOCK_DIVIDER_4);

 GPIO_setAsOutputPin(GPIO_PORT_P1, GPIO_PIN0);
 GPIO_setOutputLowOnPin(GPIO_PORT_P1, GPIO_PIN0);

 GPIO_setAsInputPinWithPullUpResistor(GPIO_PORT_P1, GPIO_PIN1);
 GPIO_clearInterruptFlag(GPIO_PORT_P1, GPIO_PIN1);
 GPIO_enableInterrupt(GPIO_PORT_P1, GPIO_PIN1);
 Interrupt_enableInterrupt(INT_PORT1);
 Interrupt_enableSleepOnIsrExit();
 Interrupt_enableMaster();

 PCM_setPowerState(PCM_LPM0_LDO_VCORE0);
}

void PORT1_IRQHandler(void){
 uint32_t status = GPIO_getEnabledInterruptStatus(GPIO_PORT_P1);
 GPIO_clearInterruptFlag(GPIO_PORT_P1, status);

 if (status & GPIO_PIN1){
  GPIO_setOutputHighOnPin(GPIO_PORT_P1, GPIO_PIN0);
  SysCtl_setWDTTimeoutResetType(SYSCTL_SOFT_RESET);
  WDT_A_initWatchdogTimer(WDT_A_CLOCKSOURCE_ACLK,
     WDT_A_CLOCKITERATIONS_32K);
  WDT_A_startTimer();}
}
```

7.5.1 System Timer Registers

There are four 32-bit registers defined to control the SysTick module. We will explain STCSR, STRVR, and STCVR in this book. For further information on the remaining registers, please see [4]. A structure named SysTick is used to control the registers. The registers of interest are named CTRL, LOAD, and VAL in this structure.

The STCSR is used to control the SysTick. Entries of this register are tabulated in Table 7.18. Here, COUNTFLAG bit is the flag of the SysTick. When the counter reaches zero, this bit is set. The constant for this bit is SysTick_CTRL_COUNTFLAG_Msk. The CLKSOURCE bit is used to source SysTick from the MCLK by the constant SysTick_CTRL_CLKSOURCE_Msk. The TICKINT bit is used to enable the SysTick interrupt by the constant SysTick_CTRL_TICKINT_Msk. The ENABLE bit is used to enable SysTick. Constant for this bit is SysTick_CTRL_ENABLE_Msk.

The STRVR is used to specify the count value for the SysTick. It can take any value between 1 and 0x00FFFFFF ($2^{24} - 1$). When the SysTick counts to 0, this value is reloaded again. The STCVR is used to get the current count value of SysTick. Any value written to this register clears the count value.

Listing 7.4 Usage of the WDT_A Module in Timer Mode, Using DriverLib Functions

```c
#include <ti/devices/msp432p4xx/driverlib/driverlib.h>

void main(void) {

  WDT_A_holdTimer();

  CS_setReferenceOscillatorFrequency(CS_REFO_32KHZ);
  CS_initClockSignal(CS_ACLK, CS_REFOCLK_SELECT, CS_CLOCK_DIVIDER_1);

  PCM_setPowerState(PCM_AM_LF_VCORE0);

  GPIO_setAsOutputPin(GPIO_PORT_P1, GPIO_PIN0);
  GPIO_setOutputLowOnPin(GPIO_PORT_P1, GPIO_PIN0);

  WDT_A_initIntervalTimer(WDT_A_CLOCKSOURCE_ACLK ,
      WDT_A_CLOCKITERATIONS_32K);
  Interrupt_enableInterrupt(INT_WDT_A);
  Interrupt_enableMaster();
  Interrupt_enableSleepOnIsrExit();
  WDT_A_startTimer();

  PCM_setPowerState(PCM_LPM3);
}

void WDT_A_IRQHandler(void) {
  GPIO_toggleOutputOnPin(GPIO_PORT_P1, GPIO_PIN0);
}
```

Bits	16	2	1	0
	COUNTFLAG	CLKSOURCE	TICKINT	ENABLE

TABLE 7.18 SysTick Control and Status Register (STCSR)

7.5.2 System Timer Usage via DriverLib Functions

The system timer can also be controlled by predefined DriverLib functions given below. The complete function list and macro definitions can be found in [11].

```
void SysTick_enableModule (void)
/*
Enables the SysTick timer by modifying CLKSOURCE and ENABLE bits of
    STCSR.
*/

void SysTick_disableModule (void)
/*
Disables the SysTick timer by modifying ENABLE bit of STCSR.
*/

void SysTick_enableInterrupt (void)
/*
```

```
Enables the SysTick timer interrupt by modifying TICKINT bit of STCSR.
    We do not need to clear the SysTick interrupt flag. The NVIC
    automatically clears it when the ISR is called.
*/

void SysTick_disableInterrupt (void)
/*
Disables the SysTick timer interrupt by modifying TICKINT bit of STCSR.
*/

void SysTick_setPeriod (uint32_t period)
/*
Sets the period of SysTick timer by modifying STRVR.
period: the period in terms of clock cycles.
*/

uint32_t SysTick_getValue (void)
/*
Returns the current SysTick counter value by reading the contents of
    STCVR.
*/
```

7.5.3 Coding Practices for the System Timer Module

We can use the SysTick to generate periodic time interrupts. We provide one example to serve this purpose using both register-level and DriverLib functions next. In both examples, LED1 connected to pin P1.0 on the MSP432 LaunchPad is used as output.

7.5.3.1 Register-Level Example

In Listing 7.5, we provide a sample code for the usage of the SysTick module. Here, LED1 toggles every 0.25 second by using SysTick. First, the AM for the system is changed to AM_LDO_VCORE1. Then, DCO frequency is set to 48 MHz and SysTick is loaded by the constant 12M to obtain a 0.25-second time interval.

7.5.3.2 DriverLib Example

We repeat the example given in Listing 7.5 using predefined DriverLib functions. We provide the corresponding C code in Listing 7.6. Here, we first set the power mode, flash wait state, and DCO frequency using functions PCM_setPowerState, FlashCtl_setWaitState, and CS_setDCOFrequency, respectively. Afterwards, we set LED1 as output. Then, we enable the SysTick module and set its period using SysTick_enableModule, and SysTick_setPeriod functions, respectively. We enable the interrupt of SysTick using the function SysTick_enableInterrupt. We toggle LED1 in the SysTick ISR.

7.6 Timer32

In MSP432, there is one Timer32 module consisting of two identical 32-bit down counters that can be configured individually. Each counter supports three different timer modes with prescale and interrupt options. The first mode is called free-running. This is the default mode in which the counter restarts after reaching zero and continues counting down from the maximum value. The second mode is called periodic timer.

Listing 7.5 Usage of the SysTick Module to Generate Periodic Interrupts, Using Registers

```c
#include "msp.h"

#define LED1 BIT0
#define TIMER_PERIOD 12000000

void main(void){

 WDT_A->CTL = WDT_A_CTL_PW | WDT_A_CTL_HOLD;

 P1->DIR = LED1;

 while (PCM->CTL1 & PCM_CTL1_PMR_BUSY);
  PCM->CTL0 = PCM_CTL0_KEY_VAL | PCM_CTL0_AMR_1;
  while (PCM->CTL1 & PCM_CTL1_PMR_BUSY);

 FLCTL->BANK0_RDCTL = (FLCTL->BANK0_RDCTL & ~(
     FLCTL_BANK0_RDCTL_WAIT_MASK)) | FLCTL_BANK0_RDCTL_WAIT_1;
 FLCTL->BANK1_RDCTL = (FLCTL->BANK0_RDCTL & ~(
     FLCTL_BANK1_RDCTL_WAIT_MASK)) | FLCTL_BANK1_RDCTL_WAIT_1;

 CS->KEY = CS_KEY_VAL;
 CS->CTL0 |= CS_CTL0_DCORSEL_5;
 CS->KEY = 0;

 SysTick->CTRL |= SysTick_CTRL_CLKSOURCE_Msk | SysTick_CTRL_ENABLE_Msk;
 SysTick->LOAD = TIMER_PERIOD;
 SysTick->VAL = 0;
 SysTick->CTRL |= SysTick_CTRL_TICKINT_Msk;

 __enable_irq();

 while(1);
}
void SysTick_Handler(void){
 P1->OUT ^= LED1;
}
```

In this mode, the counter generates an interrupt at constant intervals, reloading the original value after reaching zero. The third mode is called one-shot timer. In this mode, the counter generates an interrupt once. When the counter reaches zero, it halts until restarted by the user. To note here, the Timer32 module uses MCLK.

7.6.1 Timer32 Registers

Fourteen registers are available in the MSP432 microcontroller to control the two Timer32 counters. We will only focus on T32LOADx, T32VALUEx, T32CONTROLx, and T32INTCLRx registers in this book. For further information on the remaining registers, please see [4]. Here, x can get the value 1 or 2 to differentiate the Timer32 counters. A structure named TIMER32_x is used to control the mentioned registers. Again, x can get the value of 1 or 2. The registers of interest are named LOAD, VALUE, CONTROL, and INTCLR in this structure.

Listing 7.6 Usage of the SysTick Module to Generate Periodic Interrupts, Using DriverLib Functions

```
#include <ti/devices/msp432p4xx/driverlib/driverlib.h>

#define TIMER_PERIOD 12000000

void main(void){

 WDT_A_holdTimer();
 PCM_setPowerState(PCM_AM_LDO_VCORE1);
 FlashCtl_setWaitState(FLASH_BANK0, 2);
 FlashCtl_setWaitState(FLASH_BANK1, 2);
 CS_setDCOFrequency(CS_48MHZ);

 GPIO_setAsOutputPin(GPIO_PORT_P1, GPIO_PIN0);

 SysTick_enableModule();
 SysTick_setPeriod(12000000);
 SysTick_enableInterrupt();
 Interrupt_enableMaster();

 while(1);
}
void SysTick_Handler(void){
 GPIO_toggleOutputOnPin(GPIO_PORT_P1, GPIO_PIN0);
}
```

The T32LOADx register is used to set the count value for the Timer32 module. It can take any value between 1 and 0xFFFF ($2^{16} - 1$) or 0xFFFFFFFF ($2^{32} - 1$) based on the chosen counter size. When the Timer32 counts to 0, this value is reloaded again. The T32VALUEx register is used to get the current value of the Timer32 counter.

The T32CONTROLx register is used to control the Timer32 module. Entries of this register are tabulated in Table 7.19. Here, the ENABLE bit is used to enable or disable the Timer32 module. When this bit is set, the Timer32 module is enabled. When it is reset, the module is disabled. Constant for this bit is TIMER32_CONTROL_ENABLE. The MODE bit is used to select the Timer32 mode. When this bit is set, the free-running mode is selected. When it is reset, the periodic mode is selected. Constant for this bit is TIMER32_CONTROL_MODE. The IE bit is used to enable or disable the interrupt for the Timer32 module. When this bit is set, the Timer32 interrupt is enabled. When it is reset, the interrupt is disabled. Constant for this bit is TIMER32_CONTROL_IE. PRESCALE bits are used to divide the Timer32 clock. Constants for these bits are TIMER32_CONTROL_PRESCALE_x where x can get values 0, 1, or 2 to divide the clock frequency by 1, 16, or 256, respectively. The SIZE bit is used to choose the counter size for the Timer32 module. When this bit is set, 32-bit counter is used. When it is reset, 16-bit counter is used. Constant for this bit is TIMER32_CONTROL_SIZE. The ONESHOT bit is used to set the counter mode for the Timer32 module. When this bit is set, one-shot mode is selected. When it is reset, wrapping mode (restart after reaching zero) is selected. Constant for this bit is TIMER32_CONTROL_ONESHOT.

The T32INTCLRx register is used to clear the interrupt of the Timer32 module. Any value written to this register clears the interrupt.

Bits	7	6	5	3-2	1	0
	ENABLE	MODE	IE	PRESCALE	SIZE	ONESHOT

TABLE 7.19 Timer32 Control Register x (T32CONTROLx)

7.6.2 Timer32 Usage via DriverLib Functions

The Timer32 module can also be controlled with predefined DriverLib functions. We provide functions of interest below. Here, `timer` is the Timer32 base to be initialized. It can be either TIMER32_0_BASE or TIMER32_1_BASE. The complete function list and macro definitions can be found in [11].

```
void Timer32_initModule (uint32_t timer, uint32_t preScaler, uint32_t
    resolution, uint32_t mode)
/*
Initializes the Timer32 module by modifying the T32CONTROLx registers.
preScaler: the clock divider value. It can be one of
    TIMER32_PRESCALER_1, TIMER32_PRESCALER_16 or TIMER32_PRESCALER_256.
resolution: the Timer32 count resolution. It can be one of
    TIMER32_16BIT or TIMER32_32BIT.
mode: It is the Timer32 count mode. It can be one of
    TIMER32_FREE_RUN_MODE or TIMER32_PERIODIC_MODE.
*/

void Timer32_setCount (uint32_t timer, uint32_t count)
/*
Sets the timer count to be counted before overflow by modifying the
    T32LOADx register.
count: the timer count.
*/

void Timer32_startTimer (uint32_t timer, bool oneShot)
/*
Starts the Timer32 counting by modifying the ENABLE bit of the
    T32CONTROLx register.
oneShot: If this value is at logic level 1, then the Timer32 module
    stops after first overflow, else it runs in continuous mode.
*/

void Timer32_haltTimer (uint32_t timer)
/*
Pauses the Timer32 counting by modifying the ENABLE bit of the
    T32CONTROLx register.
*/

void Timer32_enableInterrupt (uint32_t timer)
/*
Enables the Timer32 interrupt by modifying the IE bit of the
    T32CONTROLx register.
*/

void Timer32_disableInterrupt (uint32_t timer)
/*
Disables the Timer32 interrupt by modifying the IE bit of the
    T32CONTROLx register.
```

```
*/
void Timer32_clearInterruptFlag (uint32_t timer)
/*
Clears the the Timer32 interrupt flag by modifying the T32INTCLRx
    register.
*/
```

7.6.3 Coding Practices for the Timer32 Module

We provide a sample code on the usage of the Timer32 module in this section. We provide both register-level and DriverLib functional form of the code next. In both forms, LED1 connected to pin P1.0 on the MSP432 LaunchPad is set as output.

7.6.3.1 Register-Level Example

We provide the sample C code on the usage of the Timer32 module in Listing 7.7. Here, LED1 toggles every second by using the Timer32 module interrupts. To do so, first the AM of the system is changed to AM_LDO_VCORE1. Then, HFTX is enabled to source the MCLK with 48-MHz clock frequency. The Timer32 module frequency divider is selected as 16. The timer is loaded with constant 3M to obtain a 1-second periodic time interval. The system is put in LPM0_LDO_VCORE1 mode and waits for the Timer32 interrupt. When an interrupt comes, LED1 toggles.

7.6.3.2 DriverLib Example

We repeat the example in Listing 7.7 using predefined DriverLib functions. We provide the corresponding C code in Listing 7.8. Here, we first set the power mode and flash wait states. Then, we set the crystal pins as peripheral output pins. We set the external crystal frequency and start the HFTX using `CS_setExternalClockSourceFrequency` and `CS_startHFXT` functions. Afterwards, we enable the HFXT clock source so that MCLK is supplied from 48-MHz clock source. We set LED1 pin as output. Then, we initialize the Timer32 module using `Timer32_initModule` function. We set the timer period and enable the interrupt using `Timer32_setCount` and `Timer32_enableInterrupt` functions. Finally, we start the Timer32 to count using `Timer32_startTimer` function and put system on sleep. We toggle LED1 in the Timer32 ISR function.

7.7 Timer_A

There are four identical Timer_A modules in the MSP432 microcontroller. These are named TA0, TA1, TA2, and TA3. We provide the block diagram of one of the Timer_A modules in Fig. 7.2. As can be seen in this figure, the module has a 16-bit counter that can be fed by four different clocks.

Each Timer_A module can work in four different operating modes. These are the stop, up, continuous, and up/down modes. In the stop mode, the timer of the Timer_A module stops counting while saving the present count value. The Timer_A is initially in this mode to save power. In the up mode, the timer counts up until it reaches a predefined value. Then, it restarts from zero again as shown in Fig. 7.3a. In the continuous mode, the timer counts up until it reaches 0xFFFF (or 65535 in decimal), then restarts from zero again as shown in Fig. 7.3b. In the up/down mode, the timer counts up until it reaches a predefined value. Then, counting is inverted and the timer counts down to zero as shown in Fig. 7.3c. We will explain the four operating modes in detail in Sec. 7.7.1.

Listing 7.7 Usage of the Timer32 Module to Generate Periodic Time Interrupts, Using Registers

```c
#include "msp.h"

#define LED1 BIT0
#define TIMER_PERIOD 3000000

void main(void){

  WDT_A->CTL = WDT_A_CTL_PW | WDT_A_CTL_HOLD;

  P1->DIR = LED1;

  TIMER32_1->CONTROL = TIMER32_CONTROL_SIZE | TIMER32_CONTROL_MODE |
      TIMER32_CONTROL_PRESCALE_1;
  TIMER32_1->LOAD = TIMER_PERIOD;

  while (PCM->CTL1 & PCM_CTL1_PMR_BUSY);
  PCM->CTL0 = PCM_CTL0_KEY_VAL | PCM_CTL0_AMR_1;
  while (PCM->CTL1 & PCM_CTL1_PMR_BUSY);

  FLCTL->BANK0_RDCTL = (FLCTL->BANK0_RDCTL & ~(FLCTL_BANK0_RDCTL_WAIT_
    MASK)) | FLCTL_BANK0_RDCTL_WAIT_1;
  FLCTL->BANK1_RDCTL = (FLCTL->BANK0_RDCTL & ~(FLCTL_BANK1_RDCTL_WAIT_
    MASK)) | FLCTL_BANK1_RDCTL_WAIT_1;

  PJ->SEL0 |= BIT2 | BIT3;
  PJ->SEL1 &= ~(BIT2 | BIT3);

  CS->KEY = CS_KEY_VAL ;
  CS->CTL2 |= CS_CTL2_HFXT_EN | CS_CTL2_HFXTFREQ_6;
  while(CS->IFG & CS_IFG_HFXTIFG)
   CS->CLRIFG |= CS_CLRIFG_CLR_HFXTIFG;
  CS->CTL1 &= ~(CS_CTL1_SELM_MASK | CS_CTL1_DIVM_MASK);
  CS->CTL1 |= CS_CTL1_SELM_5;
  CS->KEY = 0;

  TIMER32_1->CONTROL |= TIMER32_CONTROL_ENABLE | TIMER32_CONTROL_IE;

  __enable_irq();
  NVIC->ISER[0] = 1 << ((T32_INT1_IRQn) & 31);

  SCB->SCR |= SCB_SCR_SLEEPONEXIT_Msk;
  __sleep();
}
void T32_INT1_IRQHandler(void){
  TIMER32_1->INTCLR = 0;
  P1->OUT ^= LED1;
}
```

The Timer_A module also has seven independent capture/compare blocks as can be seen in Fig. 7.2. The purpose of the capture block is to link the changes in the input signal with time (clock) values. The purpose of the compare block is to generate interrupts at specific time intervals. This can be used to form pulse width modulation (PWM) signals. We will explain in detail in Sec. 7.7.1 how the capture/compare blocks can be used.

Power Management and Timing Operations

Listing 7.8 Usage of the Timer32 Module to Generate Periodic Time Interrupts, Using DriverLib Functions

```
#include <ti/devices/msp432p4xx/driverlib/driverlib.h>

#define TIMER_PERIOD 3000000

void main(void){

WDT_A_holdTimer();
PCM_setPowerState(PCM_AM_LDO_VCORE1);
FlashCtl_setWaitState(FLASH_BANK0, 2);
FlashCtl_setWaitState(FLASH_BANK1, 2);

GPIO_setAsPeripheralModuleFunctionOutputPin(GPIO_PORT_PJ, GPIO_PIN3 |
    GPIO_PIN2, GPIO_PRIMARY_MODULE_FUNCTION);
CS_setExternalClockSourceFrequency(32000,CS_48MHZ);
CS_startHFXT(false);
CS_initClockSignal(CS_MCLK, CS_HFXTCLK_SELECT, CS_CLOCK_DIVIDER_1);

GPIO_setAsOutputPin(GPIO_PORT_P1, GPIO_PIN0);
GPIO_setOutputLowOnPin(GPIO_PORT_P1, GPIO_PIN0);

Timer32_initModule(TIMER32_0_BASE, TIMER32_PRESCALER_16,
    TIMER32_32BIT, TIMER32_PERIODIC_MODE);
Timer32_setCount(TIMER32_0_BASE,TIMER_PERIOD);

Interrupt_enableSleepOnIsrExit();
Interrupt_enableInterrupt(TIMER32_0_INTERRUPT);
Timer32_enableInterrupt(TIMER32_0_BASE);
Interrupt_enableMaster();

Timer32_startTimer(TIMER32_0_BASE, false);

PCM_setPowerState(PCM_LPM0_LDO_VCORE1);
}

void T32_INT1_IRQHandler(void){
Timer32_clearInterruptFlag(TIMER32_0_BASE);
GPIO_toggleOutputOnPin(GPIO_PORT_P1, GPIO_PIN0);
}
```

7.7.1 Timer_A Registers

Eighteen registers are available in the MSP432 microcontroller to control each Timer_A module. These are called TAxCTL, TAxR, TAxIV, TAxEX0, TAxCCTLy, and TAxCCRy. Here, x can get values from 0 to 3 to represent four Timer_A modules and y can get values from 0 to 6 to represent seven different capture/compare blocks for each Timer_A module. A structure named TIMER_Ax is used to control these registers. The registers of interest are named CTL, R, IV, EX0, CCTL[y], and CCR[y] in this structure. Here x can get values from 0 to 3 and y can get values from 0 to 6.

The TAxR register is the counter of the Timer_A module. It is the core of the timer block. Timer count results are kept in this 16-bit register.

The TAxCTL register is used to control the Timer_A module. Entries of this register are tabulated in Table 7.20.

132 Chapter Seven

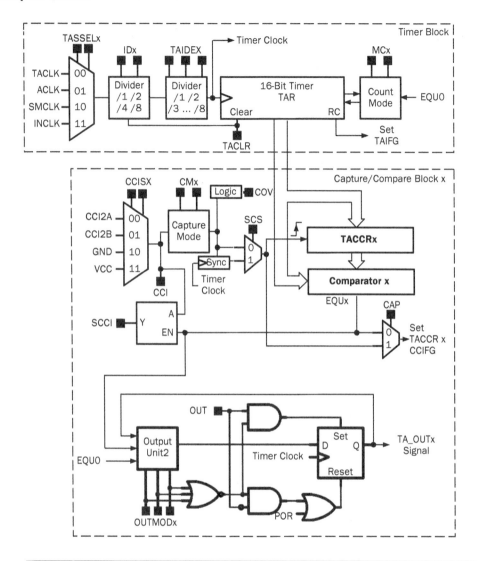

FIGURE 7.2 Block diagram of the Timer_A module.

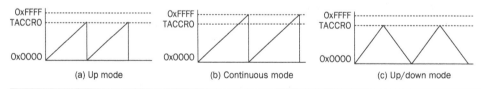

FIGURE 7.3 Timer_A module, count modes.

In Table 7.20, TASSEL bits are used to select the clock for the Timer_A module. Constants for these bits are TIMER_A_CTL_TASSEL_x, where x can get values from 0 to 3. Corresponding clock selections for these constants are TACLK, ACLK, SMCLK, and INCLK (inverse of the TACLK), respectively. ID bits are used for frequency division.

Bits	9-8	7-6	5-4	2	1	0
	TASSEL	ID	MC	TACLR	TAIE	TAIFG

TABLE 7.20 Timer_A Control Register (TAxCTL)

Constants for these bits are TIMER_A_CTL_ID_x, where x can get values from 0 to 3. Corresponding frequency division values for these constants are 1, 2, 4, and 8, respectively. MC bits are used to select the counting mode for the Timer_A module. Constants for these bits are TIMER_A_CTL_MC_x, where x can get values from 0 to 3. Corresponding mode selections for these constants are stop, up, continuous, and up/down modes, respectively.

In the stop mode, the timer stops counting and TAxR retains its value to continue later when this mode is selected. The Timer_A module is initially in this mode to save power.

In the up mode, the timer counts up until it reaches the value in TAxCCR0. Then, it restarts from zero again as shown in Fig. 7.3a. The TAIFG bit is set when the TAxR value changes from TAxCCR0 to zero. Also, the capture/compare interrupt flag (CCIFG) bit is set when the TAxR value changes from TAxCCR0-1 to TAxCCR0. The timer period for this mode can be calculated as period = (TAxCRR0+1)/ f_CLK where f_CLK stands for the frequency of the timer clock.

In the continuous mode, the timer counts up until it reaches 0xFFFF (or 65535 in decimal), then restarts from zero again as shown in Fig. 7.3b. The TAIFG bit is set when the TAxR value changes from 0xFFFF to zero. The time period for this mode can be calculated as period = 65536/ f_CLK where f_CLK stands for the frequency of the timer clock. The continuous mode is generally used for generating output with different frequencies or independent time intervals. In this mode, eight different output frequencies or time intervals can be produced by using seven TAxCCR and TAxR register entries.

In the up/down mode, the timer first counts up until it reaches the value in TAxCCR0. Then, counting is inverted and the timer counts down from TAxCCR0 to zero as shown in Fig. 7.3c. The TAIFG bit is set when the TAxR value changes from 1 to 0 in counting down. Also, the CCIFG bit is set when the TAxR value changes from TAxCCR0-1 to TAxCCR0 in counting up. The timer period for this mode can be calculated as period = (2×TAxCRR0)/ f_CLK where f_CLK stands for the frequency of the timer clock.

In Table 7.20, the TACLR bit is used to clear the Timer_A module. When this bit is set, the TAxR, clock divider, and count directions are reset. Resetting the clock divider and count directions does not mean resetting ID and MC bits. Resetting the clock divider means current prescaler counter is set to 0. Resetting the count direction means if TAxR is in counting down part in up/down mode, it is reset to counting up part. Constant for this bit is TIMER_A_CTL_CLR. The TAIE bit is used to enable the Timer_A interrupt. Constant for this bit is TIMER_A_CTL_IE. The TAIFG bit is the interrupt flag for Timer_A. The constant for this bit is TIMER_A_CTL_IFG.

The TAxCCRy register is used to hold data for the comparison of the timer value in the TAxR in compare mode. In capture mode, the TAxR value is copied to this register when a capture is performed.

The TAxCCTLy register is used to control the capture/compare block. Entries of the TAxCCTLy register are tabulated in Table 7.21. Before explaining this register, let's look at the capture and compare operations in detail.

The purpose of the capture operation is to link the changes in the input signal with TAxR values. We should first set the CAP bit of the TAxCCTLy register to use this mode. Then, the input signal source should be selected by CCIS bits of the same register. The capture edge type (rising or falling) of this selected signal is set by CM bits of the TAxCCTLy register. When a capture occurs, the value in the TAxR register is copied to the related TAxCCRy register. The capture/compare interrupt flag (CCIFG) is set to indicate that capturing is done. Also, the timer ISR is called if the CCIE bit is set. Then, the time interval between the two time instants can be calculated by these captured TAxCCRy values.

There are synchronization and overflow issues to be considered in the capture mode. If the input changes its state at the same time as the timer clock, this may cause a race condition when the TAxR value is copied to the TAxCCRy. The SCS bit of the TAxCCTLy should be set and the input should be synchronized with the timer clock to prevent this. Also, another capture may occur before the first one is processed. When this happens, the COV bit is set to indicate that an overflow occurred. Therefore, the COV bit must be cleared by software to catch subsequent overflows.

The purpose of the compare operation is to generate interrupts at specific time intervals. This can be used to form pulse width modulation (PWM) signals. The interrupt time intervals or the frequency of the PWM can be adjusted by the TAxCCRy register. When the timer counts up (until the value in the TAxR reaches the TAxCCRy value), the internal signal EQUx (which can be seen at the output of the comparator) is set. Afterwards, the interrupt flag CCIFG is set and the EQUx signal triggers (by changing the output signal OUT) according to the output mode selected. Also, the input signal of the compare block CCI is latched into the SCCI bit.

As the capture and compare operations are explained, let's focus on the entries of the TAxCCTLy register in Table 7.21. Here, CM bits are used to select the edge sensitivity in the capture mode. Constants for these bits are TIMER_A_CCTLN_CM_x, where x can get values from 0 to 3. Corresponding edge sensitivity for these constants are no capture, capture on rising edge, capture on falling edge, and capture on both edges, respectively. CCIS bits are used to select the capture/compare input. Constants for these bits are TIMER_A_CCTLN_CCIS_x, where x can get values from 0 to 3. Corresponding inputs for these constants are CCIxA, CCIxB, GND, and V_CC, respectively. The SCS bit is used to synchronize the timer clock and the capture signal (to eliminate any race condition). Constant for this bit is TIMER_A_CCTLN_SCS. The SCCI bit is used to observe the synchronized input. Constant for this bit is TIMER_A_CCTLN_SCCI. The CAP bit is used to select the capture or compare mode. When this bit is reset, the compare mode is selected. When it is set, the capture mode is selected. Constant for this bit is TIMER_A_CCTLN_CAP. OUTMOD bits are used to select the output mode for the compare operation. Constants for these bits are TIMER_A_CCTLN_OUTMOD_x, where x can get values from 0 to 7. Corresponding output modes for these constants are OUT bit value, set, toggle/reset, set/reset, toggle, reset, toggle/set, and reset/set, respectively. We will explain these modes in detail in Chap. 8 in relation to the PWM signal generation. The CCIE bit is used to enable the capture/compare interrupt. Constant for this bit is TIMER_A_CCTLN_CCIE. The CCI bit is used to observe the capture/compare input. Constant for this bit is TIMER_A_CCTLN_CCI. The OUT bit (when in TIMER_A_CCTLN_OUTMOD_0) directly controls the output. When this bit is reset, the output is logic level 0. When it is set, the output is logic level 1. Constant for this bit is TIMER_A_CCTLN_OUT. The COV bit indicates whether a capture overflow has

Bits	15-14	13-12	11	10	8	7-5	4	3	2	1	0
	CM	CCIS	SCS	SCCI	CAP	OUTMOD	CCIE	CCI	OUT	COV	CCIFG

TABLE 7.21 Timer_A Capture/Compare Control Register (TAxCCTLy)

TAIV Content	Interrupt Source	Interrupt Flag	Interrupt Priority
0x00	No interrupt pending	–	–
0x02	Capture/Compare 1	TAxCCR1 CCIFG	Highest
0x04	Capture/Compare 2	TAxCCR2 CCIFG	
0x06	Capture/Compare 3	TAxCCR3 CCIFG	
0x08	Capture/Compare 4	TAxCCR4 CCIFG	
0x0A	Capture/Compare 5	TAxCCR5 CCIFG	
0x0C	Capture/Compare 6	TAxCCR6 CCIFG	
0x0E	Timer overflow	TAxCTL TAIFG	Lowest

TABLE 7.22 TAIV Register

occurred or not. It should be cleared by software to observe a new overflow. Constant for this bit is TIMER_A_CCTLN_COV. The CCIFG is the capture/compare interrupt flag. Constant for this bit is TIMER_A_CCTLN_CCIFG.

The TAxIV register is the interrupt vector used for timer overflow and capture/compare interrupts (except capture/compare 0 interrupt). The content of this register is given in Table 7.22.

The TAxEX0 register is the additional divider for the Timer_A clock. TAIDEX bits of this register are used for selecting the frequency divider value. Constants for this register are TIMER_A_EX0_TAIDEX_x, where x can get values from 0 to 7. Corresponding frequency divider values for these constants are 1 to 8, respectively.

7.7.2 Timer_A Usage via DriverLib Functions

The Timer_A module can be controlled via predefined DriverLib functions. We provide the functions of interest below. Here, timer is the Timer_A base to be initialized. It can be one of TIMER_A0_BASE, TIMER_A1_BASE, TIMER_A2_BASE, or TIMER_A3_BASE. The complete function list and macro definitions can be found in [11].

```
void Timer_A_startCounter (uint32_t timer, uint_fast16_t timerMode)
/*
Starts the Timer_A0 counting by modifying MC bits of the TAxCTL
    register. It assumes Timer_A is already configured with configure
    functions in one of timer modes.
timerMode: the timer mode. It can be one of TIMER_A_CONTINUOUS_MODE,
    TIMER_A_UPDOWN_MODE or TIMER_A_UP_MODE.
*/

void Timer_A_configureContinuousMode (uint32_t timer, const
    Timer_A_ContinuousModeConfig *config)
/*
Configures the Timer_A0 in continuous mode by modifying the TAxCTL
    register.
```

```
    config: the config structure. It contains clock, clock divider,
       interrupt enable, and clear values. Please see the DriverLib user
       guide for macro definitions.
*/

void Timer_A_configureUpMode (uint32_t timer, const
       Timer_A_UpModeConfig *config)
/*
Configures the Timer_A0 in up mode by modifying TAxCTL and TAxCCTLy
       registers.
config: the config structure. Please see the DriverLib user guide for
       macro definitions.
*/

void Timer_A_configureUpDownMode (uint32_t timer, const
       Timer_A_UpDownModeConfig *config)
/*
Configures the Timer_A0 in up-down mode by modifying TAxCTL and
       TAxCCTLy registers.
config: the config structure. Please see the DriverLib user guide for
       macro definitions.
*/

void Timer_A_initCapture (uint32_t timer, const
       Timer_A_CaptureModeConfig *config)
/*
Configures the Timer_A0 in capture mode by modifying the TAxCCTLy
       register.
config: the config structure. Please see the DriverLib user guide for
       macro definitions.
*/

void Timer_A_initCompare (uint32_t timer, const
       Timer_A_CompareModeConfig *config)
/*
Configures the Timer_A0 in compare mode by modifying the TAxCCTLy
       register.
config: the config structure. Please see the DriverLib user guide for
       macro definitions.
*/

void Timer_A_clearTimer (uint32_t timer)
/*
Resets the Timer_A0 counter, clock divider, and count direction by
       modifying TACLR bit of the TAxCTL register.
*/

uint_fast16_t Timer_A_getCaptureCompareCount (uint32_t timer,
       uint_fast16_t captureCompareRegister)
/*
Returns the capture compare count current value by reading content of
       the TAxCCRy register.
captureCompareRegister: the capture compare register being used. It can
       be one of register 0, 1, 2, 3, 4, 5 or 6. Please see the DriverLib
       user guide for macro definitions.
```

```
*/

void Timer_A_stopTimer (uint32_t timer)
/*
Stops the Timer_A0 counting by modifying MC bits of the TAxCTL
    register.
*/

void Timer_A_setCompareValue (uint32_t timer, uint_fast16_t
    compareRegister, uint_fast16_t compareValue)
/*
Set the compare value by modifying content of the TAxCCRy register.
captureCompareRegister: the capture compare register being used. It can
    be one of register 0, 1, 2, 3, 4, 5 or 6. Please see the DriverLib
    user guide for macro definitions.
compareValue: the value to be compared with timer count.
*/

uint16_t Timer_A_getCounterValue (uint32_t timer)
/*
Returns the Timer_A0 current counter value by reading the contents of
    the TAxR register.
*/

void Timer_A_clearInterruptFlag (uint32_t timer)
/*
Clears the Timer_A0 overflow interrupt flag by modifying TAIFG bits of
    the TAxCTL register.
*/

void Timer_A_clearCaptureCompareInterrupt (uint32_t timer,
    uint_fast16_t captureCompareRegister)
/*
Clears the Timer_A0 capture compare interrupt flag by modifying CCIFG
    bits of the TAxCCTLy register.
captureCompareRegister: the capture compare register being used. It can
    be one of register 0, 1, 2, 3, 4, 5 or 6. Please see the DriverLib
    user guide for macro definitions.
*/

void Timer_A_enableInterrupt (uint32_t timer)
/*
Enables the Timer_A0 overflow interrupts by modifying TAIE bits of the
    TAxCTL register.
*/

void Timer_A_disableInterrupt (uint32_t timer)
/*
Disables the Timer_A0 overflow interrupts by modifying TAIE bits of the
    TAxCTL register.
*/

void Timer_A_enableCaptureCompareInterrupt (uint32_t timer,
    uint_fast16_t captureCompareRegister)
/*
```

*Enables the Timer_A0 capture compare interrupts by modifying CCIE bits
    of the TAxCCTLy register.
captureCompareRegister: the capture compare register being used. It can
    be one of register 0, 1, 2, 3, 4, 5 or 6. Please see the DriverLib
    user guide for macro definitions.
/

void Timer_A_disableCaptureCompareInterrupt   (uint32_t timer,
    uint_fast16_t captureCompareRegister)
/
*Disables the Timer_A0 capture compare interrupts by modifying CCIE bits
    of the TAxCCTLy register.
captureCompareRegister: the capture compare register being used. It can
    be one of register 0, 1, 2, 3, 4, 5 or 6. Please see the DriverLib
    user guide for macro definitions.
/

7.7.3 Coding Practices for the Timer_A Module

In this section, we provide several examples based on register-level and DriverLib functions on the usage of the Timer_A module. In these, we use LED1 and green color of LED2 connected to pins P1.0 and P2.1 on the MSP432 LaunchPad as output.

7.7.3.1 *Register-Level Examples*

We provide an example on the usage of timer interrupts in the Timer_A module in Listing 7.9. Here, LED1 toggles every 0.5 second by using the Timer_A0 overflow interrupt and the green color of LED2 toggles every second by using the Timer_A1 capture/compare 0 interrupt. For the Timer_A0, clock is selected as ACLK (supplied by 32768 Hz LFTX) and counter mode is selected as up/down. For the Timer_A1, clock is selected SMCLK (supplied by 128 kHz REFO) and counter mode is selected as up. Also, the system is put in LPM0_LDO_VCORE0 mode and waits for an interrupt when not operating on LEDs.

We provide an example on the usage of the capture mode in the Timer_A module in Listing 7.10. Here, the frequency of the ACLK is measured using capture mode. For this C code to work, pin P4.2 (ACLK output) should be connected to pin P2.4 (TA0.CCI1A input) on the MSP432 LaunchPad. The ACLK is supplied by VLO and the Timer_A module is supplied by 12 MHz DCO. Ten successive measurements are saved in the CaptureFreq array. As can be seen, we can obtain true values after the first entry of this array. Also, the system is put in LPM0_LDO_VCORE0 mode and waits for a capture interrupt when not operating.

Finally, we provide an example on the usage of the Timer_A module in up/down mode with overflow and capture/compare 0 interrupts enabled. We provide the corresponding C code in Listing 7.11. Here, the clock is selected as ACLK (supplied by REFO with 128 kHz). Then, the system is put in LPM0_LDO_VCORE0 mode and waits for an interrupt. Every second, LED1 toggles in the overflow ISR. The green color of LED2 toggles in the capture/compare ISR. As can be seen in this example, LED1 toggles at every second. However, the green color of LED2 toggles every second with a 0.5-second offset. Do not forget that in up/down mode, the first capture/compare interrupt occurs at the half of period value. Afterwards, consecutive interrupts occur periodically. This should be handled carefully using capture/compare interrupts in the up/down mode.

Listing 7.9 Usage of the Timer Interrupts in the Timer_A Module, Using Registers

```c
#include "msp.h"

#define LED1 BIT0
#define LED2 BIT1
#define TIMER_PERIOD_0 8192
#define TIMER_PERIOD_1 32000

void main(void){

  WDT_A->CTL = WDT_A_CTL_PW | WDT_A_CTL_HOLD;

  P1->DIR = LED1;
  P2->DIR = LED2;
  PJ->SEL0 |= BIT0 | BIT1;

  CS->KEY = CS_KEY_VAL;
  CS->CTL2 |= CS_CTL2_LFXT_EN;
  while(CS->IFG & CS_IFG_LFXTIFG)
    CS->CLRIFG |= CS_CLRIFG_CLR_LFXTIFG;
  CS->CTL1 |= CS_CTL1_SELA_0;
  CS->CLKEN |= CS_CLKEN_REFOFSEL;
  CS->CTL1 &= ~(CS_CTL1_SELS_MASK | CS_CTL1_DIVS_MASK);
  CS->CTL1 |= CS_CTL1_SELS_2;
  CS->KEY = 0;

  TIMER_A0->CCR[0] = TIMER_PERIOD_0;
  TIMER_A0->CTL = TIMER_A_CTL_TASSEL_1 | TIMER_A_CTL_MC_3 | TIMER_A_CTL_CLR
      | TIMER_A_CTL_IE;
  TIMER_A1->CCTL[0] = TIMER_A_CCTLN_CCIE;
  TIMER_A1->CCR[0] = TIMER_PERIOD_1 - 1;
  TIMER_A1->CTL = TIMER_A_CTL_TASSEL_2 | TIMER_A_CTL_MC_1 | TIMER_A_CTL_ID_2
      | TIMER_A_CTL_CLR;

  NVIC->ISER[0] = 1 << ((TA0_N_IRQn) & 31);
  NVIC->ISER[0] = 1 << ((TA1_0_IRQn) & 31);
  __enable_irq();

  SCB->SCR |= SCB_SCR_SLEEPONEXIT_Msk;
  __sleep();
}

void TA0_N_IRQHandler(void){
  TIMER_A0->CTL &= ~TIMER_A_CTL_IFG;
  P1->OUT ^= LED1;
}

void TA1_0_IRQHandler(void){
  TIMER_A1->CCTL[0] &= ~TIMER_A_CCTLN_CCIFG;
  P2->OUT ^= LED2;
}
```

7.7.3.2 DriverLib Examples

We repeat the example in Listing 7.9 using predefined DriverLib functions. We provide the corresponding C code in Listing 7.12. Here, we first set the clock and oscillator frequency. Then, we set LED1 and green color of LED2 as output. Afterwards, we set the Timer_A0 module in up/down mode and Timer_A1 module in up mode by using the functions `Timer_A_configureUpDownMode` and `Timer_A_configureUpMode`. We then enable the interrupts and start the timer modules using the function `Timer_A_`

Listing 7.10 Usage of the Capture Mode in the Timer_A Module, Using Registers

```c
#include "msp.h"

int i = 0;
volatile uint16_t CaptureValues[2] = {0};
volatile float CaptureFreq[10];

void main(void){

  WDT_A->CTL = WDT_A_CTL_PW | WDT_A_CTL_HOLD;

  P2->SEL0 |= BIT4;
  P2->DIR &= ~BIT4;
  P4->SEL0 |= BIT2;
  P4->DIR |= BIT2;

  CS->KEY = CS_KEY_VAL;
  CS->CTL1 |= CS_CTL1_SELA_1;
  CS->CTL0 &= ~ CS_CTL0_DCORSEL_MASK;
  CS->CTL0 |= CS_CTL0_DCORSEL_3;
  CS->CTL1 &= ~(CS_CTL1_SELS_MASK | CS_CTL1_DIVS_MASK);
  CS->CTL1 |= CS_CTL1_SELS_3;
  CS->KEY = 0;

  TIMER_A0->CCTL[1] = TIMER_A_CCTLN_CM_1 | TIMER_A_CCTLN_CCIS_0 |
      TIMER_A_CCTLN_CCIE | TIMER_A_CCTLN_CAP | TIMER_A_CCTLN_SCS;
  TIMER_A0->CTL |= TIMER_A_CTL_TASSEL_2 | TIMER_A_CTL_MC_2 |
      TIMER_A_CTL_CLR;

  NVIC->ISER[0] = 1 << ((TA0_N_IRQn) & 31);
  __enable_irq();

  SCB->SCR |= SCB_SCR_SLEEPONEXIT_Msk;
  __sleep();
}
void TA0_N_IRQHandler(void){
  TIMER_A0->CCTL[1] &= ~(TIMER_A_CCTLN_CCIFG);
  if(i<10){
    CaptureValues[1] = TIMER_A0->CCR[1];
    CaptureFreq[i++] = (float) 12000000/(CaptureValues[1]-CaptureValues
        [0]);
    CaptureValues[0] = CaptureValues[1];}
}
```

startCounter. In the ISR, LEDs are toggled. As the final step, interrupt flags are cleared using the functions `Timer_A_clearInterruptFlag` and `Timer_A_clearCaptureCompareInterrupt`.

We next repeat the example in Listing 7.10 using predefined DriverLib functions. We provide the corresponding C code in Listing 7.13. Here, we first set the clock and oscillator frequency. We configure the Timer_A0 module using the function `Timer_A_configureContinuousMode`. Hence, it works in continuous mode. Afterwards, we

Listing 7.11 Usage of the Timer_A Module in Up/Down Mode, Using Registers

```c
#include "msp.h"

#define LED1 BIT0
#define LED2 BIT1
#define TIMER_PERIOD 64000

void main(void){

  WDT_A->CTL = WDT_A_CTL_PW | WDT_A_CTL_HOLD;

  P1->DIR = LED1;
  P2->DIR = LED2;
  P1->OUT = 0x00;
  P2->OUT = 0x00;
  PJ->SEL0 |= BIT0 | BIT1;

  CS->KEY = CS_KEY_VAL;
  CS->CLKEN |= CS_CLKEN_REFOFSEL;
  CS->CTL1 |= CS_CTL1_SELA_2;
  CS->KEY = 0;

  TIMER_A0->CCR[0] = TIMER_PERIOD;
  TIMER_A0->CCTL[0] = TIMER_A_CCTLN_CCIE;
  TIMER_A0->CTL = TIMER_A_CTL_TASSEL_1 | TIMER_A_CTL_MC_3 | TIMER_A_CTL_CLR
      | TIMER_A_CTL_IE;

  NVIC->ISER[0] = 1 << ((TA0_N_IRQn) & 31);
  NVIC->ISER[0] = 1 << ((TA0_0_IRQn) & 31);
  __enable_irq();

  SCB->SCR |= SCB_SCR_SLEEPONEXIT_Msk;
  __sleep();
}
void TA0_N_IRQHandler(void){
  TIMER_A0->CTL &= ~TIMER_A_CTL_IFG;
  P1->OUT ^= LED1;
}

void TA0_0_IRQHandler(void) {
  TIMER_A0->CCTL[0] &= ~TIMER_A_CCTLN_CCIFG;
  P2->OUT ^= LED2;
}
```

initialize the capture mode using `Timer_A_initCapture` function. Finally, we enable the interrupts and start the timer module. We clear the CCIFG and read capture values using the function `Timer_A_getCaptureCompareCount` in the ISR.

We finally repeat the example in Listing 7.11 using DriverLib functions. We provide the corresponding C code in Listing 7.14. Here, different from the previous example in Listing 7.13, we change the configuration structure for the Timer_A0 module. We enable the overflow and capture/compare interrupts of the Timer_A0 module. We set the ACLK source as 128-kHz REFO clock. We also set LED1 and green color of LED2 as output. Then, we initialize and start the Timer_A0 module and toggle LEDs in the ISR.

Listing 7.12 Usage of the Timer Interrupts in the Timer_A Module, Using DriverLib Functions

```c
#include <ti/devices/msp432p4xx/driverlib/driverlib.h>

#define TIMER_PERIOD_0 8192
#define TIMER_PERIOD_1 32000

const Timer_A_UpDownModeConfig upDownConfig =
{
 TIMER_A_CLOCKSOURCE_ACLK,
 TIMER_A_CLOCKSOURCE_DIVIDER_1,
 TIMER_PERIOD_0,
 TIMER_A_TAIE_INTERRUPT_ENABLE,
 TIMER_A_CCIE_CCR0_INTERRUPT_DISABLE,
 TIMER_A_DO_CLEAR
};

const Timer_A_UpModeConfig upConfig =
{
 TIMER_A_CLOCKSOURCE_SMCLK,
 TIMER_A_CLOCKSOURCE_DIVIDER_1,
 TIMER_PERIOD_1,
 TIMER_A_TAIE_INTERRUPT_DISABLE,
 TIMER_A_CCIE_CCR0_INTERRUPT_ENABLE,
 TIMER_A_DO_CLEAR
};

void main(void){

 WDT_A_holdTimer();
 PCM_setPowerState(PCM_AM_LF_VCORE0);
 GPIO_setAsPeripheralModuleFunctionOutputPin(GPIO_PORT_PJ, GPIO_PIN0 | GPIO_PIN1,
     GPIO_PRIMARY_MODULE_FUNCTION);
 CS_setReferenceOscillatorFrequency(CS_REFO_32KHZ);
 CS_setExternalClockSourceFrequency(32768,48000000);
 CS_startLFXT(CS_LFXT_DRIVE3);
 CS_initClockSignal(CS_ACLK, CS_LFXTCLK_SELECT, CS_CLOCK_DIVIDER_1);
 CS_initClockSignal(CS_SMCLK, CS_REFOCLK_SELECT, CS_CLOCK_DIVIDER_1);

 GPIO_setAsOutputPin(GPIO_PORT_P1, GPIO_PIN0);
 GPIO_setAsOutputPin(GPIO_PORT_P2, GPIO_PIN1);
 GPIO_setOutputLowOnPin(GPIO_PORT_P1, GPIO_PIN0);
 GPIO_setOutputLowOnPin(GPIO_PORT_P2, GPIO_PIN1);

 Timer_A_configureUpDownMode(TIMER_A0_BASE, &upDownConfig);
 Timer_A_configureUpMode(TIMER_A1_BASE, &upConfig);

 Interrupt_enableSleepOnIsrExit();
 Interrupt_enableInterrupt(INT_TA0_N);
 Interrupt_enableInterrupt(INT_TA1_0);
 Interrupt_enableMaster();

 Timer_A_startCounter(TIMER_A0_BASE, TIMER_A_UPDOWN_MODE);
 Timer_A_startCounter(TIMER_A1_BASE, TIMER_A_UP_MODE);

 PCM_setPowerState(PCM_LPM0_LDO_VCORE1);
}

void TA0_N_IRQHandler(void){
 Timer_A_clearInterruptFlag(TIMER_A0_BASE);
 GPIO_toggleOutputOnPin(GPIO_PORT_P1, GPIO_PIN0);
}

void TA1_0_IRQHandler(void){
 Timer_A_clearCaptureCompareInterrupt(TIMER_A1_BASE,
     TIMER_A_CAPTURECOMPARE_REGISTER_0);
 GPIO_toggleOutputOnPin(GPIO_PORT_P2, GPIO_PIN1);
}
```

Listing 7.13 Usage of the Capture Mode in the Timer_A Module, Using DriverLib Functions

```c
#include <ti/devices/msp432p4xx/driverlib/driverlib.h>

const Timer_A_ContinuousModeConfig continuousModeConfig =
{
 TIMER_A_CLOCKSOURCE_SMCLK,        // ACLK Clock Source
 TIMER_A_CLOCKSOURCE_DIVIDER_1,    // ACLK/1 = 32.768khz
 TIMER_A_TAIE_INTERRUPT_ENABLE,    // Enable Overflow ISR
 TIMER_A_DO_CLEAR                  // Clear Counter
};

const Timer_A_CaptureModeConfig captureModeConfig =
{
 TIMER_A_CAPTURECOMPARE_REGISTER_1,         // CC Register 2
 TIMER_A_CAPTUREMODE_RISING_EDGE,           // Rising Edge
 TIMER_A_CAPTURE_INPUTSELECT_CCIxB,         // CCIxB Input Select
 TIMER_A_CAPTURE_SYNCHRONOUS,               // Synchronized Capture
 TIMER_A_CAPTURECOMPARE_INTERRUPT_ENABLE,   // Enable interrupt
 TIMER_A_OUTPUTMODE_OUTBITVALUE             // Output bit value
};

int i = 0;
volatile uint16_t CaptureValues[2] = {0};
volatile float CaptureFreq[10];

void main(void){

 WDT_A_holdTimer();
 CS_initClockSignal(CS_SMCLK, CS_DCOCLK_SELECT, CS_CLOCK_DIVIDER_1);
 CS_initClockSignal(CS_ACLK, CS_VLOCLK_SELECT, CS_CLOCK_DIVIDER_1);
 CS_setDCOFrequency(CS_12MHZ);

 GPIO_setAsPeripheralModuleFunctionOutputPin(GPIO_PORT_P4, GPIO_PIN2,
     GPIO_PRIMARY_MODULE_FUNCTION);
 GPIO_setAsPeripheralModuleFunctionInputPin(GPIO_PORT_P2, GPIO_PIN4,
     GPIO_PRIMARY_MODULE_FUNCTION);

 Timer_A_configureContinuousMode(TIMER_A0_BASE, &continuousModeConfig);
 Timer_A_initCapture(TIMER_A0_BASE, &captureModeConfig);

 Interrupt_enableSleepOnIsrExit();
 Interrupt_enableInterrupt(INT_TA0_N);
 Interrupt_enableMaster();

 Timer_A_startCounter(TIMER_A0_BASE, TIMER_A_CONTINUOUS_MODE);

 PCM_setPowerState(PCM_LPM0_LDO_VCORE1);
}
void TA0_N_IRQHandler(void){
 Timer_A_clearCaptureCompareInterrupt(TIMER_A0_BASE,
     TIMER_A_CAPTURECOMPARE_REGISTER_1);
 if(i<10){
  CaptureValues[1] = Timer_A_getCaptureCompareCount(TIMER_A0_BASE,
      TIMER_A_CAPTURECOMPARE_REGISTER_1);
  CaptureFreq[i++] = (float) 12000000/(CaptureValues[1]-CaptureValues[0]);
  CaptureValues[0] = CaptureValues[1];}
}
```

Chapter Seven

Listing 7.14 Usage of the Timer_A Module in Up/Down Mode, Using DriverLib Functions

```c
#include <ti/devices/msp432p4xx/driverlib/driverlib.h>

#define TIMER_PERIOD 64000

const Timer_A_UpDownModeConfig upDownConfig =
{
 TIMER_A_CLOCKSOURCE_ACLK,
 TIMER_A_CLOCKSOURCE_DIVIDER_1,
 TIMER_PERIOD,
 TIMER_A_TAIE_INTERRUPT_ENABLE,
 TIMER_A_CCIE_CCR0_INTERRUPT_ENABLE,
 TIMER_A_DO_CLEAR
};

void main(void){

 WDT_A_holdTimer();

 PCM_setPowerState(PCM_AM_LF_VCORE0);
 CS_setReferenceOscillatorFrequency(CS_REFO_128KHZ);
 CS_initClockSignal(CS_ACLK, CS_REFOCLK_SELECT, CS_CLOCK_DIVIDER_1);

 GPIO_setAsOutputPin(GPIO_PORT_P1, GPIO_PIN0);
 GPIO_setAsOutputPin(GPIO_PORT_P2, GPIO_PIN1);
 GPIO_setOutputLowOnPin(GPIO_PORT_P1, GPIO_PIN0);
 GPIO_setOutputLowOnPin(GPIO_PORT_P2, GPIO_PIN1);

 Timer_A_configureUpDownMode(TIMER_A0_BASE, &upDownConfig);

 Interrupt_enableSleepOnIsrExit();
 Interrupt_enableInterrupt(INT_TA0_N);
 Interrupt_enableInterrupt(INT_TA0_0);
 Interrupt_enableMaster();

 Timer_A_startCounter(TIMER_A0_BASE, TIMER_A_UPDOWN_MODE);

 PCM_setPowerState(PCM_LPM0_LDO_VCORE1);
}

void TA0_N_IRQHandler(void){
 Timer_A_clearInterruptFlag(TIMER_A0_BASE);
 GPIO_toggleOutputOnPin(GPIO_PORT_P1, GPIO_PIN0);
}

void TA0_0_IRQHandler(void){
 Timer_A_clearCaptureCompareInterrupt(TIMER_A0_BASE,
     TIMER_A_CAPTURECOMPARE_REGISTER_0);
 GPIO_toggleOutputOnPin(GPIO_PORT_P2, GPIO_PIN1);
}
```

Listing 7.15 Usage of the RTC_C Module

```c
#include <ti/devices/msp432p4xx/driverlib/driverlib.h>

/* Initial time is 20.03.2017 Monday 21:10:32 */
const RTC_C_Calendar initalTime = { 32, 10, 21, 0, 20, 3, 2017 };
static volatile RTC_C_Calendar currentTime;

void main(void){

 WDT_A_holdTimer();
 GPIO_setAsOutputPin(GPIO_PORT_P1, GPIO_PIN0);
 GPIO_setAsOutputPin(GPIO_PORT_P2, GPIO_PIN0 | GPIO_PIN1);
 GPIO_setOutputLowOnPin(GPIO_PORT_P1, GPIO_PIN0);
 GPIO_setOutputLowOnPin(GPIO_PORT_P2, GPIO_PIN0 | GPIO_PIN1);

 PCM_setPowerState(PCM_AM_LF_VCORE0);
 GPIO_setAsPeripheralModuleFunctionOutputPin(GPIO_PORT_PJ, GPIO_PIN0 |
     GPIO_PIN1, GPIO_PRIMARY_MODULE_FUNCTION);
 CS_setExternalClockSourceFrequency(32768,48000000);
 CS_startLFXT(CS_LFXT_DRIVE3);
 CS_initClockSignal(CS_BCLK, CS_LFXTCLK_SELECT, CS_CLOCK_DIVIDER_1);

 RTC_C_initCalendar(&initalTime, RTC_C_FORMAT_BINARY);
 RTC_C_setCalendarAlarm(12, 21, RTC_C_ALARMCONDITION_OFF,
     RTC_C_ALARMCONDITION_OFF);
 RTC_C_setCalendarEvent(RTC_C_CALENDAREVENT_MINUTECHANGE);
 RTC_C_clearInterruptFlag(RTC_C_CLOCK_READ_READY_INTERRUPT |
     RTC_C_TIME_EVENT_INTERRUPT | RTC_C_CLOCK_ALARM_INTERRUPT);
 RTC_C_enableInterrupt(RTC_C_CLOCK_READ_READY_INTERRUPT |
     RTC_C_TIME_EVENT_INTERRUPT | RTC_C_CLOCK_ALARM_INTERRUPT);
 RTC_C_startClock();

 Interrupt_enableInterrupt(INT_RTC_C);
 Interrupt_enableSleepOnIsrExit();
 Interrupt_enableMaster();

 PCM_setPowerState(PCM_LPM0_LDO_VCORE1);
}

void RTC_C_IRQHandler(void){
 uint32_t status;

 status = RTC_C_getEnabledInterruptStatus();
 RTC_C_clearInterruptFlag(status);

 if (status & RTC_C_CLOCK_READ_READY_INTERRUPT)
  GPIO_toggleOutputOnPin(GPIO_PORT_P2, GPIO_PIN0);

 if (status & RTC_C_TIME_EVENT_INTERRUPT){
  currentTime = RTC_C_getCalendarTime();
  GPIO_toggleOutputOnPin(GPIO_PORT_P2, GPIO_PIN1);}

 if (status & RTC_C_CLOCK_ALARM_INTERRUPT)
  GPIO_setOutputHighOnPin(GPIO_PORT_P1, GPIO_PIN0);
}
```

7.8 Real-Time Clock

There is an integrated real-time clock module (RTC_C) in the MSP432 microcontroller. It keeps the time value precisely by using a 15-bit counter. RTC uses BLCK with the frequency 32768 Hz as suggested by the user manual [4]. In each clock cycle, counter is incremented by one. Hence, when counter overflows at 2^{15}, exactly 1 second passes. Besides the precise time tracking property, this module can keep seconds, minutes, hours, day of week, day of month, month, and year (including leap year correction). It can also generate alarm at a given time or time intervals. The RTC_C module also offers offset-calibration and temperature drift compensation that corrects errors due to crystal offsets and temperature change.

The RTC_C module can be controlled by its dedicated registers. However, the excessive number of registers make this option less feasible. Therefore, we do not cover this option in this book. For more information on register-level usage of the RTC_C module, please see [4].

7.8.1 Real-Time Clock Usage via DriverLib Functions

We provide DriverLib functions related to the RTC_C module usage below. The complete function list and macro definitions can be found in [11].

```
void RTC_C_startClock (void)
/*
Starts the RTC_C counting.
*/

void RTC_C_holdClock (void)
/*
Stops the RTC_C counting.
*/

void RTC_C_initCalendar (const RTC_C_Calendar *calendarTime,
    uint_fast16_t formatSelect)
/*
Initializes the RTC_C in calender mode.

*calendarTime: the calender structure holding the initial time in terms
    of {seconds, minutes, hours, dayOfWeek, dayOfmonth, year}.
formatSelect: format type of RTC_C registers. It can be one of
    RTC_C_FORMAT_BINARY or RTC_C_FORMAT_BCD.
*/

RTC_C_Calendar RTC_C_getCalendarTime (void)
/*
Returns the current calender time.
*/

void RTC_C_setCalendarAlarm (uint_fast8_t minutesAlarm, uint_fast8_t
    hoursAlarm, uint_fast8_t dayOfWeekAlarm, uint_fast8_t
    dayOfmonthAlarm)
/*
Sets calender alarm.
```

*minutesAlarm: minute of the alarm time. Default value is
    RTC_C_ALARMCONDITION_OFF.
hoursAlarm: hour of the alarm time. Default value is
    RTC_C_ALARMCONDITION_OFF.
dayOfWeekAlarm: day of week of the alarm time. Default value is
    RTC_C_ALARMCONDITION_OFF.
dayOfmonthAlarm: day of month of the alarm time. Default value is
    RTC_C_ALARMCONDITION_OFF.
/

void RTC_C_setCalendarEvent (uint_fast16_t eventSelect)
/*
*Sets calender interrupt type. It is different from the calender alarm.
eventSelect: the interrupt type. It can be one of
    RTC_C_CALENDAREVENT_MINUTECHANGE, RTC_C_CALENDAREVENT_HOURCHANGE,
    RTC_C_CALENDAREVENT_NOON or RTC_C_CALENDAREVENT_MIDNIGHT.
/

void RTC_C_enableInterrupt (uint8_t interruptMask)
/*
*Enables the RTC_C interrupts.
interruptMask: the interrupt to be enabled. It can be logical OR of
    RTC_C_TIME_EVENT_INTERRUPT, RTC_C_CLOCK_ALARM_INTERRUPT,
    RTC_C_CLOCK_READ_READY_INTERRUPT, RTC_C_PRESCALE_TIMER0_INTERRUPT,
    RTC_C_PRESCALE_TIMER1_INTERRUPT, RTC_C_OSCILLATOR_FAULT_INTERRUPT.
/

void RTC_C_disableInterrupt (uint8_t interruptMask)
/*
*Disables the RTC_C interrupts.
interruptMask: the interrupt to be disabled. It can be logical OR of
    RTC_C_TIME_EVENT_INTERRUPT, RTC_C_CLOCK_ALARM_INTERRUPT,
    RTC_C_CLOCK_READ_READY_INTERRUPT, RTC_C_PRESCALE_TIMER0_INTERRUPT,
    RTC_C_PRESCALE_TIMER1_INTERRUPT, RTC_C_OSCILLATOR_FAULT_INTERRUPT.
/

void RTC_C_clearInterruptFlag (uint_fast8_t interruptFlagMask)
/*
*Clears the RTC_C interrupt flags.
interruptFlagMask: the interrupt to be cleared. It can be logical OR of
    RTC_C_TIME_EVENT_INTERRUPT, RTC_C_CLOCK_ALARM_INTERRUPT,
    RTC_C_CLOCK_READ_READY_INTERRUPT, RTC_C_PRESCALE_TIMER0_INTERRUPT,
    RTC_C_PRESCALE_TIMER1_INTERRUPT, RTC_C_OSCILLATOR_FAULT_INTERRUPT.
/

uint_fast8_t RTC_C_getEnabledInterruptStatus(**void**)
/*
*Returns the interrupt status flag masked with enabled interrupt mask.
/

7.8.2 Coding Practices for the Real-Time Clock Module

We provide a sample code on the usage of the RTC_C module in Listing 7.15. Here, we set the initial date as 20.03.2017 Monday 21:10:32 and set an alarm on 21:12 at the

148 Chapter Seven

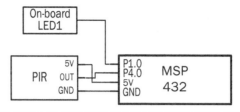

FIGURE 7.4 Layout of the energy saver application.

same date. LED1 connected to pin P1.0 on the MSP432 LaunchPad turns on when alarm occurs. The red color of LED2 connected to pin P2.0 on the MSP432 LaunchPad toggles every second when RTC ticks. The green color of LED2 connected to pin P2.1 on the MSP432 LaunchPad toggles every minute by using the minute change interrupt. The RTC_C module is supplied by BLCK as explained previously. Also, the system is put in LPM0_LDO_VCORE0 mode and it waits for RTC_C interrupts. In the ISR, interrupt status flags are checked to handle interrupts coming from different interrupt sources.

7.9 Application: Energy Saver System

The aim of this application is to learn how to use the LPMs, timers, and timer interrupts on the MSP432 microcontroller. As a real-life application, we will design an energy saver system for smart homes. In this section, we provide the equipment list, layout of the circuit, and system design specifications.

7.9.1 Equipment List

Besides LED1 connected to pin P1.0 on the MSP432 LaunchPad, we will use a PIR sensor for detecting movements in a room. A PIR sensor module gives output logic level 1 if it detects a movement in its range. If there is no movement, then it gives output logic level 0 after a time delay.

7.9.2 Layout

The layout of the energy saver system is given in Fig. 7.4.

7.9.3 System Design Specifications

The design of the energy saver system will have three main blocks, which are discussed below.

- **Block 1**: If there is any movement, we will turn on LED1 and will wait in a suitable LPM to save energy.
- **Block 2**: We will turn off LED1 if there is no movement for more than 20 seconds. Then, we will wait for a suitable LPM to save energy. Do not forget to include the delay time of the PIR sensor module in timing calculations.
- **Block 3**: We will not turn on LED1 during daytime (between 08:00 am and 06:00 pm) even a movement is detected.

We could not add the C code of the project here due to space limitations. However, we have provided the complete CCS project for this application in the companion website for the reader. We strongly suggest implementing it.

7.10 Summary

This chapter is on power management and time-based operations. Therefore, we started with the power control module and associated power supply and clock system. Using these, we considered the multiple power modes of MSP432 so that it can work under different conditions such as AMs and LPMs. This is extremely important in IoT operations in which the microcontroller depends on either battery or energy harvesting systems as power source. Power management is interwoven with the clock system of the microcontroller. Therefore, we next introduced the clock system and associated timer modules of the microcontroller. While doing this, we explored the watchdog timer, system timer, Timer32, Timer_A, and real-time clock modules in MSP432. Finally, we provided the design of the energy saver application by using the concepts introduced in this chapter.

7.11 Exercises

7.1 Repeat Problem 6.1 by adding suitable low-power and active modes. Set the CPU clock to 1 MHz using DCO.

7.2 Repeat Problem 6.2 by adding suitable low-power and active modes. Set the CPU clock to 32 kHz using LFXT.

7.3 Repeat Problem 6.3 by adding suitable low-power and active modes. Set the CPU clock to 48 MHz using DCO.

7.4 Repeat Problem 6.4 by adding suitable low-power and active modes. Set the CPU clock to 48 MHz using HFXT.

7.5 Repeat Problem 6.5 by adding suitable low-power and active modes. Set the CPU clock to 128 kHz using REFO.

7.6 Repeat Problem 6.6 by adding suitable low-power and active modes. Set the CPU clock to 10 kHz using VLO.

7.7 Write a C program to eliminate the switch bouncing problem.
   a. Use SysTick timer for this purpose.
   b. Add suitable low power and active modes. Set the CPU clock to 12 MHz using DCO.
   c. Add 100-millisecond protection for switch bouncing.

7.8 Write a C program to eliminate hang-up possibility while running the program.
   a. Use the WDT_A module timer for this purpose.
   b. Use SMCLK as clock as explained in Sec. 7.2.2. Also, use the DCO as clock source as explained in Sec. 7.2.1.
   c. Set the watchdog interval to four seconds.

7.9 Write a C program to count heartbeats in beat per minute.
   a. The heartbeat signal is generated by a digital heartbeat sensor that gives a pulse output for each heartbeat.
   b. Use the Timer_A capture mode to capture the heartbeats.

7.10 Write a program in C language such that MSP432 is in low-power mode most of the times. Initially the green color of LED2 connected to pin P2.1 on the MSP432

LaunchPad is turned on and the red color of LED2 connected to pin P2.0 on the MSP432 LaunchPad is turned off. The CPU wakes up in periodic time intervals of
   a. 10 seconds
   b. 1 minute
   c. 1 hour
   d. 1 day

As the CPU wakes up, the LEDs toggle. Then, the CPU will go to the low-power mode again. Use Timer_A in operation.

7.11 Repeat Problem 7.10 by using Timer32.

7.12 Repeat Problem 7.10 by using RTC.

7.13 Add the button S1 connected to pin P1.1 on the MSP432 LaunchPad to Problem 7.10. When this button is pressed, the timer will reset itself and the LEDs will go to their initial state.

7.14 Repeat Problem 7.13 by using RTC.

7.15 We can design a coming-home system for cars using the timer modules. The aim of the coming-home system is keeping the headlights open for a certain amount of time after the driver leaves the car. Hence, the driver can find his or her way safely. The system works as follows. When the car stops, the driver pushes the button S1 connected to pin P1.1 on the MSP432 LaunchPad twice within 2 seconds. Then, the headlights turn on and stay in this position for 20 seconds. Afterwards, they turn of automatically. LED1 connected to pin P1.0 on the MSP432 LaunchPad can be taken instead of headlights in our application.

CHAPTER 8
Mixed Signal Systems

There are specific modules to convert analog signals to digital form in microcontrollers. They perform sampling in time and quantization in amplitude. These modules are generally called analog-to-digital converter (ADC). The ADC module in MSP432 is specifically named as ADC14. There are also two comparator modules that can be taken as one-bit ADC in MSP432. They are specifically named COMP_E0 and COMP_E1. ADC14 and COMP_E modules need reference voltage level for operation. MSP432 has a reference module, called REF_A, for this purpose. In this chapter, we will explain the analog-to-digital conversion operation through REF_A, ADC14, and COMP_E modules. For digital-to-analog conversion, there may be a dedicated module in the microcontroller that performs interpolation between digital samples. Such modules are generally called digital-to-analog converter (DAC). Unfortunately, MSP432 does not have a DAC module. Therefore, we will use the pulse width modulation (PWM) operation to obtain the approximate analog representation of a given digital signal. We will also provide two real-life applications to explain the concepts introduced in this chapter.

8.1 Analog and Digital Signals

A value changing with time or another dependent variable can be taken as a signal. There are two signal types: analog and digital. By definition, an analog signal can have its amplitude with infinite precision. It can also be defined for any time value. A digital signal can represent samples of the analog signal in time. Moreover, its amplitude values are also quantized. This means that the amplitude values are represented by only certain numbers. More information on analog and digital signals (in terms of theory) can be found in [14, 15].

The digital signal is the sampled and quantized form of analog signal. Therefore, it contains less information compared to its analog counterpart. This may seem a disadvantage for the digital signal representation. However, this is not the case in practice. Analog signals are hard to store and process. They are also prone to noise. Besides, the system to process an analog signal is most of the times static. On the other hand, digital signals are very robust to noise. The system to process a digital signal can be a code block. Hence, all the system parameters (or the system itself) can easily be modified by changing a code block. That is why most recent systems are in digital form. In this book, we take MSP432 as the digital system.

8.2 The Reference Module

The ADC14 and COMP_E modules need reference voltage for operation. In MSP432, the REF_A module generates critical reference voltage levels for this purpose. These voltage levels should be constant even if the power supply voltage or temperature value fluctuates. Hence, a special circuit topology, called bandgap, is used in MSP432. To be more specific, the REF_A module consists of a factory-trimmed low-power bandgap reference voltage generator with high power supply reject ratio (PSRR), low temperature coefficient, and high accuracy. The output of the bandgap reference voltage generator should be buffered before connecting it to other modules, so that its output is not affected. There are two separate buffer stages connected to output of the bandgap reference voltage generator in MSP432. The first stage is the one with user selectable output that can generate 1.2, 1.45, and 2.5 V to be fed to the ADC_14 module. The second buffer stage is the one with constant output to fed the rest of the system. The reference voltage can also be switched to a pin for external use.

The REF_A module can be used in either static or sampled mode. The bandgap reference voltage generator is active all the time in static mode. This way, it can generate reference voltage levels with high accuracy that is suitable for ADC and comparator applications. Output of the bandgap reference voltage generator is sampled with the VLO clock in sampled mode. Besides sampling intervals, the module runs in low-power mode. Hence, energy consumption of the device decreases. However, the accuracy of the generated reference voltage level also decreases. This type of operation is more suitable for applications such as driving an LCD.

The REF_A module can also be run in either burst or continuous mode. The REF_A module becomes active when triggered by the ADC14 module in burst mode. As the ADC14 module does its job and analog-to-digital conversion ends, the REF_A module is disabled. Hence, power consumption of the device decreases. On the other hand, the REF_A module is active all the time in continuous mode. Therefore, the reference voltage level can be fed to output through a pin of the MSP432 microcontroller.

We can summarize the above operations as follows. If the user selects the static and continuous modes together while running the REF_A module, the device dissipates the highest energy. The accuracy of the obtained reference voltage levels is also high. Selecting the static and burst modes leads to less power dissipation. The accuracy of the obtained reference voltage levels is also high. However, since the REF_A module is triggered by the ADC14 module in this mode, other modules can only use the generated reference voltage levels when the ADC14 module is active. If the user selects the sampled and continuous modes together, power dissipation drops further by a sacrifice in the accuracy of the generated reference voltage levels. The last combination is using the sampled and burst modes together. Although this combination is allowed by the ADC14 module, the accuracy of the generated reference voltage levels is not suitable for conversion operations. Hence, this mode should not be used in practice.

The REF_A module has an integrated temperature sensor to trim its output with respect to temperature changes. This sensor can also be used by other modules, such as ADC14, to measure the internal temperature of the MSP432 microcontroller. The temperature sensor is enabled by default. It can be disabled by software to save power.

Bits	13	12	11	10	9	8	7
	REFBGRDY	REFGENRDY	BGMODE	REFGENBUSY	REFBGACT	REFGENACT	REFBGOT
Bits	6	5-4		3	1	0	
	REFGENOT	REFVSEL		REFTCOFF	REFOUT	REFON	

TABLE 8.1 REF_A Module Control Register 0 (REFCTL0)

8.2.1 Reference Module Registers

There is one 16-bit register, REFCTL0, to control the REF_A module in MSP432. A structure named REF_A is used to control this register. In this structure, the mentioned register is called CTL0.

The REFCTL0 register is used to configure and control the REF_A module. The entries of this register are tabulated in Table 8.1. Here, the REFGENRDY bit (with the constant REF_A_CTL0_GENRDY) is used for checking the reference voltage output. If this bit is set, the reference voltage output is ready to be used. The BGMODE bit (with constant REF_A_CTL0_BGOT) is used for selecting bandgap mode for the REF_A module. If this bit is set, sampled mode is selected. Otherwise, static mode is selected. The REFGENBUSY bit (with constant REF_A_CTL0_GENBUSY) is used for checking the reference generator. If this bit is set, the reference generator is busy. The REFVSEL bits are used for selecting the reference voltage level. Constants for these bits are REF_A_CTL0_VSEL_X where X can get values 0, 1, and 3 to generate 1.2, 1.45, and 2.5 V reference voltage levels, respectively. The REFTCOFF bit (with the constant REF_A_CTL0_TCOFF) is used for enabling or disabling the temperature sensor. If this bit is set, the temperature sensor is disabled to save power. The REFOUT bit (with the constant REF_A_CTL0_OUT) is used for enabling or disabling the reference output. If this bit is set, the reference output is enabled and can be used externally. The REFON bit (with the constant REF_A_CTL0_ON) is used for enabling or disabling the reference. If this bit is set, reference is enabled. The REFVSEL, REFTCOFF, REFOUT, and REFON bits can only be configured when the REFGENBUSY bit is reset. For further information on the remaining bits of the REFCTL0 register, please see [4].

8.2.2 Reference Module Usage via DriverLib Functions

The REF_A module can also be controlled with predefined DriverLib functions. Below, we provide the functions of interest. The complete function list and macro definitions can be found in [11].

```
void REF_A_setReferenceVoltage(uint_fast8_t referenceVoltageSelect)
/*
Sets the reference voltage by modifying the REFVSEL bit of the REFCTL0
    register.
referenceVoltageSelect: the voltage to be set. It can be one of
    REF_A_VREF1_2V, REF_A_VREF1_45V, or REF_A_VREF2_5V.
*/

void REF_A_enableReferenceVoltageOutput(void)
/*
Enables the output generated reference voltage in an analog pin by
    modifying the REFOUT bit of the REFCTL0 register.
```

```
*/

void REF_A_disableReferenceVoltageOutput(void)
/*
Disables the output generated reference voltage in an analog pin by
    modifying the REFOUT bit of the REFCTL0 register.
*/

void REF_A_enableReferenceVoltage(void)
/*
Enables the generated reference voltage used by other peripherals by
    modifying the REFON bit of the REFCTL0 register.
*/

void REF_A_disableReferenceVoltage(void)
/*
Disables the generated reference voltage used by other peripherals by
    modifying the REFON bit of the REFCTL0 register.
*/

bool REF_A_isRefGenBusy(void)
/*
Returns logic level 1 if the reference generator is busy by checking
    the REFGENBUSY bit of the REFCTL0 register.
*/

bool REF_A_isRefGenActive(void)
/*
Returns logic level 1 if the reference generator is active by checking
    the REFGENRDY bit of the REFCTL0 register.
*/

void REF_A_disableTempSensor(void)
/*
Disables the internal temperature sensor by modifying the REFTCOFF bit
    of the REFCTL0 register.
*/

void REF_A_enableTempSensor(void)
/*
Enables the internal temperature sensor by modifying the REFTCOFF bit
    of the REFCTL0 register.
*/
```

8.2.3 Coding Practices for the Reference Module

We provide two examples on how to use the REF_A module in MSP432. In both examples, we use LED1 connected to pin P1.0 on the MSP432 LaunchPad as output.

8.2.3.1 Register-Level Example

We first consider an example in register level to feed voltage from the REF_A module to LED1. We provide the corresponding C code in Listing 8.1. Here, the reference voltage is set to 2.5 V and internal temperature sensor is disabled to save power. For this program to work, the jumper labeled "Red P1.0" above LED1 on MSP432 LaunchPad must be

Listing 8.1 Powering LED1 from the REF_A Module, Using Registers

```c
#include "msp.h"

void main(void){

  WDT_A->CTL = WDT_A_CTL_PW | WDT_A_CTL_HOLD;

  P5->SEL0 |= BIT6; // Configure P5.6 to output VREF
  P5->SEL1 |= BIT6; // Configure P5.6 to output VREF

  while(REF_A->CTL0 & REF_A_CTL0_GENBUSY); // Wait until REFGENBUSY bit
      is reset
// Set reference voltage to 2.5 V, enable reference, enable reference
    output, disable temp. sensor
  REF_A->CTL0 |= REF_A_CTL0_VSEL_3 | REF_A_CTL0_ON | REF_A_CTL0_OUT |
      REF_A_CTL0_TCOFF;
  while(!(REF_A->CTL0 & REF_A_CTL0_GENRDY)); // Wait until reference
      voltage output is ready

  while (1);
}
```

removed. The pin P5.6 (VREF output) should be connected to the bottom pin of this jumper to power LED1.

8.2.3.2 DriverLib Example

We next repeat the example in Listing 8.1 using DriverLib functions. We provide the corresponding C code in Listing 8.2. Here, we first set the pin P5.6 as reference voltage output using the GPIO_setAsPeripheralModuleFunctionOutputPin function. Then, we wait until the reference voltage generator is ready using the REF_A_is RefGenBusy function. Afterwards, we set the reference voltage to 2.5 V, enable the reference voltage on the output pin, and enable the REF_A module using REF_A_setReferenceVoltage, REF_A_enableReferenceVoltageOutput, and REF_A_enableReferenceVoltage functions, respectively. Finally, we disable the temperature sensor to save power using the REF_A_disableTempSensor function and wait until the reference voltage generator output is settled using the REF_A_isRefGen Active function.

8.3 The Comparator Module

There are two identical comparator modules in the MSP432 microcontroller. These are named COMP_E0 and COMP_E1. Each module has two inputs as positive (V+) and negative (V-). One of these inputs can be used for the reference voltage (either external or internal). The other is used for the input voltage. The COMP_E module compares these two values. Let's assume that the input voltage is fed to the positive input and the reference voltage is fed to the negative input. If the input voltage is higher than the reference, then the comparator output will be logic level 1. Otherwise, it will be logic level 0. In other words, the comparator output is just one bit. Hence, this operation can be taken as one-bit analog-to-digital conversion.

Listing 8.2 Powering LED1 from the REF_A Module, Using DriverLib Functions

```
#include <ti/devices/msp432p4xx/driverlib/driverlib.h>

void main(void) {

WDT_A_holdTimer();

GPIO_setAsPeripheralModuleFunctionOutputPin(GPIO_PORT_P5, GPIO_PIN6,
    GPIO_TERTIARY_MODULE_FUNCTION);

while(REF_A_isRefGenBusy()); // Wait until REFGENBUSY bit is reset

// Set reference voltage to 2.5 V, enable reference, enable reference
    output, disable temp. sensor
REF_A_setReferenceVoltage(REF_A_VREF2_5V);
REF_A_enableReferenceVoltageOutput();
REF_A_enableReferenceVoltage();
REF_A_disableTempSensor();

while(!REF_A_isRefGenActive()); // Wait until reference voltage output
    is ready

while (1);
}
```

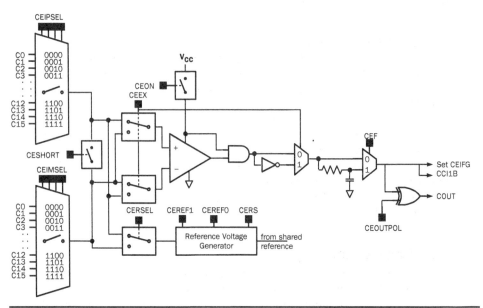

FIGURE 8.1 Block diagram of the COMP_E module.

A block diagram of one of the COMP_E modules is given in Fig. 8.1. As can be seen in this figure, the COMP_E module has 16 input channels connected to two multiplexers. Comparator inputs can be selected through these multiplexers. Moreover, input selection can be reverted using switches. Hence, the user can eliminate the comparator input offset voltage if desired. To note here, reverting input terminals also inverts the output

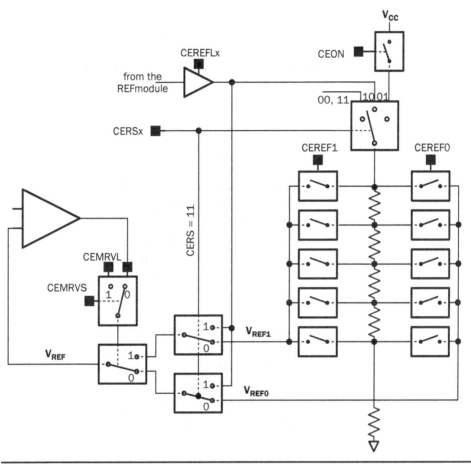

FIGURE 8.2 Block diagram of the reference voltage generator block.

polarity. The COMP_E module has a user selectable output filter (composed by a resistor and capacitor) to filter unwanted parasitic effects. The delay in this filter can be adjusted. Also, additional output inverting can be applied to output if desired.

The reference voltage generator block in Fig. 8.1 is used to generate desired reference voltage to the COMP_E module. This block can be seen in detail in Fig. 8.2. Here, internal reference voltage generated from the REF_A module or analog power supply (AV_{CC}) of the chip can be used as reference input. Then, this voltage value can be further divided by the resistor ladder shown at the bottom right part of Fig. 8.2.

8.3.1 Comparator Module Registers

There are six 16-bit registers to control the two COMP_E modules in MSP432. These are CExCTL0, CExCTL1, CExCTL2, CExCTL3, CExINT, and CExIV, where x can get values 0 or 1. Two structures named COMP_E0 and COMP_E1 are used to control these registers. The mentioned registers are named CTL0, CTL1, CTL2, CTL3, INT, and IV within these structures.

The CExCTL0 register is used to enable positive or negative inputs of the COMP_E module and select desired input channels for them. Entries of this register are tabulated

Bits	15	11-8	7	3-0
	CEIMEN	CEIMSEL	CEIPEN	CEIPSEL

TABLE 8.2 COMP_E Module Control Register 0 (CExCTL0)

Pin	Usage Area
P8.1	COMP_E0 input 0 (C0.0)
P8.0	COMP_E0 input 1 (C0.1)
P7.7	COMP_E0 input 2 (C0.2)
P7.6	COMP_E0 input 3 (C0.3)
P7.5	COMP_E0 input 4 (C0.4)
P7.4	COMP_E0 input 5 (C0.5)
P10.5	COMP_E0 input 6 (C0.6)
P10.4	COMP_E0 input 7 (C0.7)
P6.7	COMP_E1 input 0 (C1.0)
P6.6	COMP_E1 input 1 (C1.1)
P6.5	COMP_E1 input 2 (C1.2)
P6.4	COMP_E1 input 3 (C1.3)
P6.3	COMP_E1 input 4 (C1.4)
P6.2	COMP_E1 input 5 (C1.5)
P5.7	COMP_E1 input 6 (C1.6)
P5.6	COMP_E1 input 7 (C1.7)

TABLE 8.3 Pin Usage of the COMP_E Modules

in Table 8.2. Here, the CEIMEN bit (with the constant value COMP_E_CTL0_IMEN) is used to enable or disable the channel input for the negative terminal. If this bit is set, selected analog input channel for the negative terminal is enabled. The CEIMSEL bits are used for selecting the channel input for the negative terminal. Constants for these bits are COMP_E_CTL0_IMSEL_X, where X can get values between 0 and 15 to select channels from 0 to 15, respectively. The CEIPEN bit (with the constant value COMP_E_CTL0_IPEN) is used to enable or disable the channel input for the positive terminal. If this bit is set, selected analog input channel for the positive terminal is enabled. The CEIPSEL bits are used for selecting channel input for the positive terminal. Constants for these bits are COMP_E_CTL0_IPSEL_X, where X can get values between 0 and 15 to select channels from 0 to 15, respectively. The pin usage of the COMP_E modules is as in Table 8.3.

The CExCTL1 register is used to modify the COMP_E module. Entries of the CExCTL1 register are tabulated in Table 8.4. Here, the CEON bit is used to enable or

Mixed Signal Systems   159

Bits	12	11	10	9-8		7-6
	CEMRVS	CEMRVL	CEON	CEPWRMD		CEFDLY
Bits	5	4	3	2	1	0
	CEEX	CESHORT	CEIES	CEF	CEOUTPOL	CEOUT

TABLE 8.4 COMP_E Module Control Register 1 (CExCTL1)

disable the comparator. If this bit is set with the constant COMP_E_CTL1_ON, the comparator is turned on. CEPWRMD bits are used for selecting the power mode for the comparator. Constants for these bits are COMP_E_CTL1_PWRMD_X, where X can get values between 0 and 2 to select high-speed, normal, and ultra-low power modes, respectively. CEFDLY bits are used for selecting the delay value for the output filter. Constants for these bits are COMP_E_CTL1_FDLY_X, where X can get values between 0 and 3 to select 450, 900, 1800, and 3600-nanosecond delay values, respectively. The CEEX bit is used to exchange the comparator inputs and invert the comparator output. Constant for this bit is COMP_E_CTL1_EX. The CESHORT bit is used to short-circuit the positive and negative inputs of the comparator. When this bit is set with the constant COMP_E_CTL1_SHORT, inputs are shorted. The CEIES bit is used to select the edge type for the comparator output interrupt. When this bit is set with the constant COMP_E_CTL1_IES, falling edge is selected for CEIFG and rising edge is selected for CEIIFG. Otherwise, the opposite setting is applied. The CEF bit is used to enable the output filter of the comparator module. When this bit is set with the constant COMP_E_CTL1_F, the comparator output is filtered. The CEOUTPOL bit is used to select the output polarity of the comparator. When this bit is set with the constant COMP_E_CTL1_OUTPOL, output is inverted. The CEOUT bit keeps the comparators output. For further information on the remaining bits of the CExCTL1 register, please see [4].

The CExCTL2 register is used to control the reference voltage. Hence, the reference voltage source, level, accuracy, and division are configured through this register. Entries of the CExCTL2 register are tabulated in Table 8.5. Here, the CEREFACC bit is used to select the reference accuracy mode. When this bit is reset with the constant COMP_E_CTL2_REFACC, static mode is selected. When it is set, the sampled mode is selected. The CEREFL bits are used for selecting the reference voltage level. Constants for these bits are COMP_E_CTL2_CEREFL_X, where X can get values between 0 and 3 to select no reference, 1.2-V reference, 2.0-V reference, and 2.5-V reference voltage, respectively. The CEREF1 bits are used for configuring the reference resistor tap 1 block which is used to divide the reference voltage when the CEOUT bit is set. Constants for these bits are COMP_E_CTL2_REF1_X, where X can get values between 0 and 31 to multiply the reference voltage with X/32, respectively. CERS bits are used for selecting the reference source for the comparator module. Constants for these bits are COMP_E_CTL2_RS_X, where X can get values between 0 and 3 to select reference sources 0 (no current is drawn by reference circuitry), 1 (V_{CC} is applied to resistor ladder), 2 (reference voltage is applied to resistor ladder), and 3 (resistor ladder is off and reference voltage is directly applied to comparator), respectively. The CEREFACC bit is used to select the terminal for the final reference voltage to be applied to. When this bit is reset with the constant COMP_E_CTL2_RSEL, the final reference voltage is applied to the positive terminal if the CEEX bit is reset and to the negative terminal if the CEEX bit is set. Otherwise, the final reference voltage is applied to the negative terminal if the CEEX bit is reset and

Bits	15	14-13	12-8	7-6	5	4-0
	CEREFACC	CEREFL	CEREF1	CERS	CERSEL	CEREF0

TABLE 8.5 COMP_E Module Control Register 2 (CExCTL2)

Bits	15	14	13	12	11	10	9	8
	CEPD15	CEPD14	CEPD13	CEPD12	CEPD11	CEPD10	CEPD9	CEPD8
Bits	7	6	5	4	3	2	1	0
	CEPD7	CEPD6	CEPD5	CEPD4	CEPD3	CEPD2	CEPD1	CEPD0

TABLE 8.6 COMP_E Module Control Register 3 (CExCTL3)

Bits	12	9	8	4	1	0
	CERDYIE	CEIIE	CEIE	CERDYIFG	CEIIFG	CEIFG

TABLE 8.7 Comp_E Interrupt Control Register (CExINT)

to the positive terminal if the CEEX bit is set. CEREF0 bits are used for configuring the reference resistor tap 0 block which is used to divide the reference voltage when CEOUT bit is reset. Constants for these bits are COMP_E_CTL2_REF0_X, where X can get values between 0 and 31 to multiply the reference voltage with X/32, respectively. CEREF0 and CEREF1 bits are used together to generate a hysteresis reference. The voltage level constructed by CEREF1 bits is the lower limit and the voltage level constructed by CEREF0 bits is the upper limit of this hysteresis reference. In this structure, CEOUT bit is set high when input exceeds upper limit and CEOUT bit is reset when input drops below lower limit.

The CExCTL3 register is used to control the input buffer for the selected input channel. Entries of this register are tabulated in Table 8.6. The CEPDx bit in the table are used for enabling or disabling the input buffer for the input channel x. If the selected CEPDx bit is set, the input buffer for the related channel is disabled. Otherwise, the input buffer is enabled. Constants for these bits are COMP_E_CTL3_PDx, where x can get values between 0 and 15 to select the input channel.

The CExINT register is used to control the COMP_E module interrupt. Entries of this register are tabulated in Table 8.7. In this register, the CERDYIE bit is used to enable or disable the comparator-ready interrupt. If this bit is set with the constant COMP_E_INT_RDYIE, the interrupt is enabled. Otherwise, the interrupt is disabled. The CEIIE bit is used to enable or disable the interrupt for the comparator output with inverted polarity. If this bit is set with the constant COMP_E_INT_IIE, interrupt is enabled. The CEIE bit is used to enable or disable the interrupt for comparator output. If this bit is set with the constant COMP_E_INT_IE, the interrupt is enabled. The CERDYIFG bit is the interrupt flag for comparator-ready interrupt. This bit is set when the comparator is ready to be used. It has to be cleared by software. Constant for this bit is COMP_E_INT_RDYIFG. The CEIIFG bit is the interrupt flag for the comparator output interrupt with inverted polarity. CEIES bit defines the transition of the output by setting this bit. Constant for this bit is COMP_E_INT_IIFG. The CEIFG bit is the interrupt flag for the comparator output

Content	Interrupt Source	Interrupt Flag	Interrupt Priority
0x00	No interrupt pending	–	–
0x02	CEOUT interrupt	CEIFG	Highest
0x04	CEOUT interrupt with inverted polarity	CEIIFG	
0x0A	Comparator-ready interrupt	CERDYIFG	Lowest

TABLE 8.8 CExIV Register

interrupt. CEIES bit defines the transition of the output by setting this bit. Constant for this bit is COMP_E_INT_IFG.

The CExIV register is the interrupt vector used for comparator interrupts. The content of this register is given in Table 8.8.

8.3.2 Comparator Module Usage via DriverLib Functions

The COMP_E modules can also be controlled with DriverLib functions given below. Here, comparator is the comparator module to be configured. It can be one of COMP_E0_BASE or COMP_E1_BASE. The complete function list and macro definitions can be found in [11].

```
bool COMP_E_initModule(uint32_t comparator, const COMP_E_Config
    *config)
/*
Configures the comparator module by modifying CExCTL0, CExCTL1,
    CExCTL2, and CExCTL3 registers.
config: is the config structure. It contains the positive and negative
    terminal connection, output filter, output inversion, and power
    mode selection. Please see the DriverLib user guide for macro
    definitions.
*/

void COMP_E_setReferenceVoltage(uint32_t comparator, uint_fast16_t
    supplyVoltageReferenceBase, uint_fast16_t
    lowerLimitSupplyVoltageFractionOf32, uint_fast16_t
    upperLimitSupplyVoltageFractionOf32)
/*
Configures the reference module and sets the reference voltage.
The voltage is determined by the equation: Vbase * (Numerator / 32). If
    the upper and lower limit voltage numerators are equal, then a
    static reference is defined. If they are different, then a
    hysteresis effect is generated. It uses the CExCTL2 registers.
supplyVoltageReferenceBase: the voltage reference base. It can be one
    of COMP_E_REFERENCE_AMPLIFIER_DISABLED, COMP_E_VREFBASE1_2V,
    COMP_E_VREFBASE2_0V or COMP_E_VREFBASE2_5V.
lowerLimitSupplyVoltageFractionOf32: the numerator for lower limit of
    the reference voltage.
upperLimitSupplyVoltageFractionOf32: the numerator for upper limit of
    the reference voltage.
*/

void COMP_E_setPowerMode(uint32_t comparator, uint_fast16_t powerMode)
```

```
/*
Sets the power mode of the comparator module by modifying CExCTL1
    registers.
powerMode: the power mode to be set. It can be one of
    COMP_E_HIGH_SPEED_MODE, COMP_E_NORMAL_MODE or
    COMP_E_ULTRA_LOW_POWER_MODE.
*/

void COMP_E_enableModule(uint32_t comparator)
/*
Enables the comparator module by modifying the CEON bit of the CExCTL1
    register.
*/

void COMP_E_disableModule(uint32_t comparator)
/*
Disables the comparator module by modifying the CEON bit of the CExCTL1
    register.
*/

uint8_t COMP_E_outputValue(uint32_t comparator)
/*
Returns the output value of the comparator module by checking the CEOUT
    bit of the CExCTL1 register.
*/

void COMP_E_enableInterrupt(uint32_t comparator, uint_fast16_t mask)
/*
Enables the comparator interrupts.
mask: interrupt to be enabled. It can be logical OR form of
    COMP_E_OUTPUT_INTERRUPT, COMP_E_INVERTED_POLARITY_INTERRUPT, and
    COMP_E_READY_INTERRUPT.
*/

void COMP_E_disableInterrupt(uint32_t comparator, uint_fast16_t mask)
/*
Disables the comparator interrupts.
mask: interrupt to be disabled. It can be logical OR form of
    COMP_E_OUTPUT_INTERRUPT, COMP_E_INVERTED_POLARITY_INTERRUPT, and
    COMP_E_READY_INTERRUPT.
*/

void COMP_E_clearInterruptFlag(uint32_t comparator, uint_fast16_t mask)
/*
Clears the comparator interrupts flags.
mask: interrupt to be cleared. It can be logical OR form of
    COMP_E_OUTPUT_INTERRUPT, COMP_E_INVERTED_POLARITY_INTERRUPT, and
    COMP_E_READY_INTERRUPT.
*/

uint_fast16_t COMP_E_getEnabledInterruptStatus(uint32_t comparator)
/*
Returns the interrupt status flag masked with enabled interrupt mask.
*/
```

Listing 8.3 Comparing an Input Voltage with a Reference Value, Using Registers

```c
#include "msp.h"

#define LED2_RED BIT0
#define LED2_GREEN BIT1

void main(void){

 WDT_A->CTL = WDT_A_CTL_PW | WDT_A_CTL_HOLD;

 P2->DIR = LED2_RED | LED2_GREEN;
 P2->OUT = 0x00;
 P6->SEL0 |= BIT4; // Configure P6.4 for comparator input C1.3
 P6->SEL1 |= BIT4; // Configure P6.4 for comparator input C1.3

// Input channel 3 is selected for V+ terminal
 COMP_E1->CTL0 = COMP_E_CTL0_IPEN |COMP_E_CTL0_IPSEL_3;
// Select high-speed mode, enable output filter with 3600 ns delay
 COMP_E1->CTL1 = COMP_E_CTL1_PWRMD_0 | COMP_E_CTL1_F|
     COMP_E_CTL1_FDLY_3;
// Vcc/2 is selected as reference voltage for both VREF0 and VREF1 and
    applied to V- terminal
 COMP_E1->CTL2 = COMP_E_CTL2_RS_1| COMP_E_CTL2_RSEL |
     COMP_E_CTL2_CEREFL_0 | COMP_E_CTL2_REF0_16 | COMP_E_CTL2_REF1_16;
 COMP_E1->CTL3 = COMP_E_CTL3_PD3; // Disable input buffer for channel 3
 COMP_E1->CTL1 |= COMP_E_CTL1_ON; // Turn on the comparator

 while(1){
  if(COMP_E1->CTL1 & CEOUT){ // If comparator output is set
   P2->OUT |= LED2_RED;
   P2->OUT &= ~LED2_GREEN;}
  else { // If comparator output is reset
   P2->OUT |= LED2_GREEN;
   P2->OUT &= ~LED2_RED;}
}}
```

8.3.3 Coding Practices for the Comparator Module

We provide four examples on how to use the COMP_E module in this section. In the first and third examples, we use the red and green colors of LED2 connected to pins P2.0 and P2.1 on the MSP432 LaunchPad as output. In the second and fourth examples, we use LED1 connected to pin P1.0 on the MSP432 LaunchPad as output.

8.3.3.1 Register-Level Examples

We first provide an example where the red and green colors of LED2 are controlled by the comparator output. We provide the C code in Listing 8.3. Here, the comparator input is selected as pin P6.4 (connected to input channel 3 of the COMP_E1 module). The comparator input is connected to the positive terminal. The reference voltage connected to the negative terminal is set to $V_{CC}/2$ V. Red color of LED2 is turned on when the comparator output is at logic level 1. Otherwise, the green color of LED2 is turned on.

We next provide an example where LED1 is controlled by the comparator output interrupt. We provide the C code in Listing 8.4. Here, the comparator input is selected as pin P6.4 (connected to input channel 3 of the COMP_E1 module). The comparator

Listing 8.4 Usage of the Comparator Interrupt, Using Registers

```c
#include "msp.h"

#define LED1 BIT0

void main(void) {

  WDT_A->CTL = WDT_A_CTL_PW | WDT_A_CTL_HOLD;

  P1->DIR = LED1;
  P1->OUT = 0x00;
  P6->SEL0 |= BIT4; // Configure P6.4 for comparator input C1.3
  P6->SEL1 |= BIT4; // Configure P6.4 for comparator input C1.3

  // Input channel 3 is selected for V+ terminal
  COMP_E1->CTL0 = COMP_E_CTL0_IPEN |COMP_E_CTL0_IPSEL_3;
  // Select high-speed mode, enable output filter with 3600 ns delay
  COMP_E1->CTL1 = COMP_E_CTL1_PWRMD_0 | COMP_E_CTL1_F|
     COMP_E_CTL1_FDLY_3;
  // Vcc/2 is selected as reference voltage for both VREF0 and VREF1 and
     applied to V- terminal
  COMP_E1->CTL2 = COMP_E_CTL2_RS_1| COMP_E_CTL2_RSEL |
     COMP_E_CTL2_CEREFL_0 | COMP_E_CTL2_REF0_31 | COMP_E_CTL2_REF1_31;
  COMP_E1->CTL3 = COMP_E_CTL3_PD3; // Disable input buffer for channel 3
  COMP_E1->INT = COMP_E_INT_IE; // Enable the comparator output
     interrupt
  COMP_E1->CTL1 |= COMP_E_CTL1_ON; // Turn on the comparator

  NVIC->ISER[0] = 0x00000080; // COMP_E1 interrupt is enabled in NVIC
  __enable_interrupts(); // All interrupts are enabled

  SCB->SCR |= SCB_SCR_SLEEPONEXIT_Msk;
  __sleep();
}

void COMP_E1_IRQHandler(void) {
  COMP_E1->INT &= ~COMP_E_INT_IFG; // Clear comparator output interrupt
     flag
  P1->OUT ^= LED1;
}
```

input is connected to the positive terminal. The reference voltage connected to the negative terminal is set to $V_{CC}/2$ V. The comparator output interrupt edge is selected as low to high. The system is set to LPM0_LDO_VCORE0 mode and it waits for an interrupt. Every time the comparator output changes from low to high, an interrupt is generated and LED1 is toggled in the ISR.

8.3.3.2 DriverLib Examples

We repeat the example in Listing 8.3 using DriverLib functions. We provide the corresponding C code in Listing 8.5. Here, we first set the red and green colors of LED2 as output. Then, we set the pin P6.4 (connected to input channel 3 of the COMP_E1 module) as input using the function `GPIO_setAsPeripheralModuleFunctionInputPin`.

Listing 8.5 Comparing the Input Voltage with a Reference Value, Using DriverLib Functions

```c
#include <ti/devices/msp432p4xx/driverlib/driverlib.h>

const COMP_E_Config compConfig =
{
 COMP_E_VREF,
 COMP_E_INPUT3,
 COMP_E_FILTEROUTPUT_DLYLVL4,
 COMP_E_NORMALOUTPUTPOLARITY
};

void main(void){

 WDT_A_holdTimer();

 GPIO_setAsOutputPin(GPIO_PORT_P2, GPIO_PIN0 | GPIO_PIN1);
 GPIO_setOutputLowOnPin(GPIO_PORT_P2, GPIO_PIN0 | GPIO_PIN1);
 GPIO_setAsPeripheralModuleFunctionInputPin(GPIO_PORT_P6, GPIO_PIN4,
     GPIO_TERTIARY_MODULE_FUNCTION);

 COMP_E_initModule(COMP_E1_BASE, &compConfig);
 COMP_E_setPowerMode(COMP_E1_BASE, COMP_E_HIGH_SPEED_MODE);
 COMP_E_setReferenceVoltage(COMP_E1_BASE,
     COMP_E_REFERENCE_AMPLIFIER_DISABLED, 16, 16);
 COMP_E_disableInputBuffer(COMP_E1_BASE, COMP_E_INPUT3);
 COMP_E_enableModule(COMP_E1_BASE);

 while(1){
  if(COMP_E_outputValue(COMP_E1_BASE) == COMP_E_HIGH){
   GPIO_setOutputLowOnPin(GPIO_PORT_P2, GPIO_PIN0);
   GPIO_setOutputHighOnPin(GPIO_PORT_P2, GPIO_PIN1);}
  else{
   GPIO_setOutputLowOnPin(GPIO_PORT_P2, GPIO_PIN1);
   GPIO_setOutputHighOnPin(GPIO_PORT_P2, GPIO_PIN0);}}
}
```

Afterwards, we initialize the COMP_E1 module by the function COMP_E_initModule. Here, we set the configuration structure such that the comparator input is connected to the positive terminal and the reference voltage is connected to the negative terminal. Output has normal polarity. We also enable the comparator output filter with 3600-nanosecond delay. Then, we set the power mode as high speed and the reference voltage as $V_{CC}/2$ V, and disable the output buffer using COMP_E_setPowerMode, COMP_E_setReferenceVoltage, and COMP_E_disableInputBuffer functions. Finally, we enable the COMP_E1 module using the COMP_E_enableModule function. In a while loop, red color of LED2 is turned on when the comparator output is set to logic level 1. Otherwise, the green color of LED2 is turned on. To do so, we use the COMP_E_outputValue function to check the output of the comparator module.

We repeat the example in Listing 8.4 using DriverLib functions. We provide the corresponding C code in Listing 8.6. Here, different from the previous example in Listing 8.5,

Listing 8.6 Usage of the Comparator Interrupt, Using DriverLib Functions

```c
#include <ti/devices/msp432p4xx/driverlib/driverlib.h>

const COMP_E_Config compConfig =
{
 COMP_E_INPUT3,
 COMP_E_VREF,
 COMP_E_FILTEROUTPUT_DLYLVL4,
 COMP_E_NORMALOUTPUTPOLARITY
};

void main(void) {

 WDT_A_holdTimer();

 GPIO_setAsOutputPin(GPIO_PORT_P1, GPIO_PIN0);
 GPIO_setOutputLowOnPin(GPIO_PORT_P1, GPIO_PIN0);
 GPIO_setAsPeripheralModuleFunctionInputPin(GPIO_PORT_P6, GPIO_PIN4,
    GPIO_TERTIARY_MODULE_FUNCTION);

 COMP_E_initModule(COMP_E1_BASE, &compConfig);
 COMP_E_setPowerMode(COMP_E1_BASE, COMP_E_HIGH_SPEED_MODE);
 COMP_E_setReferenceVoltage(COMP_E1_BASE,
    COMP_E_REFERENCE_AMPLIFIER_DISABLED, 16, 16);
 COMP_E_disableInputBuffer(COMP_E1_BASE, COMP_E_INPUT3);
 COMP_E_enableInterrupt(COMP_E1_BASE, COMP_E_OUTPUT_INTERRUPT);

 Interrupt_enableSleepOnIsrExit();
 Interrupt_enableInterrupt(INT_COMP_E1);
 Interrupt_enableMaster();

 COMP_E_enableModule(COMP_E1_BASE);

 PCM_setPowerState(PCM_LPM0_LDO_VCORE1);
}
void COMP_E1_IRQHandler(void) {
 COMP_E_clearInterruptFlag(COMP_E1_BASE, COMP_E_OUTPUT_INTERRUPT_FLAG);
 GPIO_toggleOutputOnPin(GPIO_PORT_P1, GPIO_PIN0);
}
```

we enable the COMP_E output interrupt using the `COMP_E_enableInterrupt` function. We clear the interrupt flag using `COMP_E_clearInterruptFlag` function and toggle LED1 in the ISR.

8.4 Analog-to-Digital Conversion

There are several analog-to-digital conversion methods each having its advantages and disadvantages [16]. The ADC14 module in MSP432 uses the successive approximation register (SAR) conversion method. Therefore, we will only deal with it in this section. Then, we will focus on the properties of the ADC14 module.

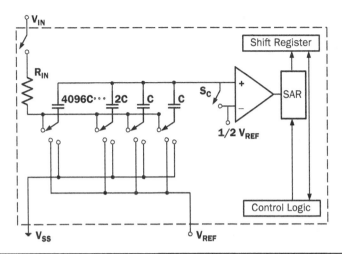

FIGURE 8.3 The circuit diagram of the SAR converter.

8.4.1 Successive Approximation Register Converter

As the name implies, the SAR converter works iteratively in obtaining the digital form of the analog signal. The MSB of the digital form is obtained first in iteration. Then, the remaining lower order bits are obtained step by step till the LSB.

The SAR circuitry is as in Fig. 8.3. In this figure, V_{IN} stands for the analog voltage value to be converted to digital form. The working principle of the given SAR circuitry is as follows. Initially, all the capacitors will be discharged to the offset voltage of the comparator. As the analog signal is fed to input, it will be kept at that value by a sample-and-hold circuit. Then, this voltage is applied to all capacitors. Since each capacitor has a different capacitance (in powers of two), they will be charged accordingly. These values are compared with the reference voltage. Based on the comparison, the bit value (either zero or one) is generated and saved in the shift register. Then, the reference voltage (V_{REF}) is updated and the conversion continues until the desired accuracy is obtained.

To explain how the SAR conversion works, we simulate it with the C code in Listing 8.7. Here, the constant `bitsize` represents the bitsize of the digital form to be obtained. The variable `Vin` stands for the analog voltage to be converted to digital form. The variable `Vref` stands for the reference voltage. The array `bits` holds the digital form obtained. Finally, the variable `quantized` holds the quantized approximate form of the analog input voltage. We can observe the `bits` and `quantized` variables from the Expressions window in CCS.

Let's give an example on the usage of this simulation program. Assume that we take the reference voltage (`Vref`) as 3 (V). We set the bitsize to 14 (bits). Assume the input voltage (`Vin`) to be 1.9 (V). We will get the digital representation `0x2888` as we run the simulation program. As can be observed from the Expressions window, the input voltage is approximated by 1.89990234 V. Hence, this simulation program gives insight on the working principles of the SAR conversion.

8.4.2 The ADC14 Module

The ADC14 module in MSP432 supports up to 1 MSPS 14-bit conversion. The result can be saved in 32 independent conversion registers controlled by 32 control registers.

Listing 8.7 Simulation Program for the SAR Conversion

```c
#include "msp.h"

#define bitsize 14

float Vref = 3.6;
float Vin = 3.3;
float thresh;
float quantized = 0;
int count;
int bitval;
int bits[bitsize];

void main (void){

WDT_A->CTL = WDT_A_CTL_PW | WDT_A_CTL_HOLD;

Vref /= 2;
thresh = Vref;

for(count=0; count<bitsize; count++){
 Vref /= 2;
 if (Vin >= thresh){
  bitval = 1;
  thresh += Vref;}
 else {
  bitval = 0;
  thresh -= Vref;}

 bits[count] = bitval;
 quantized += 2*Vref*bitval;}

while(1);
}
```

Therefore, each conversion register is controlled by its own control register. Using both conversion and control registers, 32 independent samples can be converted without any CPU intervention. The ADC14 module supports software selectable sample-and-hold-time, internal or external selectable reference voltage, selectable single and differential input channel, internal temperature sensor channel, selectable clock sources, and four (single channel, repeat single channel, sequence channel, and repeat sequence channel) conversion modes.

The ADC14 module uses two selectable voltage reference values (V_{R+} and V_{R-}) for conversion that define upper and lower limits. These reference voltages can be supplied by the REF_A module explained in Sec. 8.2 or by an external source. When the user decides on using an external source, he or she should make sure that the reference voltage is between 0 and 3.3 V. Otherwise, MSP432 may be damaged.

The ADC14 module can be used in either single channel or differential channel input settings. We will extensively use the first option throughout the book. For more information on the second option, please see [17].

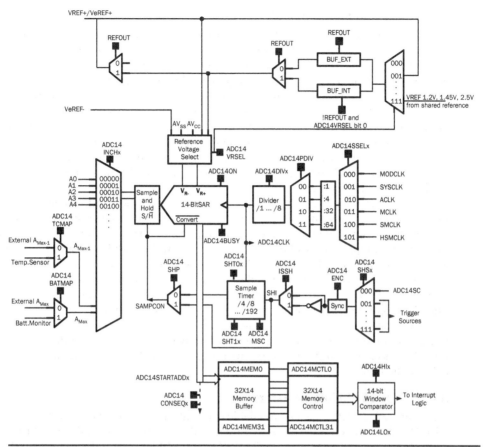

FIGURE 8.4 Block diagram of the ADC14 module.

For a single channel input (V_{in+}), the digital output value (N) obtained by the ADC14 module can be calculated as

$$N = \frac{V_{in+} - V_{R-}}{V_{R+} - V_{R+}} \times 16384 \qquad (8.1)$$

where N is the converted digital value. The full scale value (0x3FFF) is obtained when V_{R+} or higher voltage is applied to the analog input. Similarly, zero value is obtained when V_{R-} or a lower voltage level is applied to the analog input.

A block diagram of the ADC14 module is as in Fig. 8.4. As can be seen in this figure, the ADC14 module can sample various internal and external input channels using sample and hold circuit. The 14-bit ADC core can use the internal variable reference voltage generated by the REF_A module, external reference voltage, or supply voltage as its reference voltage. It can be triggered by various sources. The ADC core can be clocked by MODCLK, SYSCLK, ACLK, MCLK, SMCLK, and HSMCLK. The ADC14CLK clock can be generated by dividing these clocks. The conversion result will be stored in the configured conversion register(s) depending on the conversion mode selected.

8.4.3 The Window Comparator

The ADC14 module has a window comparator that can be used to measure the input signal continuously. When the value of this signal falls between or out of predefined threshold levels, the corresponding interrupt will be generated. The user can set the low and high threshold levels. This gives high flexibility in applications.

Assume that we want to continuously check whether an input signal falls between two threshold levels. Without using the window comparator, the ADC14 module should be running continuously sampling the input signal. For each sampled signal, the CPU should be activated to check whether signal value falls between the two threshold levels. When the window comparator is used, the comparison operation is also performed within the module. Therefore, the CPU will not be activated till an interrupt is generated by the window comparator.

Let's consider an actual example where the usage of the window comparator simplifies operations. Assume that we want to process an audio signal in the environment whenever the signal level gets higher than a threshold. We can form a structure using the window comparator as follows. We can set the low threshold level such that when the audio signal value is below it, no ADC operation is done. When the audio signal value gets higher than the low threshold, then the window comparator triggers an interrupt. This interrupt starts the analog-to-digital conversion operation via the ADC14 module. Therefore, only the signal of interest is sampled via this operation. Then, the sampled signal can be processed. The main advantage of this setup is that the CPU will be in low-power mode for low-level signals. It will only be activated for the signals to be processed. Hence, the energy consumption of the system will be low compared to using the CPU all the times.

8.4.4 ADC14 Module Registers

There are 77 32-bit registers available to control the ADC14 module. These are ADC14CTL0, ADC14CTL1, ADC14LO0, ADC14HI0, ADC14LO1, ADC14HI1, ADC14MCTLx, ADC14MEMx, ADC14IER0, ADC14IER1, ADC14IFGR0, ADC14IFGR1, ADC14CLRIFGR0, ADC14CLRIFGR1, and ADC14IV where x can get values from 0 to 31. A structure named ADC14 is used to control these registers. Within this structure, the mentioned registers are named CTL0, CTL1, LO0, HI0, LO1, HI1, MCTL[x], MEM[x], IER0, IER1, IFGR0, IFGR1, CLRIFGR0, CLRIFGR1, and IV. Again, x can get values from 0 to 31.

The ADC14CTL0 register is used to configure and control the main properties of the ADC14 module. Entries of this register are tabulated in Table 8.9. Here, ADC14PDIV bits are used to select the predivider for the ADC14 clock. Constants for these bits are ADC14_CTL0_PDIV_x, where x can get values from 0 to 3. Corresponding predivider values are 1, 4, 32, and 64, respectively. ADC14SHSx bits are used for selecting sample-and-hold source for the ADC14 module. Constants for these bits are

Bits	31-10	29-27	26	25	24-22	21-19	18-17
	ADC14PDIV	ADC14SHSx	ADC14SHP	ADC14ISSH	ADC14DIVx	ADC14SSELx	ADC14CONSEQx
Bits	16	15-12	11-8	7	4	1	0
	ADC14BUSY	ADC14SHT1x	ADC14SHT0x	ADC14MSC	ADC14ON	ADC14ENC	ADC14SC

TABLE 8.9 ADC14 Module Control Register 0 (ADC14CTL0)

ADC14_CTL0_SHS_x, where x can get values from 0 to 7. Corresponding sample-and-hold sources are ADC14SC bit, Timer_A0 output 1, Timer_A0 output 2, Timer_A1 output 1, Timer_A1 output 2, Timer_A2 output 1, Timer_A2 output 2, and Timer_A3 output 1, respectively. The ADC14SHP bit is used to select the source of the sampling signal. When this bit is set with the constant ADC14_CTL0_SHP, the sampling signal is sourced from the output of the sampling timer. When the bit is reset, the sampling signal is sourced from the input sample signal directly. The ADC14ISSH bit is used to invert the sample input signal. When this bit is set, with the constant ADC14_CTL0_ISSH, the sample input signal is inverted. ADC14DIVx bits are used to select the ADC14 clock divider. Constants for these bits are ADC14_CTL0_DIV_x, where x can get values from 0 to 7. Corresponding divider values are 1, 2, 3, 4, 5, 6, 7, and 8, respectively. ADC14SSELx bits are used to select the clock for the ADC14 module. Constants for these bits are ADC14_CTL0_SSEL_x, where x can get values from 0 to 5. Corresponding clocks are MODCLK, SYSCLK, ACLK, MCLK, SMCLK, and HSMCLK, respectively. ADC14CONSEQx bits are used to select the conversion mode. Constants for these bits are ADC14_CTL0_CONSEQ_x, where x can get values from 0 to 3. Corresponding conversion modes are single-channel single-conversion, multi-channel single-conversion, single-channel repeated-conversion, and multi-channel repeated-conversion, respectively. The ADC14BUSY bit is used to observe the ADC operation activity. When this bit is set by the ADC14 module, it indicates that an active sampling or conversion process is going on. Constant for this bit is ADC14_CTL0_BUSY. ADC14SHT1x bits are used to define sample-and-hold time in terms of ADC14 clock cycles for the ADC14MEMx registers from 8 to 23. Constants for these bits are ADC14_CTL0_SHT1_x, where x can get values from 0 to 7. Corresponding ADC14 clock cycle values are calculated by the formula 4×2^x. ADC14SHT0x bits are used to define sample-and-hold time in terms of ADC14 clock cycles for the ADC14MEMx registers from 0 to 7 and 24 to 31. Constants for these bits are ADC14_CTL0_SHT0_x, where x can get values from 0 to 7. Corresponding ADC14 clock cycle values are calculated by 4×2^x. The ADC14MSC bit allows multiple sample and conversion operations for valid sampling modes. When this bit is set, with the constant ADC14_CTL0_MSC, the sampling and conversion operations after the first one are performed automatically. The ADC14ON bit is used to enable the ADC14 module. Constant for this bit is ADC14_CTL0_ON. The ADC14ENC bit enables the conversion. Constant for this bit is ADC14_CTL0_ENC. The ADC14SC bit starts the analog-to-digital conversion. Constant for this bit is ADC14_CTL0_SC.

The ADC14CTL1 register is mainly used to control the resolution and power consumption of the ADC14 module. Also, some special input channels can be activated through this register. Entries of the ADC14CTL1 register are tabulated in Table 8.10. The ADC14TCMAP bit should be used to enable the internal temperature sensor for the ADC14 module. When this bit is set, with the constant ADC14_CTL1_TCMAP, the internal temperature sensor channel is enabled and can be used via channel A22. The ADC14BATMAP bit is used to monitor the voltage level of the analog power supply

Bits	23	22	5-4	2	1-0
	ADC14TCMAP	ADC14BATMAP	ADC14RES	ADC14REFBURST	ADC14PWRMD

TABLE 8.10 ADC14 Module Control Register 1 (ADC14CTL1)

Bits	15	14	11-8	7	4-0
	ADC14WINCTH	ADC14WINC	ADC14VRSEL	ADC14EOS	ADC14INCHx

TABLE 8.11 ADC14 Module Conversion Memory Control Register x (ADC14MCTLx)

Constant	V_{R+}	V_{R-}
ADC14_MCTLN_VRSEL_0	AV_{CC}	AV_{SS}
ADC14_MCTLN_VRSEL_1	V_{REF} buffered	AV_{SS}
ADC14_MCTLN_VRSEL_14	V_{eREF+}	V_{eREF-}
ADC14_MCTLN_VRSEL_15	V_{eREF+} buffered	V_{eREF-}

TABLE 8.12 Reference Values Used in the ADC14 Module

(AV_{CC}) of the microcontroller. It is especially needed when the microcontroller is supplied by external battery. When this bit is set, with constant ADC14_CTL1_BATMAP, internal $1/2 \times AV_{CC}$ channel is enabled and can be used via channel A23. Half of the AV_{CC} voltage can be obtained from this channel. ADC14RES bits are used to select the resolution of the ADC14 module. Constants for these bits are ADC14_CTL1_RES_x, where x can get values from 0 to 3. Corresponding resolution values are 8, 10, 12, and 14 bits, respectively. The ADC14REFBURST bit is used to select on-time of reference buffer. When this bit is set, with the constant ADC14_CTL1_REFBURST, the ADC14 reference buffer is only enabled during the sample and conversion process. When this bit is reset, the buffer will be enabled continuously. ADC14PWRMD bits are used to select power modes for the ADC14 module. Constants for these bits are ADC14_CTL1_PWRMD_x, where x can get values 0 or 2. When ADC14_CTL1_PWRMD_0 is selected, the ADC14 module is used in regular power mode with up to 1 MSPS sample rate. Also, any resolution can be used in this mode. When ADC14_CTL1_PWRMD_2 is selected, the ADC14 module is used in low-power mode with up to 200 kSPS sample rate. Only 8-, 10-, or 12-bit resolution can be used in this mode. When this mode is used, power consumption of the ADC14 module is less than half of the regular mode. For further information on the remaining bits of this register, please see [4].

ADC14LO0 and ADC14LO1 registers are used to set low threshold value for window comparator. On the other hand, ADC14HI0 and ADC14HI1 registers are used to set high threshold value for the window comparator.

The ADC14MCTLx register is mainly used to select ADC references and input channels. It also controls the window comparator. Entries of this register are tabulated in Table 8.11. Here, ADC14WINCTH is used to select window comparator registers 0 or 1. Constant for this bit is ADC14_MCTLN_WINCTH. If this bit is reset, ADC14LO0 and ADC14HI0 registers are used. If it is set, ADC14LO1 and ADC14HI1 registers are used. ADC14WINC bit is used to enable window comparator. Constant for this bit is ADC14_MCTLN_WINC. ADC14VRSEL bits are used to select the voltage reference values. Constants for these bits are ADC14_MCTLN_VRSEL_x, where x can get values 0, 1, 14, and 15. Corresponding reference values are given in Table 8.12. Here, AV_{CC} is the analog power supply of the ADC14 module with the nominal value 3 V. AV_{SS} is the analog ground supply voltage of the ADC14 module with the nominal value 0 V. V_{REF} is the internal positive reference voltage that is generated by the REF_A module and shared with other modules. V_{eREF+} is the external positive reference voltage for the ADC14 module that can be obtained from pin P5.6. V_{eREF-} is the external negative reference voltage

Pin	Usage Area
P5.5	ADC14 analog input A0
P5.4	ADC14 analog input A1
P5.3	ADC14 analog input A2
P5.2	ADC14 analog input A3
P5.1	ADC14 analog input A4
P5.0	ADC14 analog input A5
P4.7	ADC14 analog input A6
P4.6	ADC14 analog input A7
P4.5	ADC14 analog input A8
P4.4	ADC14 analog input A9
P4.3	ADC14 analog input A10
P4.2	ADC14 analog input A11
P4.1	ADC14 analog input A12
P4.0	ADC14 analog input A13
P6.1	ADC14 analog input A14
P6.0	ADC14 analog input A15
P9.1	ADC14 analog input A16
P9.0	ADC14 analog input A17
P8.7	ADC14 analog input A18
P8.6	ADC14 analog input A19
P8.5	ADC14 analog input A20
P8.4	ADC14 analog input A21
P8.3	ADC14 analog input A22
P8.2	ADC14 analog input A23
P5.6	ADC14 positive reference voltage V_{REF+}/V_{eREF+}
P5.7	ADC14 negative reference voltage V_{REF-}/V_{eREF-}

TABLE 8.13 Pin Usage of the ADC14 Module

for the ADC14 module that can be obtained from pin P5.7. It is advised to connect V_{eREF-} to AV_{SS}.

The ADC14EOS bit is used to stop conversion in multi-channel conversion mode. When this bit is set, with the constant ADC14_MCTLN_EOS, related input channel is specified as the last input channel to be converted. ADC14INCHx bits are used for selecting the input channels for the ADC14 module. Constants for these bits are ADC14_MCTLN_INCH_x, where x can get values from 0 to 31 to select the input channel from A0 to A31, respectively. The pin usage of the ADC14 module is as in Table 8.13. For further information on the remaining bits of the ADC14MCTLx register, please see [4].

ADC14MEMx registers are used for keeping the conversion results, where x can get values from 0 to 31. This means that there are 32 different 32-bit memory locations to keep conversion results. For single-channel conversion, the result is written to the ADC14MEM0 register. For multi-channel conversion, the first result is written to the

Bits	3	2	1
	ADC14HIIE	ADC14LOIE	ADC14INIE

TABLE 8.14 ADC14 Module Interrupt Enable Register 1 (ADC14IER1)

Bits	3	2	1
	ADC14HIIFG	ADC14LOIFG	ADC14INIFG

TABLE 8.15 ADC14 Module Interrupt Flag Register 1 (ADC14IFGR1)

ADC14MEM0 register, then other ADC14MEMx registers are used for the remaining channels.

The ADC14IER0 register is used to enable the ADC interrupt for the selected ADC14MEMx register. Constants for this register are ADC14_IER0_IEx, where x can get values from 0 to 31. These constants enable interrupts for the ADC14MEMx register from 0 to 31, respectively. An interrupt is generated when a conversion result is written to the ADC14MEMx register. To note here, the interrupt should have been enabled for that ADC14MEMx register.

The ADC14IER1 register is used to enable the ADC interrupt for the window comparator. Entries of this register are tabulated in Table 8.14. Here, ADC14HIIE is used to enable the interrupt for exceeding the high threshold of window comparator. Constant for this bit is ADC14_IER1_HIIE. ADC14LOIE is used to enable the interrupt for falling below the low threshold of window comparator. Constant for this bit is ADC14_IER1_LOIE. ADC14INIE is used to enable the interrupt for staying between the thresholds of window comparator. Constant for this bit is ADC14_IER1_INIE. For further information on the remaining bits of the ADC14IER1 register, please see [4].

The ADC14IFGR0 register is the interrupt flag for the ADC14 interrupt. Constants for this register are ADC14_IFGR0_IFGx, where x can get values from 0 to 31. When a conversion result is written to an ADC14MEMx register, related interrupt flag is set.

The ADC14IFGR1 register is the interrupt flag for the window comparator. Entries of this register are tabulated in Table 8.15. Here, ADC14HIIFG is the interrupt flag for exceeding the high threshold of window comparator. Constant for this bit is ADC14_IFGR1_HIIFG. ADC14LOIE is the interrupt flag for falling below the low threshold of window comparator. Constant for this bit is ADC14_IFGR1_LOIFG. ADC14INIE is the interrupt flag for staying between the thresholds of window comparator. Constant for this bit is ADC14_IFGR1_INIFG. For further information on the remaining bits of the ADC14IFGR1 register, please see [4].

The ADC14CLRIFGR0 register is used for clearing the selected interrupt flags. Constants for this register are ADC14_CLRIFGR0_CLRIFGx, where x can get values from 0 to 31. These constants clear interrupt flags from 0 to 31, respectively.

The ADC14CLRIFGR1 register is used for clearing the interrupt flags for window comparator. Entries of this register are tabulated in Table 8.16. Here, the CLRADC14HIIFG is used to clear the ADC14HIIFG (with the constant ADC14_

Bits	3	2	1
	CLRADC14HIIFG	CLRADC14LOIFG	CLRADC14INIFG

TABLE 8.16 ADC14 Module Clear Interrupt Flag Register 1 (ADC14CLRIFGR1)

TAIV Content	Interrupt Source	Interrupt Flag	Interrupt Priority
0x00	No interrupt pending	–	–
0x02	ADC14MEMx overflow	ADC14OVIFG	Highest
0x04	Conversion time overflow	ADC14TOVIFG	
0x06	ADC14 window high interrupt flag	ADC14HIIFG	
0x08	ADC14 window low interrupt flag	ADC14LOIFG	
0x0A	ADC14 window between interrupt flag	ADC14INIFG	
0x0C	ADC14MEM0 interrupt flag	ADC14IFG0	
0x0E	ADC14MEM1 interrupt flag	ADC14IFG1	
...	...	...	
0x48	ADC14MEM30 interrupt flag	ADC14IFG30	
0x4A	ADC14MEM31 interrupt flag	ADC14IFG31	
0x4C	Local buffered reference is ready	ADC14RDYIFG	Lowest

TABLE 8.17 ADC14IV Register

CLRIFGR1_CLRHIIFG). The CLRADC14LOIFG is used to clear the ADC14LOIFG (with the constant ADC14_CLRIFGR1_CLRLOIFG). The CLRADC14INIFG is used to clear the ADC14INIFG (with the constant ADC14_CLRIFGR1_CLRINIFG). For further information on the remaining bits of the ADC14CLRIFGR1 register, please see [4].

The ADC14IV register is the interrupt vector used for all ADC14 interrupts. The contents of this register are given in Table 8.17.

8.4.5 ADC14 Module Usage via DriverLib Functions

The ADC14 module can also be controlled via predefined DriverLib functions. We provide functions of interest below. The complete function list and macro definitions can be found in [11].

```
void ADC14_enableModule(void)
/*
Enables the ADC14 module by modifying the ADC14ON bit of the ADC14CTL0
    register.
*/

bool ADC14_disableModule(void)
/*
Disables the ADC14 module by modifying the ADC14ON bit of the ADC14CTL0
    register.
```

```
*/

bool ADC14_initModule(uint32_t clockSource, uint32_t clockPredivider,
    uint32_t clockDivider, uint32_t internalChannelMask)
/*
Configures and initializes the ADC14 module by modifying the ADC14CTL0
    and ADC14CTL1 registers.
clockSource: clock to feed the ADC14 module. It can be one of MODCLK,
    SYSCLK, ACLK, MCLK, SMCLK, and HSMCLK. Please see DriverLib user
    guide for macro definitions.
clockPredivider:  pre-divider for the clock. It can be one of 1, 4, 32,
    or 64. Please see DriverLib user guide for macro definitions.
clockDivider: clock divider for the pre-divided clock. It can be one of
    1, 2, 3, 4, 5, 6, 7, or 8. Please see DriverLib user guide for
    macro definitions.
internalChannelMask: the channel mapping for the ADC14 module. It can
    be logical OR form of the internal channel 0, 1, 2, 3, temperature
    sensor, and battery channels. No internal channel routing can be
    selected if desired. Please see DriverLib user guide for macro
    definitions.
*/

void ADC14_setResolution(uint32_t resolution)
/*
Sets the sampling resolution of the ADC14 module by modifying ADC14RES
    bits of the ADC14CTL1 register.
resolution: resolution to be set. It can be one of ADC_8BIT, ADC_10BIT,
    ADC_12BIT or ADC_14BIT.
*/

bool ADC14_setSampleHoldTrigger(uint32_t source, bool invertSignal)
/*
Sets the trigger for the ADC14 module by modifying the ADC14SHSx bits
    of the ADC14CTL0 register.
source: the trigger source. It can be one of software, 1, 2, 3, 4, 5,
    6, or 7 values. Please see DriverLib user guide for macro
    definitions.
invertSignal: when set to logic level 1, the source triggers the ADC at
    falling edge.
*/

bool ADC14_setSampleHoldTime(uint32_t firstPulseWidth, uint32_t
    secondPulseWidth)
/*
Sets the sample-and-hold-time by modifying the ADC14SHT1x or ADC14SHT0x
    bits of ADC14CTL0 register. There are two configuration parameters
    to control the sample-and-hold-time for different memory
    locations.
firstPulseWidth: first sample and hold duration parameter for memory
    locations ADC_MEMORY_0 through ADC_MEMORY_7 and ADC_MEMORY_24
    through ADC_MEMORY_31 in terms of ADC clock cycles. It can be one
    of 4, 8, 16, 32, 64, 96, 128 or 192. Please see DriverLib user
    guide for macro definitions.
```

*secondPulseWidth: second sample and hold duration parameter for memory
    locations ADC_MEMORY_8 through ADC_MEMORY_23 in terms of ADC clock
    cycles. Possible values are the same as firstPulseWidth. Please see
    DriverLib user guide for macro definitions.*
*/

bool ADC14_configureMultiSequenceMode(uint32_t memoryStart, uint32_t
    memoryEnd, bool repeatMode)
/*
*Configures the ADC14 module in multi channel (scan) mode by modifying
    the ADC14CTL0 and ADC14CTL1 registers. The results are saved into
    ADC memory locations.*
*memoryStart: start address of the memory locations. It can be one of
    ADC_MEM0 through ADC_MEM31.*
*memoryEnd: end address of the memory locations. It can be one of
    ADC_MEM0 through ADC_MEM31.*
*repeatMode: configures the ADC14 module to repeat conversion
    automatically after first conversion. It can be at logic level 1
    or 0.*
*/

bool ADC14_configureSingleSampleMode(uint32_t memoryDestination, bool
    repeatMode)
/*
*Configures the ADC14 module in single channel mode by modifying
    ADC14CTL0 and ADC14CTL1 registers. The result is saved in ADC
    memory location.*
*memoryDestination: the memory location to save the result. It can be
    one of ADC_MEM0 through ADC_MEM31.*
*repeatMode: configures the ADC14 module to repeat conversion
    automatically after the first conversion. It can be at logic level
    1 or 0.*
*/

bool ADC14_enableConversion(**void**)
/*
*Enables the configured conversion mode by modifying the ADC14ENC bit of
    the ADC14CTL0 register. Note that it does not trigger the
    conversion.*
*/

void ADC14_disableConversion(**void**)
/*
*Stops and disables the configured conversion mode by modifying the
    ADC14ENC bit of the ADC14CTL0 register.*
*/

bool ADC14_isBusy(**void**)
/*
*Returns logic level 1 if the ADC14 module is active and conversion is
    in progress by checking the ADC14BUSY bit of the ADC14CTL0
    register.*
*/

```c
bool ADC14_configureConversionMemory(uint32_t memorySelect, uint32_t
    refSelect, uint32_t channelSelect, bool differntialMode)
/*
Configures the conversion memory locations by individually or in group
    by modifying ADC14MCTLx registers.
memorySelect: memory location to be configured. It can be logical OR
    form of ADC_MEM0 through ADC_MEM31.
refSelect: voltage reference for conversion. It can be one of power
    supply voltages, internal or external reference voltages. Please
    see DriverLib user guide for macro definitions.
channelSelect: the channel to be sampled. It can be one of ADC_INPUT_A0
    through ADC_INPUT_A31.
differntialMode: sets the conversion in single or differential mode. It
    can be one of ADC_NONDIFFERENTIAL_INPUTS or
    ADC_DIFFERENTIAL_INPUTS. If differential mode is enabled, then the
    configured channel is combined with its neighbor channel to set a
    differential channel. For example if A0 or A1 channel is
    configured, then conversation takes place between A0 and A1
        channels.
*/

bool ADC14_enableComparatorWindow(uint32_t memorySelect, uint32_t
    windowSelect)
/*
Enables the window comparator by modifying the ADC14MCTLx register.
memorySelect: the selected memory location. It can be logical OR form
    of ADC_MEM0 through ADC_MEM31.
windowSelect: memory location to store the converted samples. It can be
     one of ADC_COMP_WINDOW0 or ADC_COMP_WINDOW1 values.
*/

bool ADC14_disableComparatorWindow(uint32_t memorySelect)
/*
Disables the window comparator by modifying the ADC14MCTLx register.
memorySelect: the selected memory location. It can be logical OR form
    of ADC_MEM0 through ADC_MEM31.
*/

bool ADC14_setComparatorWindowValue(uint32_t window, int16_t low,
    int16_t high)
/*
Sets the lower and upper threshold levels of window comparator by
    modifying ADC14LOx and ADC14HIx registers.
window: memory location to store the converted samples. It can be one
    of ADC_COMP_WINDOW0 or ADC_COMP_WINDOW1 values.
low: lower threshold value.
high: upper threshold value.
*/

uint_fast16_t ADC14_getResult(uint32_t memorySelect)
/*
Returns the conversion result from the selected memory location by
    reading the contents of individual ADC14MEMx registers.
memorySelect: the selected memory location. It can be one of ADC_MEM0
    through ADC_MEM31.
```

```
*/

void ADC14_getMultiSequenceResult(uint16_t *res)
/*
Returns the conversion results from configured multi-sequence memory
    addresses by reading the contents of ADC14MEMx registers.
res: the array pointer to save conversion results.
*/

void ADC14_getResultArray(uint32_t memoryStart, uint32_t memoryEnd,
    uint16_t *res)
/*
Returns the conversion results from the selected memory addresses by
    reading the contents of ADC14MEMx registers.
memoryStart: start address of the memory locations. It can be one of
    ADC_MEM0 through ADC_MEM31.
memoryEnd: end address of the memory locations. It can be one of
    ADC_MEM0 through ADC_MEM31.
res: the array pointer to save conversion results.
*/

bool ADC14_enableSampleTimer(uint32_t multiSampleConvert)
/*
Enables the timer trigger for sampling by modifying ADC14SHP and
    ADC14MSC bits of the ADC14CTL0 register. After each conversion, ADC
    can be automatically or manually triggered to convert next sample.
multiSampleConvert: the iteration method to convert next sample. It can
    be one of ADC_MANUAL_ITERATION or ADC_AUTOMATIC_ITERATION.
*/

bool ADC14_disableSampleTimer(void)
/*
Enables the timer trigger for sampling by modifying ADC14SHP bit of the
    ADC14CTL0 register.
*/

void ADC14_enableInterrupt(uint_fast64_t mask)
/*
Enables the ADC interrupt by modifying the ADC14IER0 register.
mask: the interrupt type to be enabled. It can be logical OR form of
    ADC_INT0 through ADC_INT31, ADC_IN_INT, ADC_LO_INT, ADC_HI_INT,
    ADC_OV_INT, ADC_TOV_INT or ADC_RDY_INT.
*/

void ADC14_disableInterrupt(uint_fast64_t mask)
/*
Disables the ADC interrupts by modifying the ADC14IER0 register.
mask: the interrupt type to be disabled. It can be logical Or form of
    ADC_INT0 through ADC_INT31, ADC_IN_INT, ADC_LO_INT, ADC_HI_INT,
    ADC_OV_INT, ADC_TOV_INT or ADC_RDY_INT values.
*/

uint_fast64_t ADC14_getEnabledInterruptStatus(void)
/*
```

*Returns the status of enabled interrupt flags by reading the ADC14IER0
    and ADC14IFGR0 registers.
*/

void ADC14_clearInterruptFlag(uint_fast64_t mask)
/*
Clears ADC interrupts by modifying the ADC14CLRIFGR0 register.
mask: the interrupt type to be cleared. It can be logical OR form of
    ADC_INT0 through ADC_INT31, ADC_IN_INT, ADC_LO_INT, ADC_HI_INT,
    ADC_OV_INT, ADC_TOV_INT or ADC_RDY_INT.
*/

8.4.6 ADC14 Module Usage via Energia Function

Energia can also be used to control the ADC14 module in MSP432. We provide the related function below. More information on Energia functions can be found in [12].

int analogRead(pin);
/*
Configures and reads analog value from specified pin by modifying and
    reading ADC14CTL0, ADC14CTL1, ADC14MCTLx, and ADC14MEMx registers.
pin: selected pin to be configured as analog input.
*/

8.4.7 Coding Practices for the ADC14 Module

We provide example codes on the usage of the ADC14 module in this section. We consider register-level, DriverLib, and Energia function usages through these examples. We use the input channels A0 (connected to pin P5.5) and A1 (connected to pin P5.4) on the MSP432 LaunchPad in some operations. We also use LED1 connected to pin P1.0 on the MSP432 LaunchPad as output.

8.4.7.1 Register-Level Examples

The first example on the usage of the ADC14 module is actually on forming a comparator. We provide the corresponding C code in Listing 8.8. Here, the input voltage level at channel A0 is checked in an infinite loop. If the converted value is above a certain level, LED1 is turned on. We used single-channel, single-conversion mode in this example using software trigger.

The second example on the usage of the ADC14 module is on measuring the internal temperature level of the MSP432 microcontroller. We provide the corresponding C code in Listing 8.9. Here, the internal temperature sensor is used with 2.5-V internal reference. We use single-channel, multiple conversion mode in this example using software trigger. The system is put in LPM0_LDO_VCORE0 mode. Hence, it waits for an ADC14 interrupt. Then, the temperature value is read and converted to degrees Celsius in the ISR.

The third example on the usage of the ADC14 module focuses on the multi-channel, multiple conversion mode. We provide the corresponding C code for this operation in Listing 8.10. Here, the ADC14 module is triggered by using the Timer_A0 output 1. The microcontroller is put into LPM0_LDO_VCORE0 mode. Hence, it waits for an ADC interrupt. The conversion results of input channels A0 and A1 are obtained in the ISR every 0.5 seconds.

Listing 8.8 Usage of the ADC14 Module as a Comparator, Using Registers

```c
#include "msp.h"

#define LED1 BIT0

void main(void){

  WDT_A->CTL = WDT_A_CTL_PW | WDT_A_CTL_HOLD;

  P1->DIR = LED1;
  P5->SEL1 |= BIT5; // Set P5.5 as ADC input A0
  P5->SEL0 |= BIT5; // Set P5.5 as ADC input A0

// Configure MCLK as 24 MHz from DCO
  CS->KEY = CS_KEY_VAL;
  CS->CTL0 &= ~ CS_CTL0_DCORSEL_MASK;
  CS->CTL0 |= CS_CTL0_DCORSEL_4;
  CS->CTL1 &= ~(CS_CTL1_SELM_MASK | CS_CTL1_DIVM_MASK);
  CS->CTL1 |= CS_CTL1_SELM_3;
  CS->KEY = 0;

  ADC14->CTL0 &= ~ADC14_CTL0_ENC; // Reset ENC bit for configuration
// Predivide ADC clock by 4, Software trigger, Sampling timer is the
    source of sampling signal, Divide ADC clock by 6, ADC clock source
    is MCLK, single channel-single conversion mode, Sample and hold
    time is 16 ADC clock cycles, ADC14 on
  ADC14->CTL0 = ADC14_CTL0_PDIV_1 | ADC14_CTL0_SHS_0 | ADC14_CTL0_SHP |
      ADC14_CTL0_DIV_5 | ADC14_CTL0_SSEL_3 | ADC14_CTL0_CONSEQ_0 |
      ADC14_CTL0_SHT0_2 | ADC14_CTL0_ON;
  ADC14->CTL1 = ADC14_CTL1_RES_3; // ADC 14-bit resolution
  ADC14->MCTL[0] = ADC14_MCTLN_VRSEL_0 | ADC14_MCTLN_INCH_0; // Vcc -
      Vss references, input channel A0 is selected
  ADC14->CTL0 |= ADC14_CTL0_ENC; // Enable conversion

  while (1){
    ADC14->CTL0 |= ADC14_CTL0_SC; // Start conversion
    while(ADC14->CTL0 & ADC14_CTL0_BUSY); // Wait until conversion is
        done
    if(ADC14->MEM[0] > 0x200) P1->OUT |= LED1;
    else P1->OUT &= ~LED1;}
}
```

8.4.7.2 DriverLib Examples

We can repeat the previous example in Listing 8.8 using DriverLib functions. We provide the C code in Listing 8.11. Here, we first set the DCO frequency as 24 MHz and LED1 as output. We set pin P5.5 as analog input using the `GPIO_setAsPeripheralModuleFunctionInputPin` function. Afterwards, we enable and initialize the ADC14 module using the `ADC14_enableModule` and `ADC14_initModule` functions. We set the ADC14 module in single sample mode using the `ADC14_configureSingleSampleMode` function. Afterwards, we configure the ADC_MEM0 as conversion memory using the `ADC14_configureConversionMemory` function. Finally, we configure the trigger

Listing 8.9 Measuring the Internal Temperature Level of the Microcontroller by the ADC14 Module, Using Registers

```c
#include "msp.h"

int TempADC;
float TempC;

int TempCal_2_5v_30C; // Variable to store the calibration value for
    2.5V reference and 30 Celsius
int TempCal_2_5v_85C; // Variable to store the calibration value for
    2.5V reference and 85 Celsius

void main(void){

  WDT_A->CTL = WDT_A_CTL_PW | WDT_A_CTL_HOLD;

// Read the ADC temperature reference calibration value
  TempCal_2_5v_30C = TLV->ADC14_REF2P5V_TS30C;
  TempCal_2_5v_85C = TLV->ADC14_REF2P5V_TS85C;

  while(REF_A->CTL0 & REF_A_CTL0_GENBUSY); // Wait until REFGENBUSY bit
      is reset
  REF_A->CTL0 |= REF_A_CTL0_VSEL_3 | REF_A_CTL0_ON; // Set reference
      voltage to 2.5 V, enable reference
  REF_A->CTL0 &= ~REF_A_CTL0_TCOFF; // Enable temperature sensor
  while(!(REF_A->CTL0 & REF_A_CTL0_GENRDY)); // Wait until reference
      voltage output is ready

  ADC14->CTL0 &= ~ADC14_CTL0_ENC; // Reset ENC bit for configuration
//Software trigger, Sampling timer is the source of sampling signal,
    single channel-repeated conversion mode, Sample and hold time is
    128 ADC clock cycles, multiple sample and conversion, ADC14 on
  ADC14->CTL0 = ADC14_CTL0_SHS_0 | ADC14_CTL0_SHP |
      ADC14_CTL0_CONSEQ_2 | ADC14_CTL0_SHT0_6 | ADC14_CTL0_MSC |
      ADC14_CTL0_ON;
  ADC14->CTL1 = ADC14_CTL1_RES_3 | ADC14_CTL1_TCMAP; // ADC 14-bit
      resolution, Enable internal temperature sensor
  ADC14->MCTL[0] = ADC14_MCTLN_VRSEL_1 | ADC14_MCTLN_INCH_22; // VREF -
      Vss references, input channel A22 is selected
  ADC14->IER0 = ADC14_IER0_IE0; // ADC interrupt enabled for ADC14->
      MEM[0]
  ADC14->CTL0 |= ADC14_CTL0_ENC|ADC14_CTL0_SC; // Enable and start
      conversion

  NVIC->ISER[0] = 0x01000000; // ADC14 interrupt is enabled in NVIC
  __enable_irq(); // All interrupts are enabled

  SCB->SCR |= SCB_SCR_SLEEPONEXIT_Msk; // Sleep on exit
  __sleep(); // enter LPM0
}

// ADC14 interrupt service routine
void ADC14_IRQHandler(void){
 if (ADC14->IFGR0 & ADC14_IFGR0_IFG0){
  TempADC = ADC14->MEM[0]; // Read temp. sensor ADC value
// Obtain temperature in Celsius
  TempC = ((((float) TempADC - TempCal_2_5v_30C) * (85 - 30)) /
      (TempCal_2_5v_85C - TempCal_2_5v_30C) + 30.0f;}
}
```

Listing 8.10 ADC Operation on multi-channels, Using Registers

```c
#include "msp.h"

uint16_t resultsBuffer[2];

void main(void){

 WDT_A->CTL = WDT_A_CTL_PW | WDT_A_CTL_HOLD;

 P5->SEL1 |= BIT5 | BIT4; // Set P5.5 as ADC input A0 and P5.4 as ADC
    input A1
 P5->SEL0 |= BIT5 | BIT4; // Set P5.5 as ADC input A0 and P5.4 as ADC
    input A1

 CS->KEY = CS_KEY_VAL;
 CS->CTL1 |= CS_CTL1_SELA_2;
 CS->KEY = 0;

 ADC14->CTL0 &= ~ADC14_CTL0_ENC; // Reset ENC bit for configuration
//Timer_A0 output 1 trigger, Sampling timer is the source of sampling
    signal, multi channel-repeated conversion mode, Sample and hold
    time is 96 ADC clock cycles, ADC14 on
 ADC14->CTL0 = ADC14_CTL0_SHS_1 | ADC14_CTL0_SHP |
    ADC14_CTL0_CONSEQ_3 | ADC14_CTL0_SHT0_5 | ADC14_CTL0_ON;
 ADC14->MCTL[0] = ADC14_MCTLN_VRSEL_0 | ADC14_MCTLN_INCH_0; // Vcc -
    Vss references, input channel A0, End of sequence
 ADC14->MCTL[1] = ADC14_MCTLN_VRSEL_0 | ADC14_MCTLN_INCH_1 |
    ADC14_MCTLN_EOS; // Vcc - Vss references, input channel A1, End of
     sequence
 ADC14->IER0 = ADC14_IER0_IE1; // ADC interrupt enabled for ADC14->MEM
    [1]

 ADC14->CTL0 |= ADC14_CTL0_ENC; // Start conversion-software trigger

 TIMER_A0->CCR[0] = 16384;  // PWM period
 TIMER_A0->CCR[1] = 16384;  // CCR1 PWM duty cycle
 TIMER_A0->CCTL[1] = TIMER_A_CCTLN_OUTMOD_3; // CCR1 set/reset
 TIMER_A0->CTL = TIMER_A_CTL_TASSEL_1 | TIMER_A_CTL_MC_1 |
    TIMER_A_CTL_CLR; // ACLK, Up mode, Clear TAR

 NVIC->ISER[0] = 0x01000000; // ADC14 interrupt is enabled in NVIC
 __enable_irq(); // All interrupts are enabled

 SCB->SCR |= SCB_SCR_SLEEPONEXIT_Msk; // Sleep on exit
 __sleep(); // enter LPM0
}

// ADC14 interrupt service routine
void ADC14_IRQHandler(void){
 if (ADC14->IFGR0 & ADC14_IFGR0_IFG1){
  resultsBuffer[0] = ADC14->MEM[0];
  resultsBuffer[1] = ADC14->MEM[1];
  __no_operation();}
}
```

Listing 8.11 Usage of the ADC14 Module as a Comparator, Using DriverLib Functions

```c
#include <ti/devices/msp432p4xx/driverlib/driverlib.h>

static volatile uint16_t curADCResult;

void main(void){

 WDT_A_holdTimer();

 CS_setDCOFrequency(CS_24MHZ);

 GPIO_setAsOutputPin(GPIO_PORT_P1, GPIO_PIN0);
 GPIO_setAsPeripheralModuleFunctionInputPin(GPIO_PORT_P5, GPIO_PIN5,
    GPIO_TERTIARY_MODULE_FUNCTION);

 ADC14_enableModule();
 ADC14_initModule(ADC_CLOCKSOURCE_MCLK, ADC_PREDIVIDER_1,
    ADC_DIVIDER_4, 0);
 ADC14_configureSingleSampleMode(ADC_MEM0, false);
 ADC14_configureConversionMemory(ADC_MEM0,
    ADC_VREFPOS_AVCC_VREFNEG_VSS, ADC_INPUT_A0, false);
 ADC14_enableSampleTimer(ADC_MANUAL_ITERATION);
 ADC14_enableConversion();

 while (1){
  ADC14_toggleConversionTrigger();
  while(ADC14_isBusy());
  curADCResult = ADC14_getResult(ADC_MEM0);
  if(curADCResult > 0x200)
   GPIO_setOutputHighOnPin(GPIO_PORT_P1, GPIO_PIN0);
  else
   GPIO_setOutputLowOnPin(GPIO_PORT_P1, GPIO_PIN0);}
}
```

as software and enable the ADC conversion using the `ADC14_enableSampleTimer` and `ADC14_enableConversion` functions. In a while loop, we trigger the ADC14 module to get a sample and wait until ADC completes the conversion using `ADC14_toggleConversionTrigger` and `ADC14_isBusy` functions. We read the conversion result using the function `ADC14_getResult` and toggle LED1 accordingly.

We next consider the example in Listing 8.9 using DriverLib functions. We provide the corresponding C code in Listing 8.12. Here, we configure the ADC14 module similar to the previous example. We set the ADC input channel as temperature sensor with reference voltage as 2.5 V. Different from the previous example, we set the ADC14 module in single-channel, multiple conversion mode using the `ADC14_enableSampleTimer` function. We enable ADC interrupts using the `ADC14_enableInterrupt` function. In the ISR, we read and clear the ADC interrupt flag using `ADC14_getEnabledInterruptStatus` and `ADC14_clearInterruptFlag` functions. Then, we read the conversion result using the `ADC14_getResult` function. We finally compute the temperature value in degrees Celsius.

Listing 8.12 Measuring the Internal Temperature Level of the Microcontroller by the ADC14 Module, Using DriverLib Functions

```c
#include <ti/devices/msp432p4xx/driverlib/driverlib.h>

volatile float tempC;
int cal30, cal85;

void main(void){

 WDT_A_holdTimer();

 cal30 = SysCtl_getTempCalibrationConstant(SYSCTL_2_5V_REF,
     SYSCTL_30_DEGREES_C);
 cal85 = SysCtl_getTempCalibrationConstant(SYSCTL_2_5V_REF,
     SYSCTL_85_DEGREES_C);

 REF_A_setReferenceVoltage(REF_A_VREF2_5V);
 REF_A_enableReferenceVoltage();
 REF_A_enableTempSensor();

 ADC14_enableModule();
 ADC14_initModule(ADC_CLOCKSOURCE_MCLK, ADC_PREDIVIDER_1,
     ADC_DIVIDER_1, ADC_TEMPSENSEMAP);
 ADC14_configureSingleSampleMode(ADC_MEM0, true);
 ADC14_configureConversionMemory(ADC_MEM0,
     ADC_VREFPOS_INTBUF_VREFNEG_VSS, ADC_INPUT_A22, false);
 ADC14_enableSampleTimer(ADC_AUTOMATIC_ITERATION);
 ADC14_enableInterrupt(ADC_INT0);

 Interrupt_enableSleepOnIsrExit();
 Interrupt_enableInterrupt(INT_ADC14);
 Interrupt_enableMaster();

 ADC14_enableConversion();
 ADC14_toggleConversionTrigger();
}
void ADC14_IRQHandler(void){
 uint64_t status;
 int32_t  TempADC;

 status = ADC14_getEnabledInterruptStatus();
 ADC14_clearInterruptFlag(status);

 if(status & ADC_INT0){
  TempADC = ADC14_getResult(ADC_MEM0);
  tempC = ((((float)TempADC - cal30) * 55) / (cal85 - cal30) + 30.0f;}
}
```

We next consider the example in Listing 8.10 using DriverLib functions. We provide the corresponding C code in Listing 8.13. Here, we set the ADC14 module in multi-channel, multiple conversion mode. We set its trigger as Timer_A0 output 1 using `ADC14_configureMultiSequenceMode` and `ADC14_setSampleHoldTrigger` functions. In the ISR, we read and clear the interrupt flag first. Then, we read multiple channel conversion results using the `ADC14_getMultiSequenceResult` function.

Listing 8.13 ADC Operation on Multiple Channels, Using DriverLib Functions

```c
#include <ti/devices/msp432p4xx/driverlib/driverlib.h>

const Timer_A_UpModeConfig upModeConfig =
{
 TIMER_A_CLOCKSOURCE_ACLK,
 TIMER_A_CLOCKSOURCE_DIVIDER_1,
 16384,
 TIMER_A_TAIE_INTERRUPT_DISABLE,
 TIMER_A_CCIE_CCR0_INTERRUPT_DISABLE,
 TIMER_A_DO_CLEAR
};

const Timer_A_CompareModeConfig compareConfig =
{
 TIMER_A_CAPTURECOMPARE_REGISTER_1,
 TIMER_A_CAPTURECOMPARE_INTERRUPT_DISABLE,
 TIMER_A_OUTPUTMODE_SET_RESET,
 16384
};

uint16_t resultsBuffer[2];

void main(void){

 WDT_A_holdTimer();

 CS_setReferenceOscillatorFrequency(CS_REFO_32KHZ);
 CS_initClockSignal(CS_ACLK, CS_REFOCLK_SELECT, CS_CLOCK_DIVIDER_1);

 Timer_A_configureUpMode(TIMER_A0_BASE, &upModeConfig);
 Timer_A_initCompare(TIMER_A0_BASE, &compareConfig);

 GPIO_setAsPeripheralModuleFunctionInputPin(GPIO_PORT_P5, GPIO_PIN4 | GPIO_PIN5,
     GPIO_TERTIARY_MODULE_FUNCTION);

 ADC14_enableModule();
 ADC14_initModule(ADC_CLOCKSOURCE_MCLK, ADC_PREDIVIDER_1, ADC_DIVIDER_1, 0);
 ADC14_configureMultiSequenceMode(ADC_MEM0, ADC_MEM1, true);
 ADC14_configureConversionMemory(ADC_MEM0, ADC_VREFPOS_AVCC_VREFNEG_VSS,
     ADC_INPUT_A0, false);
 ADC14_configureConversionMemory(ADC_MEM1, ADC_VREFPOS_AVCC_VREFNEG_VSS,
     ADC_INPUT_A1, false);

 ADC14_setSampleHoldTrigger(ADC_TRIGGER_SOURCE1, false);
 ADC14_enableSampleTimer(ADC_MANUAL_ITERATION);
 ADC14_enableInterrupt(ADC_INT1);
 ADC14_enableConversion();

 Interrupt_enableSleepOnIsrExit();
 Interrupt_enableInterrupt(INT_ADC14);
 Interrupt_enableMaster();

 Timer_A_startCounter(TIMER_A0_BASE, TIMER_A_UP_MODE);

 PCM_setPowerState(PCM_LPM0_LDO_VCORE1);
}
void ADC14_IRQHandler(void){
 uint64_t status;

 status = ADC14_getEnabledInterruptStatus();
 ADC14_clearInterruptFlag(status);

 if (status & ADC_INT1)
  ADC14_getMultiSequenceResult(resultsBuffer);
}
```

Listing 8.14 ADC Window Comparator Operation, Using DriverLib Functions

```c
#include <ti/devices/msp432p4xx/driverlib/driverlib.h>

volatile float tempC;
int cal30, cal85, comp_h, comp_l;

void main(void){

 WDT_A_holdTimer();

 cal30 = SysCtl_getTempCalibrationConstant(SYSCTL_2_5V_REF,
     SYSCTL_30_DEGREES_C);
 cal85 = SysCtl_getTempCalibrationConstant(SYSCTL_2_5V_REF,
     SYSCTL_85_DEGREES_C);
 comp_l = ((int32_t) ((20.0 - 30.0) * (cal85 - cal30)) / 55) + cal30;
 comp_h = ((int32_t) ((50.0 - 30.0) * (cal85 - cal30)) / 55) + cal30;

 GPIO_setAsOutputPin(GPIO_PORT_P1, GPIO_PIN0);

 REF_A_setReferenceVoltage(REF_A_VREF2_5V);
 REF_A_enableReferenceVoltage();
 REF_A_enableTempSensor();

 ADC14_enableModule();
 ADC14_initModule(ADC_CLOCKSOURCE_MCLK, ADC_PREDIVIDER_1,
     ADC_DIVIDER_1, ADC_TEMPSENSEMAP);
 ADC14_configureSingleSampleMode(ADC_MEM0, true);
 ADC14_configureConversionMemory(ADC_MEM0,
     ADC_VREFPOS_INTBUF_VREFNEG_VSS, ADC_INPUT_A22, false);
 ADC14_setComparatorWindowValue(ADC_COMP_WINDOW0, comp_l, comp_h);
 ADC14_enableComparatorWindow(ADC_MEM0, ADC_COMP_WINDOW0);
 ADC14_enableSampleTimer(ADC_AUTOMATIC_ITERATION);
 ADC14_clearInterruptFlag(ADC_IN_INT);
 ADC14_enableInterrupt(ADC_IN_INT);

 Interrupt_enableSleepOnIsrExit();
 Interrupt_enableInterrupt(INT_ADC14);
 Interrupt_enableMaster();

 ADC14_enableConversion();
 ADC14_toggleConversionTrigger();

 PCM_setPowerState(PCM_LPM0_LDO_VCORE1);
}

void ADC14_IRQHandler(void){
 uint64_t status;
 int32_t  TempADC;

 status = ADC14_getEnabledInterruptStatus();
 ADC14_clearInterruptFlag(status);

 if(status & ADC_INT0){
```

```
  TempADC = ADC14_getResult(ADC_MEM0);
  tempC = (((float)TempADC - cal30) * 55) / (cal85 - cal30) + 30.0f;
  ADC14_disableInterrupt(ADC_INT0);
  GPIO_setOutputLowOnPin(GPIO_PORT_P1, GPIO_PIN0);}
 else if(status & ADC_IN_INT ){
  GPIO_setOutputHighOnPin(GPIO_PORT_P1, GPIO_PIN0);
  ADC14_enableInterrupt(ADC_INT0);}
}
```

Listing 8.15 Usage of the ADC14 Module as a Comparator, Using Energia Functions

```
int analogIn = A0;
int curADCResult;

void setup(){
 pinMode(YELLOW_LED, OUTPUT);
}
void loop(){
 curADCResult = analogRead(analogIn);
 if(curADCResult > 0x200) digitalWrite(YELLOW_LED, HIGH);
 else digitalWrite(YELLOW_LED, LOW);
 delay(100);
}
```

We finally consider the window comparator usage via DriverLib functions. We provide the corresponding C code in Listing 8.14. Here, we set the ADC14 module similar to Listing 8.12. We compute the threshold levels for 20°C and 50°C, respectively. Then, we set the window comparator using the function ADC14_setComparatorWindowValue. Afterwards, we enable the window comparator for the selected memory location using the function ADC14_enableComparatorWindow. Finally, we enable the ADC interrupt for the window comparator. In the ISR, we read and clear the interrupt flag first. If the result is either greater than the lower threshold level or lower than the upper threshold level, we enable the ADC conversion interrupt and turn on LED1 connected to pin P1.0 on the MSP432 LaunchPad. Then, we read the conversion result, compute the temperature value in degrees Celsius, and turn off LED1.

8.4.7.3 Energia Example

We finally repeat the example given in Listing 8.8 using Energia functions. We provide the corresponding C code in Listing 8.15. Here, first we set LED1 as output in the setup function. Then, we configure the ADC14 module and read the conversion result using the analogRead function and toggle LED1 accordingly.

8.5 Digital-to-Analog Conversion

To convert a digital signal to analog form, a digital-to-analog converter (DAC) is needed. Unfortunately, MSP432 does not have such a module. However, we can use PWM for this purpose.

The output signal in PWM is a high-frequency digital pulse sequence. The width of the pulses change depending on the setting. As this high-frequency signal is smoothed by a low-pass filter (such as a simple RC circuit), we will get an average voltage that is approximately a DC signal. This average voltage (V_{avg}) obtained from the system will be

$$V_{avg} = \frac{t_{on}}{t_{period}} \times V_{CC} = D \times V_{CC} \quad (8.2)$$

where t_{on} is the duration the pulse will be on and t_{period} is the period of the pulse. The ratio t_{on}/t_{period} is called the duty cycle (D) of the PWM signal. By changing the duty cycle (changing t_{on} and keeping t_{period} constant), an approximate DC signal can be generated.

8.5.1 PWM Registers

The Timer_A compare mode can be used to generate PWM signals in MSP432. We briefly explained the compare mode in Sec. 7.7. Here, we will explain the output modes of Timer_A in detail. There are eight different output types in the compare mode. They are described below:

- **Out bit value** (OUTMOD_0): The OUTx bit controls the output signal.
- **Set** (OUTMOD_1): The output is set only once when the TAxR reaches the TACCRx value.
- **Toggle/Reset** (OUTMOD_2): The output is toggled when the TAxR reaches the TACCRx value. It is reset when the TAxR reaches the TACCR0 value.
- **Set/Reset** (OUTMOD_3): The output is set when the TAxR reaches the TACCRx value. It is reset when the TAxR reaches the TACCR0 value.
- **Toggle** (OUTMOD_4): The output is toggled when the TAxR reaches the TACCRx value.
- **Reset** (OUTMOD_5): The output is reset only once when the TAxR reaches the TACCRx value.
- **Toggle/Set** (OUTMOD_6): The output is toggled when the TAxR reaches the TACCRx value. It is set when the TAxR reaches the TACCR0 value.
- **Reset/Set** (OUTMOD_7): The output is reset when the TAxR reaches the TACCRx value. It is set when the TAxR reaches the TACCR0 value.

The compare mode output types are given in Fig. 8.5. In the first part of the figure, the timer is in up mode. In the second part of the figure, the timer is in continuous mode. In the third part of the figure, the timer is in up/down mode. This figure clearly shows that the compare mode can be used in PWM generation.

The arrangement for the PWM while using the Timer_A capture/compare block (in output mode 7) is as follows. The output is turned on when the TAxR value reaches 0. It is turned off when the TAxR value reaches TACCRx. This means that increasing the value in TACCRx increases the duty cycle. The period of the PWM is the same as that of the timer. Therefore, its frequency is

$$f_{PWM} = \frac{f_{CLK}}{TACCR0 + 1} \quad (8.3)$$

where f_{CLK} stands for the frequency of the timer clock. The duty cycle of the PWM signal can be calculated by

$$D = \frac{TACCRx}{TACCR0 + 1} \quad (8.4)$$

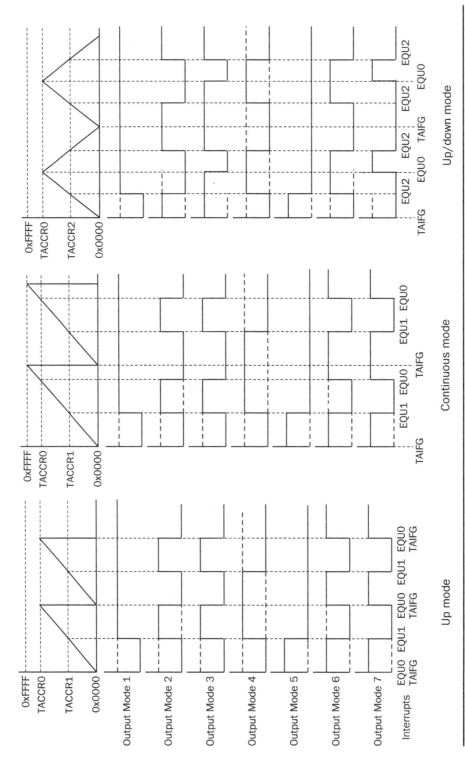

FIGURE 8.5 Compare mode outputs.

Mixed Signal Systems

Pin	Usage Area
P7.3	Timer_A0, capture: CCI0A input, compare Out0 output
P2.4	Timer_A0, capture: CCI1A input, compare Out1 output
P2.5	Timer_A0, capture: CCI2A input, compare Out2 output
P2.6	Timer_A0, capture: CCI3A input, compare Out3 output
P2.7	Timer_A0, capture: CCI4A input, compare Out4 output
P8.0	Timer_A1, capture: CCI0A input, compare Out0 output
P7.7	Timer_A1, capture: CCI1A input, compare Out1 output
P7.6	Timer_A1, capture: CCI2A input, compare Out2 output
P7.5	Timer_A1, capture: CCI3A input, compare Out3 output
P7.4	Timer_A1, capture: CCI4A input, compare Out4 output
P8.1	Timer_A2, capture: CCI0A input, compare Out0 output
P5.6	Timer_A2, capture: CCI1A input, compare Out1 output
P5.7	Timer_A2, capture: CCI2A input, compare Out2 output
P6.6	Timer_A2, capture: CCI3A input, compare Out3 output
P6.7	Timer_A2, capture: CCI4A input, compare Out4 output
P10.4	Timer_A3, capture: CCI0A input, compare Out0 output
P10.5	Timer_A3, capture: CCI1A input, compare Out1 output
P8.2	Timer_A3, capture: CCI2A input, compare Out2 output
P9.2	Timer_A3, capture: CCI3A input, compare Out3 output
P9.3	Timer_A3, capture: CCI4A input, compare Out4 output

TABLE 8.18 Capture Mode Inputs and Compare Mode Outputs for the Timer_A Module

where *TACCRx* stands for the xth Timer_A capture/compare register. To note here, the timer will be in the up mode for this equation to be valid.

Finally, the generated PWM signal can be taken out from specific pins given in Table 8.18. Do not forget to set the corresponding bit in the PxSEL0 register for PWM output.

8.5.2 PWM Usage via DriverLib Function

The PWM generation property of the Timer_A0 module can be controlled via DriverLib functions. We provide the function of interest below. The complete function list and macro definitions can be found in [11].

```
void Timer_A_generatePWM(uint32_t timer, const Timer_A_PWMConfig *
    config)
/*
Configures the Timer_A0 in PWM mode by modifying TAxCTL, TAxCCTLy, and
    TAxCCRy registers.
timer: the Timer_A0 instance to be configured. It can be one of
    TIMER_A0_BASE, TIMER_A1_BASE, TIMER_A2_BASE or TIMER_A3_BASE.
config: the configuration structure. It contains clock source, clock
    source divider, timer period, compare register, compare output
    mode, and duty cycle values. Please see DriverLib user guide for
        macro definitions.
*/
```

8.5.3 PWM via Energia Function

Energia can also be used to control the PWM operation in MSP432. We provide the related function below. More information on Energia functions can be found in [12].

```
void analogWrite(pin, value)
/*
Configures selected pin as PWM output and sets duty cycle. The
    frequency of the PWM signal is approximately 490 Hz.
pin: selected pin to be configured as analog output.
value: the duty cycle value between 0 and 255.
*/
```

8.5.4 Coding Practices for the PWM Module

We consider an example to show how the PWM signal can be of use in generating an analog signal from digital pulses. In doing this, we use registers, DriverLib, and Energia functions. We use the red color of LED2 connected to pin P2.0 on the MSP432 LaunchPad as output. For our programs to work, the jumper labeled "Red P2.0" above LED2 on MSP432 LaunchPad must be removed and pin P2.4 (PWM output) should be connected to the bottom pin of this jumper. We set the button S1 connected to the pin P1.1 on the MSP432 LaunchPad as output.

8.5.4.1 Register-Level Example

We provide the C code of the example, where the red color of LED2 is adjusted by PWM signal, in Listing 8.16. Here, the SMCLK is sourced from 3-MHz DCO and Timer_A0 module is sourced from this SMCLK in up mode. The PWM frequency is set to 3000000/3000 = 1 kHz and initial duty cycle is set to 0.8. Duty cycle is increased by 0.05 when S1 is released in the ISR.

8.5.4.2 DriverLib Example

We next repeat the example in Listing 8.16 using DriverLib functions. We provide the corresponding C code in Listing 8.17. Here, we first set pin P2.4 as PWM output. Then, we configure the Timer_A0 module using the `Timer_A_generatePWM` function. Afterwards, we set S1 as input with the pull-up resistor. We also enable the port interrupt. In the ISR, the duty cycle of the PWM signal is increased by 0.05. To do so, the Timer_A0 module is configured accordingly.

8.5.4.3 Energia Example

We finally repeat the example in Listing 8.16 using Energia functions. We provide the corresponding code in Listing 8.18.

8.6 Application: Leakage Control System

The aim of this application is to learn how to use the ADC14 module on the MSP432 microcontroller. As a real-life application, we will design a leakage control system for smart homes. In this section, we provide the equipment list, layout of the circuit, and system design specifications.

Listing 8.16 Usage of the Timer_A0 Module in PWM Signal Generation, Using Registers

```c
#include "msp.h"

#define S1 BIT1
#define DELAY 500
int i;

int period = 3000; // Period of PWM
float D = 0.8f;

void main(void){

  WDT_A->CTL = WDT_A_CTL_PW | WDT_A_CTL_HOLD;

  P2->DIR |= BIT4;
  P2->SEL0 |= BIT4; // Set P2.4 as PWM output

  P1->DIR = 0x00;
  P1->REN = S1;
  P1->OUT = S1;
  P1->IE = S1;
  P1->IES = 0x00;
  P1->IFG = 0x00;

  TIMER_A0->CCR[0] = period - 1;  // PWM period
  TIMER_A0->CCR[1] = period*D; // CCR1 PWM duty cycle
  TIMER_A0->CCTL[1] = TIMER_A_CCTLN_OUTMOD_7; // CCR1 reset/set
  TIMER_A0->CTL = TIMER_A_CTL_TASSEL_2 | TIMER_A_CTL_MC_1 |
      TIMER_A_CTL_CLR; // SMCLK, Up mode, Clear TAR

  NVIC->ISER[1] = 0x00000008; // Port P1 interrupt is enabled in NVIC
  _enable_interrupts(); // All interrupts are enabled

  SCB->SCR |= SCB_SCR_SLEEPONEXIT_Msk;
  __sleep();
}
void PORT1_IRQHandler(void){
 if(P1->IFG & S1){
  if(D >= 1.0f) D = 0.0f;
  else D += 0.05f;

  TIMER_A0->CCR[1] = period*D;} // CCR1 PWM duty cycle

 for(i=0;i<DELAY;i++){} // Delay for debounce
 P1->IFG &= ~S1;
}
```

8.6.1 Equipment List

Below, we provide the equipment list to be used in the leakage control system:

- MQ-135 air-quality sensor module
- MQ-5 gas sensor module
- Water-level sensor module

Listing 8.17 Usage of the Timer_A0 Module in PWM Signal Generation, Using DriverLib Functions

```c
#include <ti/devices/msp432p4xx/driverlib/driverlib.h>

int period = 3000; // Period of PWM
float D = 0.8f;

/* Timer_A PWM Configuration Parameter */
Timer_A_PWMConfig pwmConfig =
{
 TIMER_A_CLOCKSOURCE_SMCLK,
 TIMER_A_CLOCKSOURCE_DIVIDER_1,
 3000,
 TIMER_A_CAPTURECOMPARE_REGISTER_1,
 TIMER_A_OUTPUTMODE_RESET_SET,
 2400
};

void main(void){

 WDT_A_holdTimer();

 GPIO_setAsPeripheralModuleFunctionOutputPin(GPIO_PORT_P2, GPIO_PIN4,
     GPIO_PRIMARY_MODULE_FUNCTION);
 Timer_A_generatePWM(TIMER_A0_BASE, &pwmConfig);

 GPIO_setAsInputPinWithPullUpResistor(GPIO_PORT_P1, GPIO_PIN1);
 GPIO_clearInterruptFlag(GPIO_PORT_P1, GPIO_PIN1);
 GPIO_enableInterrupt(GPIO_PORT_P1, GPIO_PIN1);

 Interrupt_enableInterrupt(INT_PORT1);
 Interrupt_enableSleepOnIsrExit();
 Interrupt_enableMaster();

 PCM_setPowerState(PCM_LPM0_LDO_VCORE1);
}

void PORT1_IRQHandler(void){
 uint32_t status = GPIO_getEnabledInterruptStatus(GPIO_PORT_P1);
 GPIO_clearInterruptFlag(GPIO_PORT_P1, status);

 if (status & GPIO_PIN1){
  if(D >= 1.0f) D = 0.0f;
  else D += 0.05f;

  pwmConfig.dutyCycle = period*D;
  Timer_A_generatePWM(TIMER_A0_BASE, &pwmConfig);}
}
```

MQ-135 is an analog gas sensor that can detect NH_3, NOx, alcohol, benzene, smoke, and CO_2. The response of the sensor is the same for different gases. Hence, it is not suitable particularly to detect single gas. Instead, it is useful to detect overall air quality by detecting harmful indoor gases. MQ-5 is an analog gas sensor. It can detect H_2, LPG, CH_4, CO, and alcohol. It can detect gas concentrations between 200 and 10000 ppm.

Listing 8.18 Usage of the Timer_A0 Module in PWM Signal Generation, Using Energia Functions

```
int dutycycle = 0;

void setup(){
 pinMode(RED_LED, OUTPUT);
 pinMode(PUSH1, INPUT_PULLUP);
 attachInterrupt(PUSH1, pushButton, FALLING);
}

void loop(){}

void pushButton(){
 delay(100);
 dutycycle+=32;
 analogWrite(RED_LED, (dutycycle%255));
}
```

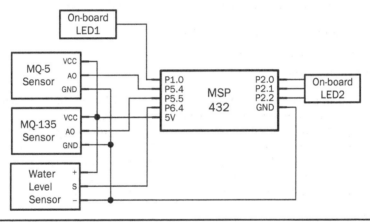

FIGURE 8.6 Layout of the leakage control system.

The water-level sensor is a simple resistive circuit that senses the water level using a series of parallel wires exposed to conductive liquid.

8.6.2 Layout

The layout of the leakage control system is given in Fig. 8.6.

8.6.3 System Design Specifications

The design of our leakage control system will have two main blocks as follows:

- **Block 1**: If the water level exceeds a certain level, then LED1 connected to P1.0 pin on the MSP432 LaunchPad will turn on to indicate there is a general alert. To specifically indicate the water leakage, the red color of LED2 connected to P2.0 pin on the MSP432 LaunchPad will turn on.
- **Block 2**: If the harmful gas level sensed over a period of time exceeds a threshold level, then LED1 connected to P1.0 pin on the MSP432 LaunchPad will turn on to indicate there is a general alert. To specifically indicate the air-quality alert,

the green color of LED2 connected to pin P2.1 on the MSP432 LaunchPad will turn on. Similarly, the blue color of LED2 connected to pin P2.2 on the MSP432 LaunchPad will turn on if there is a gas leakage.

We could not add the C code of the project here due to page limitations. However, we provided the complete CCS project for this application in the companion website for the reader. We strongly suggest implementing it.

8.7 Application: Blinds Control System

The aim of this application is to learn how to use the ADC14 module and PWM operation on the MSP432 microcontroller. As a real-life application, we will design a blinds control system for smart homes. In this section, we provide the equipment list, layout of the circuit, and system design specifications.

8.7.1 Equipment List

Below, we provide the equipment list to be used in the blinds control system:

- One light-dependent resistor (LDR)
- One continuous servo motor
- One 10-kΩ resistor

In this application, the PWM signal generated by the MSP432 will be fed to the continuous servo motor directly. While a regular servo motor only turns over a narrow range, a continuous rotation servo can spin continuously. In the continuous servo motor, 1-millisecond pulse (with the period 50 Hz) corresponds to full speed in one direction, while 2-millisecond pulse corresponds to full speed in the other direction. A 1.5-millisecond pulse should cause the servo motor to stop. Other pulse widths between one and two milliseconds rotate motor with slower speed.

8.7.2 Layout

The layout of the blinds control system is given in Fig. 8.7.

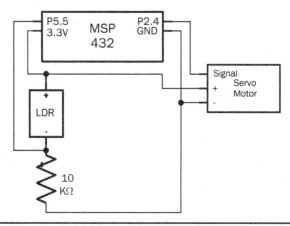

FIGURE 8.7 Layout of the blinds control system.

8.7.3 System Design Specifications

The design of the blinds control system will be as follows. There will be four threshold levels for light intensity. This will correspond to four levels for the blinds. If the light level exceeds a certain level, the motor will turn with full speed for 2 seconds to lower or raise the blinds one level.

We could not add the C code of the project here due to page limitations. However, we have provided the complete CCS project for this application in the companion website for the reader. We strongly suggest implementing it.

8.8 Summary

MSP432 can process analog signals as well. Therefore, we considered ADC and DAC operations in this chapter. In analog-to-digital conversion, we need reference voltage levels. The REF_A module in MSP432 handles this issue. Therefore, we first focused on it. Then, we explored the COMP_A module that provides binary output by comparing its input values. We provided sample codes for this module. Afterwards, we focused on the ADC14 module that provides a 14-bit digital representation of the analog signal fed to it. The ADC14 module uses the SAR method in analog-to-digital conversion. We explored the principles of this method through a simulation program. As in the COMP_A module, we provided sample C codes for the usage of the ADC14 module. Unfortunately, MSP432 does not have a DAC module. Therefore, we used PWM to obtain analog signals from digital representations. Although the obtained analog signal is an approximation, it is sufficient for most applications. If needed, we can use an external DAC module to obtain precise analog signals. Finally, we considered two real-life applications to show how ADC and PWM operations can be used in actual systems.

8.9 Exercises

8.1 Use the C code in Listing 8.7 to calculate the 14-bit SAR conversion of the analog voltage levels 0.42, 0.83, 1.65, and 2.752 V. The reference voltage will be 3 V.

8.2 Design a solar charge controller using MSP432 with the below specifications:
   a. There will be a solar panel connected to the microcontroller with a controllable switching element (such as power multiplexer or analog switch IC). If the voltage output of the solar panel is greater than 2.5 V, then it will supply the system.
   b. There will be a battery connected to the microcontroller with a controllable switching element. If the voltage level of the solar panel is lower than 2.5 V, then the battery will supply the system.
   c. If the voltage level of the battery drops below the output of the solar panel, then the third switching element will connect the battery to the solar panel.
   d. Use the red, green, and blue colors of LED2 connected to pins P2.0, P2.1, and P2.2 on the MSP432 LaunchPad to indicate the used power supply source.
   e. The controllable switching elements can be controlled via digital I/O interface.

8.3 Design a fan controller using MSP432 with the below specifications:
   a. The temperature will be monitored using the ADC14 module. Here, take 100 samples over 5 seconds period and calculate the average temperature to find a reliable value.

b. Generate a PWM signal to change the speed of the fan according to the calculated temperature value. Set the PWM frequency as 250 Hz.
c. Use the button S1, connected to pin P1.1 on the MSP432 LaunchPad, to turn on or off the fan.
d. Use the button S2 connected to pin P1.4 on the MSP432 LaunchPad to change the sensitivity of the fan speed with respect to temperature. Use three sensitivity levels such that the lower the sensitivity the lower the speed of the fan. As an example, you can set the duty cycle of the PWM signal as duty cycle = level×10×(temp-20) where the level can be 0.5, 1, or 1.5.

8.4 Design a musical tone generator using MSP432 with the below specifications:
  a. The frequencies are 261.6, 293.6, 329.6, 349.2, 392.0, 440.0, and 493.8 Hz for C4, D4, E4, F4, G4, A4, and B4 notes, respectively.
  b. Write a function that generates the note for a given duration.
  c. Play the notes sequentially as C4, D4, E4, F4, G4, A4, and B4 for two seconds each.
  Note: You can use a 3-V piezo buzzer and PWM signal with 50% duty cycle to generate a tone signal. The frequency values can be generated up to 10% error. You can use the pin P2.4 on the MSP432 LaunchPad to connect the buzzer.

CHAPTER 9
Digital Communication

Data transfer between two (or more) microcontrollers becomes a necessity for complex projects. Moreover, some peripheral devices (such as sensors and DAC modules) communicate with the microcontroller through data transfer channels. Therefore, digital communication has become an essential part of a modern microcontroller. The module responsible for digital communication in MSP432 is called enhanced universal serial communication interface (eUSCI). This module supports universal asynchronous receiver/transmitter (UART), serial peripheral interface (SPI), and inter-integrated circuit (I^2C) communication modes. We will concentrate in this chapter on the eUSCI module and communication modes it provides. Next, we start with the brief description of the eUSCI module.

9.1 Enhanced Universal Serial Communication Interface

There are eight eUSCI modules called eUSCI_Ax and eUSCI_Bx in MSP432, where x can get values from 0 to 3. eUSCI_Ax can support UART and SPI communication modes. Similarly, eUSCI_Bx can support SPI and I^2C communication modes. We will provide the general properties of the eUSCI_Ax and eUSCI_Bx modules in this section.

9.1.1 eUSCI Registers

The eUSCI module has several special function control and status registers for the UART, SPI, and I^2C communication modes. Some of these registers are specific to the communication mode. Some of them share the same name for different communication modes. All eUSCI registers are tabulated in Tables 9.1 and 9.2. In these tables, the usage area of each register is also provided.

We will explain the control and status registers in Tables 9.1 and 9.2 in detail for each communication mode in the following sections. However, receive and transmit buffer registers for eUSCI_Ax and eUSCI_Bx modules deserve specific consideration. Data to be transmitted should be written to the transmit buffer register for any communication mode. These are UCAxTXBUF and UCBxTXBUF for the eUSCI_Ax and eUSCI_Bx modules, respectively. Similarly, data received will be read from the receive buffer register. These are UCAxRXBUF and UCBxRXBUF for the eUSCI_Ax and eUSCI_Bx modules, respectively.

Register Name	Short Form	Used In
eUSCI_Ax control word register 0	UCAxCTLW0	UART, SPI
eUSCI_Ax control word register 1	UCAxCTLW1	UART
eUSCI_Ax baud rate control word register	UCAxBRW	UART, SPI
eUSCI_Ax modulation control word register	UCAxMCTLW	UART
eUSCI_Ax status register	UCAxSTATW	UART, SPI
eUSCI_Ax receive buffer register	UCAxRXBUF	UART, SPI
eUSCI_Ax transmit buffer register	UCAxTXBUF	UART, SPI
eUSCI_Ax interrupt enable register	UCAxIE	UART, SPI
eUSCI_Ax interrupt flag register	UCAxIFG	UART, SPI
eUSCI_Ax interrupt vector register	UCAxIV	UART, SPI
eUSCI_Ax auto baud rate control register	UCAxABCTL	UART
eUSCI_Ax IrDA control word register	UCAxIRCTL	UART
eUSCI_Ax IrDA receive control register	UCA0IRRCTL	UART

TABLE 9.1 eUSCI_Ax Control and Status Registers

Register	Short Form	Used In
eUSCI_Bx control word register 0	UCBxCTLW0	SPI, I^{2}C
eUSCI_Bx control word register 1	UCBxCTLW1	I^{2}C
eUSCI_Bx bit rate control word register	UCBxBRW	SPI, I^{2}C
eUSCI_Bx status register	UCBxSTATW	SPI, I^{2}C
eUSCI_Bx byte counter threshold register	UCBxTBCNT	I^{2}C
eUSCI_Bx receive buffer register	UCBxRXBUF	SPI, I^{2}C
eUSCI_Bx transmit buffer register	UCBxTXBUF	SPI, I^{2}C
eUSCI_Bx interrupt enable register	UCBxIE	SPI, I^{2}C
eUSCI_Bx interrupt flag register	UCBxIFG	SPI, I^{2}C
eUSCI_Bx interrupt vector register	UCBxIV	SPI, I^{2}C
eUSCI_Bx I^{2}C own address register 0	UCBxI2COA0	I^{2}C
eUSCI_Bx I^{2}C own address register 1	UCBxI2COA1	I^{2}C
eUSCI_Bx I^{2}C own address register 2	UCBxI2COA2	I^{2}C
eUSCI_Bx I^{2}C own address register 3	UCBxI2COA3	I^{2}C
eUSCI_Bx I^{2}C received address register	UCBxADDRX	I^{2}C
eUSCI_Bx I^{2}C address mask register	UCBxADDMASK	I^{2}C
eUSCI_Bx I^{2}C slave address register	UCBxI2CSA	I^{2}C

TABLE 9.2 eUSCI_Bx Control and Status Registers

9.1.2 eUSCI Clocks

The eUSCI module has three clocks: BRCLK, BITCLK, and BITCLK16. The BRCLK represents the selected clock for the eUSCI module. The UART mode can use UCCLK, ACLK, and SMCLK as BRCLK. UCCLK is the external clock for the UART mode. The SPI mode can use ACLK and SMCLK as BRCLK. Finally, the I^{2}C mode can use UCCLKI, ACLK, and SMCLK as BRCLK. UCCLKI is the external clock for the I^{2}C mode. The BITCLK is generated by directly dividing BRCLK in SPI and I^{2}C modes. On the other hand, BITCLK is generated by modulating BRCLK in UART mode. BITCLK is mainly used in controlling the bit transmission and reception rates. Finally, the BITCLK16 is used as the sampling clock in UART oversampling mode. We will explain all these clocks in specific communication modes in the following sections.

The register UCBxBRW is responsible to divide the clock for the eUSCI_Ax and eUSCI_Bx modules. UCBxBRW register forms the 16-bit division coefficient for the BRCLK. This is called UCBRx. UCBxBRW is specifically called baud rate control word register in the eUSCI_Ax module for UART mode. In the eUSCI_Bx and eUSCI_Ax modules, the register UCBxBRW is called the bit rate control word register for SPI mode.

9.1.3 Common Properties

The UART, SPI, and I^{2}C communication modes are initialized by the same steps. Initially, the eUSCI module must be reset to configure all related eUSCI registers. The eUSCI module must be set after this operation. Finally, if the interrupts are used in the eUSCI module, they should be enabled. We will explain these steps for each communication mode separately in the following sections.

The eUSCI module is not functional when the device is in deep sleep or shut down modes. However, the user can choose to use deep sleep or shutdown modes, or eUSCI module using FORCE_LPM_ENTRY bit of PCMCTL1 register of the PCM module. If this bit is reset, the entry to deep sleep or shutdown modes should be aborted and eUSCI module stays active. If this bit is set, the eUSCI module is stopped and device goes into the desired deep sleep or shutdown mode. The latter option is useful if the eUSCI module is transmitting or receiving data at a very slow rate and the application should enter deep sleep or shut down modes at the expense of packet data lost. Please see Chap. 7 for further details on the PCM module.

9.1.4 Pin Layout for eUSCI

The pin usage for the eUSCI module is as in Table 9.3. Do not forget to set these pins by appropriate PxSEL bits before using them. In this table, pins P1.1 and P9.5 can be used for external UCCLK input. Pins P1.5, P6.3, P8.1, and P10.1 can be used for external UCCLKI input when they are not used by the SPI mode.

9.2 Universal Asynchronous Receiver/Transmitter

Universal asynchronous receiver/transmitter (UART) is the asynchronous communication mode used between two or more devices. Being asynchronous, there is no need for a common clock in UART. Hence, connected devices can work independently. In fact, UART is the only asynchronous communication mode in MSP432. UART is simple to use compared to the synchronous communication modes to be introduced in the following sections. We will only focus on the UART mode for communication between two microcontrollers (or one microcontroller and a host computer) in this book. Also, we will not consider the enhanced UART usage with automatic baud rate detection (local interconnect network, LIN) and infrared data association (IrDA). More information on these topics can be found in [4].

Pin	Usage Area
P1.0	eUSCI_A0 slave transmit enable in SPI mode (UCA0STE)
P1.1	eUSCI_A0 clock input/output (UCA0CLK)
P1.2	eUSCI_A0 receive data input in UART mode (UCA0RXD), slave out-master in in SPI mode (UCA0SOMI)
P1.3	eUSCI_A0 transmit data output in UART mode (UCA0TXD), slave in-master out in SPI mode (UCA0SIMO)
P2.0	eUSCI_A1 slave transmit enable in SPI mode (PM_UCA1STE)
P2.1	eUSCI_A1 clock input/output (PM_UCA1CLK)
P2.2	eUSCI_A1 receive data input in UART mode (PM_UCA1RXD), slave out-master in in SPI mode (PM_UCA1SOMI)
P2.3	eUSCI_A1 transmit data output in UART mode (PM_UCA1TXD), slave in-master out in SPI mode (PM_UCA1SIMO)
P3.0	eUSCI_A2 slave transmit enable in SPI mode (PM_UCA2STE)
P3.1	eUSCI_A2 clock input/output (PM_UCA2CLK)
P3.2	eUSCI_A2 receive data input in UART mode (PM_UCA2RXD), slave out-master in in SPI mode (PM_UCA2SOMI)
P3.3	eUSCI_A2 transmit data output in UART mode (PM_UCA2TXD), slave in-master out in SPI mode (PM_UCA2SIMO)
P9.4	eUSCI_A3 slave transmit enable in SPI mode (UCA3STE)
P9.5	eUSCI_A3 clock input/output (UCA3CLK)
P9.6	eUSCI_A3 receive data input in UART mode (UCA3RXD), slave out-master in in SPI mode (UCA3SOMI)
P9.7	eUSCI_A3 transmit data output in UART mode (UCA3TXD), slave in-master out in SPI mode (UCA3SIMO)
P1.4	eUSCI_B0 slave transmit enable in SPI mode (UCB0STE)
P1.5	eUSCI_B0 clock input/output (UCB0CLK)
P1.6	eUSCI_B0 data in I^2C mode (UCB0SDA), slave in-master out in SPI mode (UCB0SIMO)
P1.7	eUSCI_B0 clock in I^2C mode (UCB0SCL), slave out-master in in SPI mode (UCB0SOMI)
P6.2	eUSCI_B1 slave transmit enable in SPI mode (UCB1STE)
P6.3	eUSCI_B1 clock input/output (UCB1CLK)
P6.4	eUSCI_B1 data in I^2C mode (UCB1SDA), slave in-master out in SPI mode (UCB1SIMO)
P6.5	eUSCI_B1 clock in I^2C mode (UCB1SCL), slave out-master in in SPI mode (UCB1SOMI)
P3.4	eUSCI_B2 slave transmit enable in SPI mode (PM_UCB2STE)
P3.5	eUSCI_B2 clock input/output (PM_UCB2CLK)
P3.6	eUSCI_B2 data in I^2C mode (PM_UCB2SDA), slave in-master out in SPI mode (PM_UCB2SIMO)
P3.7	eUSCI_B2 clock in I^2C mode (PM_UCB2SCL), slave out-master in in SPI mode (PM_UCB2SOMI)
P8.0	eUSCI_B3 slave transmit enable in SPI mode (UCB3STE)
P8.1	eUSCI_B3 clock input/output (UCB3CLK)
P6.6	eUSCI_B3 data in I^2C mode (UCB3SDA), slave in-master out in SPI mode (UCB3SIMO)
P6.7	eUSCI_B3 clock in I^2C mode (UCB3SCL), slave out-master in in SPI mode (UCB3SOMI)
P10.0	eUSCI_B3 slave transmit enable in SPI mode (UCB3STE)
P10.1	eUSCI_B3 clock input/output (UCB3CLK)
P10.2	eUSCI_B3 data in I^2C mode (UCB3SDA), slave in-master out in SPI mode (UCB3SIMO)
P10.3	eUSCI_B3 clock in I^2C mode (UCB3SCL), slave out-master in in SPI mode (UCB3SOMI)

TABLE 9.3 Pin Usage Table for the eUSCI Module

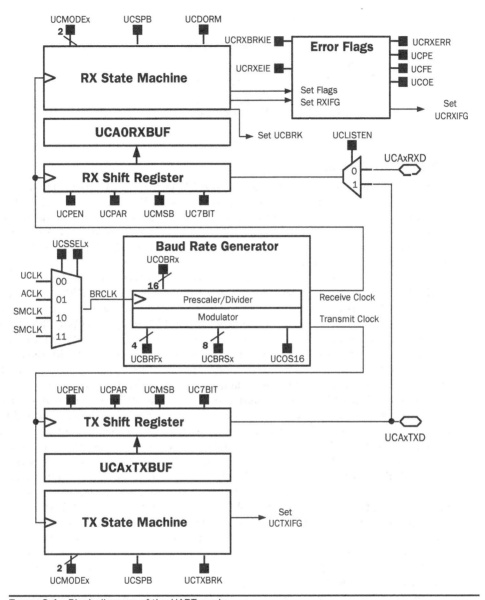

FIGURE 9.1 Block diagram of the UART mode.

Block diagram of the UART is given in Fig. 9.1. As can be seen in this figure, the UART mode has two pins to communicate with other devices. These are the receive (UCAxRXD) and transmit (UCAxTXD) pins. In this block diagram, the transmit and receive shift registers are not visible to the user. Instead, the transmit and receive buffers will be used for communication.

Bits	15	14	13	12	11
	UCPEN	UCPAR	UCMSB	UC7BIT	UCSPB
Bits	10-9	8	7-6	0	
	UCMODEx	UCSYNC	UCSSELx	UCSWRST	

TABLE 9.4 eUSCI_Ax Control Word Register 0 (UCAxCTLW0)

9.2.1 UART Registers

There are 12 16-bit registers defined to control UART. We will explain UCAxCTLW0, UCAxBRW, UCAxMCTLW, UCAxSTATW, UCAxRXBUF, UCAxTXBUF, UCAxIE, UCAxIFG, and UCAxIV registers in this book. For further information on the remaining registers, please see [4]. A structure named EUSCI_Ax is used to control UART registers. The registers of interest are named CTLW0, BRW, MCTLW, STATW, RXBUF, TXBUF, IE, IFG, and IV in this structure.

The eUSCI_Ax control word register 0 (UCAxCTLW0) is used to configure the UART mode. The entries of this register are tabulated in Table 9.4. Here, UCPEN and UCPAR bits are used for parity bit settings [18]. The UCPEN bit is used to enable the parity bit for the system. If this bit is set with the constant EUSCI_A_CTLW0_PEN, the parity bit is enabled. Otherwise, the parity bit is disabled. After the UCPEN bit is set, the UCPAR bit is used to decide on the parity type. When this bit is set with the constant EUSCI_A_CTLW0_PAR, even parity is used. Otherwise, odd parity is used. The UCMSB bit is used to choose the start bit for data transfer. When this bit is set with the constant EUSCI_A_CTLW0_MSB, the transmission starts from the MSB. Otherwise, the transmission starts from the LSB. The latter configuration (LSB first) is generally selected in the UART mode. The UC7BIT bit is used to select data length. When this bit is set with the constant EUSCI_A_CTLW0_SEVENBIT, data length is set to seven bits. Otherwise, it is set to eight bits. The UCSPB bit is used to decide on the number of stop bits. When this bit is set with the constant EUSCI_A_CTLW0_SPB, two stop bits are used. Otherwise, one stop bit is used. UCMODEx bits are used to select the asynchronous mode. Constants for these bits are EUSCI_A_CTLW0_MODE_0 (UART mode), EUSCI_A_CTLW0_MODE_1 (idle-line multiprocessor mode), EUSCI_A_CTLW0_MODE_2 (address-bit multiprocessor mode), and EUSCI_A_CTLW0_MODE_3 (automatic baud rate detection mode). The default setting is UCMODE_0 for the UART communication between two devices. UCMODE_1 and UCMODE_2 can be used for the UART communication between more than two devices. The UCSYNC bit is used to choose the asynchronous or synchronous communication mode. When this bit is reset, the asynchronous mode (UART) is selected. When it is set with the constant EUSCI_A_CTLW0_SYNC, the synchronous mode (SPI) is selected. Therefore, UCSYNC bit should be reset for the UART mode. UCSSELx bits are used to select the UART clock. Constants for these bits are EUSCI_A_CTLW0_UCSSEL_0 (for UCLK), EUSCI_A_CTLW0_UCSSEL_1 (for ACLK), and EUSCI_A_CTLW0_UCSSEL_2 (for SMCLK). The UCSWRST bit is used to reset the USCI module. When this bit is set with the constant EUSCI_A_CTLW0_SWRST, the USCI module is reset. When it is reset, the USCI module will be ready for operation. For further information on the remaining bits of this register, please see [4].

The UART mode also has a status register called UCAxSTATW that is specifically used to observe the changes in the system. The entries of this register are given

Bits	7	0
	UCLISTEN	UCBUSY

TABLE 9.5 eUSCI_Ax Status Register (UCAxSTATW)

Bits	15-8	7-4	0
	UCBRSx	UCBRFx	UCOS16

TABLE 9.6 eUSCI_Ax Modulation Control Word Register (UCAxMCTLW)

in Table 9.5. In this table, the UCLISTEN bit is used to generate an internal loop between the transmitter and receiver on the same device. When this bit is set with the constant EUSCI_A_STATW_LISTEN, the loopback is enabled. This property can be used to troubleshoot the communication codes on a single device. The UCBUSY bit shows whether the eUSCI module is busy or not. This bit is set when the transmit or receive operation is performed. It is reset when the system is inactive. The status of this bit can be checked with the constant EUSCI_A_STATW_BUSY. For further information on the remaining bits of this register, please see [4].

9.2.1.1 Baud Rate Generation

Baud rate represents the number of received or sent bits per second (bps). Desired baud rates can be generated by using the baud rate generator block in the UART mode. This block receives the selected clock (BRCLK) as input. The clock frequency can be divided by the 16-bit coefficient UCBRx (as explained in Sec. 9.1.2). The baud rate generator block also has an eUSCI_Ax modulation control word register (UCAxMCTLW) to set the modulation property. The entries of this register are given in Table 9.6. Depending on the settings and the input clock frequency, the UART baud rate generator block can be used in low- or high-frequency modes. We will discuss about these in the following paragraphs.

In Table 9.6, UCBRFx bits (with constant EUSCI_A_MCTLW_BRF_OFS) are used to select the modulation pattern for BITCLK16. For more information on these patterns, please see [4]. This is the first modulation step for the oversampling (high frequency) mode. This step is not applicable in the low-frequency mode. UCBRSx bits (with constant EUSCI_A_MCTLW_BRS_OFS) are used to select the modulation pattern for the BITCLK that has the closest frequency for the desired baud rate. This is the only modulation step for the low-frequency mode. Also, this is the second step for the oversampling mode. The UCOS16 bit is used to activate the oversampling mode. When this bit is reset, oversampling mode is disabled and the baud rate is generated by using low-frequency clocks. In this mode, high-frequency clocks can also be used. However, this is generally not recommended since it decreases the time interval for majority votes (to be explained in the following section). When the UCOS16 bit is set with the constant EUSCI_A_MCTLW_OS16, the oversampling mode is enabled.

In the UART mode, baud rate calculation formulas are given in [4]. However, typical baud rates can be generated by setting the UCOS16, UCBRx, UCBRFx, and UCBRSx values. Based on the status of the UCOS16 bit, the baud rates that can be generated are given in Tables 9.7 and 9.8.

Baud Rate	BRCLK Freq. (Hz)	UCBRx	UCBRSx
2,400	32,768	13	0xB6
4,800	32,768	6	0xEE
9,600	32,768	3	0x92
57,600	1,000,000	17	0x4A
57,600	1,048,576	18	0x11
115,200	1,000,000	8	0xD6
115,200	1,048,576	9	0x08
230,400	4,000,000	17	0x4A
230,400	4,194,304	18	0x11
460,800	8,000,000	17	0x4A
460,800	8,388,608	18	0x11

TABLE 9.7 Typical Baud Rates That Can Be Generated When the UCOS16 Bit Is Reset

9.2.1.2 UART Transmit/Receive Operations

Before focusing on the transmit and receive operations, we should mention that these operations are done on a seven- or eight-bit character basis in the UART mode. Also, the character is not sent alone. The character format for the UART mode is as in Table 9.9. In this table, D0···D6 stand for the seven data (character) bits. *D7* stands for the eight data bit. In Table 9.9, italic characters indicate that the mentioned bits are optional to use. Also, the LSB first transmission is typically used in the UART mode. As a reminder, this is achieved by resetting the UCMSB bit in the UCAxCTLW0 register.

Transmit and receive operations are simple in the UART mode. If there is no data written to the UCAxTXBUF, the baud rate generator does not provide any clock to the UART. Hence, it stays in the idle state. The transmit operation starts when the data is written to the UCAxTXBUF. Then, the baud rate generator starts working. Data within the UCAxTXBUF is moved to the transmit shift register. Meanwhile, the UCTXIFG bit in the UCAxIFG register is set to indicate that UCAx0TXBUF is ready to accept new data. Data in the transmit shift register is sent to the receiver in a serial manner. Then, UART returns to the idle state.

The receive operation starts when the falling edge of the start bit is detected. Until then, the baud rate generator does not provide any clock to the UART. Therefore, the UART stays in the idle state as in the transmit operation. The baud rate generator starts working after the falling edge of the start bit is detected. Then, the receiver checks the validity of the start bit. If the start bit is not valid, the UART goes to the idle state. Otherwise, each received signal pulse is checked by majority voting. Here, three samples are taken from the pulse. If the number of zeros is more than ones in these samples, then the receive shift register gets a logic level zero. Otherwise, it gets a logic level one. The binary data is shifted in the receive shift register. This operation continues until the stop bit is detected. The final result is transferred to the UCAxRXBUF.

Baud Rate	BRCLK Freq. (Hz)	UCBRx	UCBRSx	UCBRFx	Baud Rate	BRCLK Freq. (Hz)	UCBRx	UCBRSx	UCBRFx
1,200	32,768	1	0x25	11	38,400	16,777,216	27	0xFB	4
9,600	1,000,000	6	0x20	8	38,400	20,000,000	32	0xEE	8
9,600	1,048,576	6	0x22	13	57,600	4,000,000	4	0x55	5
9,600	4,000,000	26	0xB6	0	57,600	4,194,304	4	0xEE	8
9,600	4,194,304	27	0xFB	4	57,600	8,000,000	8	0xF7	10
9,600	8,000,000	52	0x49	1	57,600	8,388,608	9	0xB5	1
9,600	8,388,608	54	0xEE	9	57,600	12,000,000	13	0x25	0
9,600	12,000,000	78	0x00	2	57,600	16,000,000	17	0xDD	5
9,600	16,000,000	104	0xD6	2	57,600	16,777,216	18	0x44	3
9,600	16,777,216	109	0xB5	3	57,600	20,000,000	21	0x22	11
9,600	20,000,000	130	0x25	3	115,200	4,000,000	2	0xBB	2
19,200	1,000,000	3	0x02	4	115,200	4,194,304	2	0x92	4
19,200	1,048,576	3	0xAD	6	115,200	8,000,000	4	0x55	5
19,200	4,000,000	13	0x84	0	115,200	8,388,608	4	0xEE	8
19,200	4,194,304	13	0x55	10	115,200	12,000,000	6	0x20	8
19,200	8,000,000	26	0xB6	0	115,200	16,000,000	8	0xF7	10
19,200	8,388,608	27	0xFB	4	115,200	16,777,216	9	0xB5	1
19,200	12,000,000	39	0x00	1	115,200	20,000,000	10	0xAD	13
19,200	16,000,000	52	0x49	1	230,400	8,000,000	2	0xBB	2
19,200	16,777,216	54	0xEE	9	230,400	8,388,608	2	0x92	4
19,200	20,000,000	65	0xD6	1	230,400	12,000,000	3	0x02	4
38,400	1,000,000	1	0x00	10	230,400	16,000,000	4	0x55	5
38,400	1,048,576	1	0x25	11	230,400	16,777,216	4	0xEE	8
38,400	4,000,000	6	0x20	8	230,400	20,000,000	5	0xEE	6
38,400	4,194,304	6	0x22	13	460,800	12,000,000	1	0x00	10
38,400	8,000,000	13	0x84	0	460,800	16,000,000	2	0xBB	2
38,400	8,388,608	13	0x55	10	460,800	16,777,216	2	0x92	4
38,400	12,000,000	19	0x65	8	460,800	20,000,000	2	0x92	11
38,400	16,000,000	26	0xB6	0					

TABLE 9.8 Typical Baud Rates That Can Be Generated When the UCOS16 Bit Is Set

Start Bit	D0 · · D6	D7	Address Bit	Parity Bit	Stop Bit	Second Stop Bit

TABLE 9.9 UART Character Format

9.2.1.3 UART Interrupts

UART has three registers to control transmit and receive interrupts. These are eUSCI_Ax interrupt enable register (UCAxIE), eUSCI_Ax interrupt flag register (UCAxIFG), and eUSCI_Ax interrupt vector register (UCAxIV).

In the UCAxIE register, the UCTXIE and UCRXIE bits are used to enable the transmit and receive interrupts, respectively. When these bits are set, with the constants EUSCI_A_IE_TXIE and EUSCI_A_IE_RXIE for the transmit and receive operations, related interrupts are enabled. Otherwise, interrupts are disabled. Related to these, the UCTXIFG and

UCIVx Content	Interrupt Source	Interrupt Flag	Interrupt Priority
0x00	No interrupt pending	–	–
0x02	Receive buffer full	UCRXIFG	Highest
0x04	Transmit buffer empty	UCTXIFG	
0x06	Start bit received	UCSTTIFG	
0x08	Transmit complete	UCTXCPTIFG	Lowest

TABLE 9.10 UCAxIV Register

UCRXIFG bits of the UCAxIFG register are the flags for the transmit and receive interrupts, respectively. Each interrupt flag can be checked by the constant EUSCI_A_IFG_TXIFG or EUSCI_A_IFG_RXIFG depending on the transmit or receive operation.

The UCAxIV register is the interrupt vector used for transmit and receive interrupts. The contents of this register are given in Table 9.10.

The interrupt-based communication operation works in the UART mode as follows. Initially, the UCTXIE and UCRXIE bits should be set to enable transmit and receive interrupts. In the transmit operation, an interrupt is requested when the UCAxTXBUF is ready for another character. Then, the UCTXIFG flag is set. This flag is automatically cleared when a new character is written to the UCAxTXBUF. In the reception operation, an interrupt is requested when a character is loaded to the UCAxRXBUF. Then, the UCRXIFG flag is set. This flag is automatically cleared when data in UCAxRXBUF is read.

9.2.2 eUSCI Module in UART Mode via DriverLib Functions

The eUSCI module in UART mode can be controlled with DriverLib functions given below. Here, `moduleInstance` is the eUSCI base to be initialized and controlled. It can be one of EUSCI_A0_BASE, EUSCI_A1_BASE, EUSCI_A2_BASE, and EUSCI_A3_BASE values. The complete function list and macro definitions can be found in [11].

```
bool UART_initModule(uint32_t moduleInstance, const eUSCI_UART_Config *
    config)
/*
Initializes the UART module by modifying the UCAxCTL0 and UCAxCTL1
    registers.
config: the configuration structure. It contains clock, clock divider,
    first and second modulation stage register selection, parity, shift
     register direction, stop bit, uart mode and oversampling values.
    Please see the datasheet for more information.
*/

void UART_transmitData(uint32_t moduleInstance, uint_fast8_t
    transmitData)
/*
Sends a byte by modifying the UCAxTXBUF register.
transmitData: data to be transmitted.
*/

uint8_t UART_receiveData(uint32_t moduleInstance)
/*
```

Digital Communication

Returns the byte sent by UART by reading the UCAxRXBUF register.
*/

void UART_enableModule(uint32_t moduleInstance)
/*
Enables the UART mode by modifying the UCAxCTL1 register.
*/

void UART_disableModule(uint32_t moduleInstance)
/*
Disables the UART mode by modifying the UCAxCTL1 register.
*/

void UART_transmitBreak(uint32_t moduleInstance)
/*
Sends a UART break by modifying the UCAxCTL1 and UCAxTXBUF registers.
*/

void UART_enableInterrupt(uint32_t moduleInstance, uint_fast8_t mask)
/*
Enables the UART interrupts by modifying the UCAxIFG, UCAxIE, and
*   UCAxCTL1 registers.*
mask: the interrupt type to be enabled. It can be logical OR form of
*   EUSCI_A_UART_RECEIVE_INTERRUPT, EUSCI_A_UART_TRANSMIT_INTERRUPT,*
*   EUSCI_A_UART_RECEIVE_ERRONEOUSCHAR_INTERRUPT, and*
*   EUSCI_A_UART_BREAKCHAR_INTERRUPT values.*
*/

void UART_disableInterrupt(uint32_t moduleInstance, uint_fast8_t mask)
/*
Disables UART interrupts by modifying the UCAxIE and UCAxCTL1
*   registers.*
mask: the interrupt type to be disabled. It can be logical OR form of
*   EUSCI_A_UART_RECEIVE_INTERRUPT, EUSCI_A_UART_TRANSMIT_INTERRUPT,*
*   EUSCI_A_UART_RECEIVE_ERRONEOUSCHAR_INTERRUPT, and*
*   EUSCI_A_UART_BREAKCHAR_INTERRUPT values.*
*/

uint_fast8_t UART_getEnabledInterruptStatus(uint32_t moduleInstance)
/*
Returns the status of the enabled interrupt flags by reading the UCAxIE
*   and UCAxIFG registers.*
*/

void UART_clearInterruptFlag(uint32_t moduleInstance, uint_fast8_t
    mask)
/*
Clears UART interrupts by modifying the UCAxIFG register.
mask: the interrupt type to be cleared. It can be logical OR form of
*   EUSCI_A_UART_RECEIVE_INTERRUPT, EUSCI_A_UART_TRANSMIT_INTERRUPT,*
*   EUSCI_A_UART_RECEIVE_ERRONEOUSCHAR_INTERRUPT, and*
*   EUSCI_A_UART_BREAKCHAR_INTERRUPT values.*
*/

9.2.3 eUSCI Module in UART Mode via Energia Functions

Energia can also be used to control the eUSCI module in UART mode. We provide the related functions below. More information on Energia functions can be found in [12].

```
void begin(speed);
/*
Opens serial port and configures the data rate.
speed: selected baud rate.
*/

int available();
/*
Returns the number of available bytes from serial port. Serial receive
    buffer stores received bytes until read. It can hold 128 bytes.
*/

int write(val);
int write(str);
int write(buf, len);
/*
Sends data and returns the number of data sent.
val: a single byte data to be sent.
str: string data to be sent.
buf: array data to be sent.
len: the length of the array
*/

int print(val);
/*
Sends data in ASCII form. It automatically converts numbers and float
    variables to ASCII characters.
val: data to be sent.
*/

int println(val);
/*
Sends data in ASCII form followed by a carriage return character (ASCII
    0x13, or \ r) and a new line character (ASCII 0x10, or \ n). It
    automatically converts numbers, floats to ASCII characters.
val: data to be sent.
*/

int read();
/*
Receives the first byte of incoming data.
*/
```

9.2.4 Coding Practices for the UART Mode

In this section, we provide sample C codes for the UART communication mode. We will be using the red, green, and blue colors of LED2 connected to pins P2.0, P2.1, and P2.2 on the MSP432 LaunchPad as output. The button S1 connected to pin P1.1 on the MSP432 LaunchPad is used as input. Please also check Sec. 2.1.11 for the usage of the terminal program in CCS.

9.2.4.1 Register-Level Examples

In the first example, the duty cycle of a PWM signal (between 0 and 100) is obtained from the host computer using the UART mode in Listing 9.1. The connection diagram for this example is given in Fig. 9.2. The PWM signal is used to adjust the brightness of the red color of LED2. This operation is done continuously. For this program to work, the jumper labeled Red P2.0 above LED2 on MSP432 LaunchPad must be removed. The pin P2.4 (PWM output) should be connected to the bottom pin of this jumper.

Listing 9.1 The UART PWM Application, Using Registers

```c
#include "msp.h"

int convert(uint32_t ByteCounter, char *data);
void transmit(char *str);

 char enter[] = "Enter Duty Cycle\r\n";
 char enter_new[] = "Enter New Duty Cycle\r\n";
 char digits[3];
 uint8_t RXByteCounter = 0;
 int dutyCycle;

void main(void){

 WDT_A->CTL = WDT_A_CTL_PW | WDT_A_CTL_HOLD;

 P2->DIR |= BIT4;
 P2->SEL0 |= BIT4;
 P1->SEL0 |= BIT2 | BIT3; // Set P1.2 and P1.3 as UCA0RXD and UCA0TXD

// Configure SMCLK as 12 MHz from DCO
 CS->KEY = CS_KEY_VAL;
 CS->CTL0 &= ~ CS_CTL0_DCORSEL_MASK;
 CS->CTL0 |= CS_CTL0_DCORSEL_3;
 CS->CTL1 |= CS_CTL1_SELS_3;
 CS->KEY = 0;

 EUSCI_A0->CTLW0 |= EUSCI_A_CTLW0_SWRST; // Hold EUSCI_A0 module in
     reset state
 EUSCI_A0->CTLW0 |= EUSCI_A_CTLW0_UCSSEL_2; // Select SMCLK as EUSCI_A0
     clock

// Set baud-rate to 57600 from Table
 EUSCI_A0->BRW = 13;
 EUSCI_A0->MCTLW = 0x2500 | EUSCI_A_MCTLW_OS16;
 EUSCI_A0->CTLW0 &= ~EUSCI_A_CTLW0_SWRST; // Clear SWRST to resume
     operation
 EUSCI_A0->IFG &= ~EUSCI_A_IFG_RXIFG; // Clear EUSCI_A0 RX interrupt
     flag
 EUSCI_A0->IE |= EUSCI_A_IE_RXIE; // Enable EUSCI_A0 RX interrupt

 TIMER_A0->CCR[0] = 12000 - 1;  // PWM period
 TIMER_A0->CCR[1] = 0; // CCR1 PWM duty cycle
 TIMER_A0->CCTL[1] = TIMER_A_CCTLN_OUTMOD_7; // CCR1 reset/set
 TIMER_A0->CTL = TIMER_A_CTL_TASSEL_2 | TIMER_A_CTL_MC_1 |
     TIMER_A_CTL_CLR; // SMCLK, Up mode, Clear TAR
```

```c
    NVIC->ISER[0] = 0x00010000; // EUSCI_A0 interrupt is enabled in NVIC
    __enable_irq(); // All interrupts are enabled

    transmit(enter); // Transmit enter array

    while(1){
     SCB->SCR |= SCB_SCR_SLEEPONEXIT_Msk; // Sleep on exit
     __sleep(); // enter LPM0
     dutyCycle = 120 * convert(RXByteCounter, digits); // Convert duty
        cycle value
     RXByteCounter = 0; // Reset RX byte counter
     TIMER_A0->CCR[1] = dutyCycle; // Set new duty cycle
     transmit(enter_new);} // Transmit enter new array
}

void EUSCIA0_IRQHandler(void){
 char RXData;
 if(EUSCI_A0->IFG & EUSCI_A_IFG_RXIFG){
   EUSCI_A0->IFG &=~ EUSCI_A_IFG_RXIFG; // Clear EUSCI_A0 RX interrupt
       flag
   RXData = EUSCI_A0->RXBUF; // Get received data
   if((RXData != 0x0D) && (RXByteCounter <= 3)) digits[RXByteCounter++]
       = RXData;
   else SCB->SCR &= ~SCB_SCR_SLEEPONEXIT_Msk;} // Not sleep on exit
}

void transmit(char *str){
 while(*str != 0){ // Loop until null character
   while(!(EUSCI_A0->IFG & EUSCI_A_IFG_TXIFG)); // Check if the transmit
       interrupt flag is set
   EUSCI_A0->TXBUF = *str++;} // Load the transmit buffer with current
       string element
}

int convert(uint32_t ByteCounter, char *data){
 char hundreds = '0', tens = '0', ones = '0';
 if(ByteCounter == 1) ones = data[0]; // If the ByteCounter equals 1,
    take only ones digit
 else if(ByteCounter == 2){ // If the ByteCounter equals 2, take ones
    and tens digits
   ones = data[1];
   tens = data[0];}
 else{ // If the ByteCounter equals 3, take ones, tens and hundreds
      digits
   ones = data[2];
   tens = data[1];
   hundreds = data[0];}
 return (int)(((hundreds - 0x30)*100) + ((tens - 0x30)*10) + (ones -
      0x30));
}
```

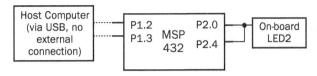

FIGURE 9.2 The connection diagram for the UART PWM application.

Listing 9.2 The Transmitter Part of the UART Communication Between Two MSP432 LaunchPads, Using Registers

```c
#include "msp.h"

#define S1 BIT1

 char TXData[10] = "RGBBGRBGGR";
 int i = 0;

void main(void){

 WDT_A->CTL = WDT_A_CTL_PW | WDT_A_CTL_HOLD;

 P1->DIR &= ~S1;
 P1->REN = S1;
 P1->OUT = S1;
 P1->IE = S1;
 P1->IES = S1;
 P1->IFG = 0x00;
 P3->SEL0 |= BIT2 | BIT3; // Set P3.2 and P3.3 as UCA2RXD and UCA2TXD

// Configure SMCLK as 12 MHz from DCO
 CS->KEY = CS_KEY_VAL;
 CS->CTL0 &= ~ CS_CTL0_DCORSEL_MASK;
 CS->CTL0 |= CS_CTL0_DCORSEL_3;
 CS->CTL1 |= CS_CTL1_SELS_3;
 CS->KEY = 0;

 EUSCI_A2->CTLW0 |= EUSCI_A_CTLW0_SWRST; // Hold eUSCI module in reset
     state
 EUSCI_A2->CTLW0 |= EUSCI_A_CTLW0_UCSSEL_2; // Select SMCLK as EUSCI_A2
     clock

// Set baud-rate to 57600 from Table
 EUSCI_A2->BRW = 13;
 EUSCI_A2->MCTLW = 0x2500 | EUSCI_A_MCTLW_OS16;
 EUSCI_A2->CTLW0 &= ~EUSCI_A_CTLW0_SWRST; // Clear SWRST to resume
     operation

 NVIC->ISER[1] = 0x00000008; // Port P1 interrupt is enabled in NVIC
 NVIC->ISER[0] = 0x00040000; // EUSCI_A2 interrupt is enabled in NVIC
 __enable_irq(); // All interrupts are enabled

 SCB->SCR |= SCB_SCR_SLEEPONEXIT_Msk; // Sleep on exit
 __sleep(); // enter LPM0
}
```

```c
void EUSCIA2_IRQHandler(void){
 uint32_t status = EUSCI_A2->IFG; // Get EUSCI_A2 interrupt flag
 EUSCI_A2->IFG &=~ EUSCI_A_IFG_TXIFG; // Clear EUSCI_A2 TX interrupt
     flag

 if(status & EUSCI_A_IFG_TXIFG){ // Check if transmit interrupt occurs
  EUSCI_A2->TXBUF = TXData[i++%10]; // Load current TXData value to
      transmit buffer
  EUSCI_A2->IE &= ~EUSCI_A_IE_TXIE;} // Disable EUSCI_A2 TX interrupt
}

void PORT1_IRQHandler(void){
 uint32_t status = P1->IFG; // Get Port1 interrupt flag
 P1->IFG &= ~S1; // Interrupt flag is cleared for P1.1
 if(status & S1) EUSCI_A2->IE |= EUSCI_A_IE_TXIE; // Enable EUSCI_A2 TX
     interrupt
}
```

Listing 9.3 The Receiver Part of the UART Communication Between Two MSP432 LaunchPads, Using Registers

```c
#include "msp.h"

#define LED2_RED BIT0
#define LED2_GREEN BIT1
#define LED2_BLUE BIT2

void main(void){

 WDT_A->CTL = WDT_A_CTL_PW | WDT_A_CTL_HOLD;

 P2->DIR = LED2_RED|LED2_GREEN|LED2_BLUE;
 P2->OUT = 0x00;
 P3->SEL0 |= BIT2 | BIT3; // Set P3.2 and P3.3 as UCA2RXD and UCA2TXD

// Configure SMCLK as 12 MHz from DCO
 CS->KEY = CS_KEY_VAL;
 CS->CTL0 &= ~ CS_CTL0_DCORSEL_MASK;
 CS->CTL0 |= CS_CTL0_DCORSEL_3;
 CS->CTL1 |= CS_CTL1_SELS_3;
 CS->KEY = 0;

 EUSCI_A2->CTLW0 |= EUSCI_A_CTLW0_SWRST; // Hold eUSCI module in reset
     state
 EUSCI_A2->CTLW0 |= EUSCI_A_CTLW0_UCSSEL_2; // Select SMCLK as EUSCI_A2
     clock

// Set baud-rate to 57600 from Table
 EUSCI_A2->BRW = 13;
 EUSCI_A2->MCTLW = 0x2500 | EUSCI_A_MCTLW_OS16;
 EUSCI_A2->CTLW0 &= ~EUSCI_A_CTLW0_SWRST; // Clear SWRST to resume
     operation
 EUSCI_A2->IFG &= ~EUSCI_A_IFG_RXIFG; // Clear EUSCI_A2 RX interrupt
     flag
```

```c
EUSCI_A2->IE |= EUSCI_A_IE_RXIE; // Enable EUSCI_A2 RX interrupt

NVIC->ISER[0] = 0x00040000; // EUSCI_A2 interrupt is enabled in NVIC
__enable_irq(); // All interrupts are enabled
SCB->SCR |= SCB_SCR_SLEEPONEXIT_Msk; // Sleep on exit
__sleep(); // enter LPM0
}

void EUSCIA2_IRQHandler(void){
  uint32_t status = EUSCI_A2->IFG; // Get EUSCI_A2 interrupt flag
  EUSCI_A2->IFG &=~ EUSCI_A_IFG_RXIFG; // Clear EUSCI_A2 RX interrupt
      flag

  if(status & EUSCI_A_IFG_RXIFG){ // Check if receive interrupt occurs
    if((EUSCI_A2->RXBUF) == 'R') P2->OUT ^= LED2_RED; // Toggle P2.0 if
        'R' is received
    else if((EUSCI_A2->RXBUF) == 'G') P2->OUT ^= LED2_GREEN; // Toggle
        P2.1 if 'G' is received
    else if((EUSCI_A2->RXBUF) == 'B') P2->OUT ^= LED2_BLUE;} // Toggle
        P2.2 if 'B' is received
}
```

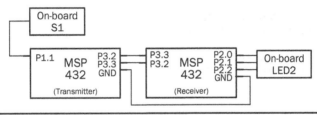

FIGURE 9.3 The connection diagram for the UART communication between two MSP432 LaunchPads.

In Listings 9.2 and 9.3, the UART mode is used to establish a digital communication between the two MSP432 LaunchPads. The connection diagram for this example is given in Fig. 9.3. The C code for the transmitter device is given in Listing 9.2. The C code for the receiver device is given in Listing 9.3. In this application, when the button S1 of the transmitter device is pressed, the transmitter sends the next character from the TXData array to control the red, green, and blue colors of LED2 on the receiver.

9.2.4.2 DriverLib Examples

We repeat the example in Listing 9.1 using DriverLib functions. We provide the corresponding C code in Listing 9.4. Here, we first set the peripheral pins, power mode, and DCO frequency. Afterwards, we initialize and enable the EUSCI_A0 module by the functions UART_initModule and UART_enableModule, respectively. Here, we set the configuration structure such that the EUSCI_A0 module uses SMCLK as its clock, MSB bit is first to send, no parity is used, and baud rate is 115200 bps. Then, we enable the EUSCI_A0 UART receive interrupt using the function UART_enableInterrupt. We clear the enabled interrupt flag, read received values using the function UART_receiveData, and set PWM duty cycle in the ISR. We use the transmit function to transmit the data using the function UART_transmitData.

Listing 9.4 The UART PWM Application, Using DriverLib Functions

```c
#include <ti/devices/msp432p4xx/driverlib/driverlib.h>

// http://software-dl.ti.com/msp430/msp430_public_sw/mcu/msp430/
    MSP430BaudRateConverter/index.html
const eUSCI_UART_Config uartConfig =
{
 EUSCI_A_UART_CLOCKSOURCE_SMCLK,// SMCLK Clock Source
 13,// BRDIV = 13
 0,// UCxBRF = 0
 37, // UCxBRS = 37
 EUSCI_A_UART_NO_PARITY, // No Parity
 EUSCI_A_UART_LSB_FIRST, // MSB First
 EUSCI_A_UART_ONE_STOP_BIT, // One stop bit
 EUSCI_A_UART_MODE, // UART mode
 EUSCI_A_UART_OVERSAMPLING_BAUDRATE_GENERATION // Oversampling
};

 int period = 12000;

Timer_A_PWMConfig pwmConfig =
{
 TIMER_A_CLOCKSOURCE_SMCLK,
 TIMER_A_CLOCKSOURCE_DIVIDER_1,
 12000,
 TIMER_A_CAPTURECOMPARE_REGISTER_1,
 TIMER_A_OUTPUTMODE_RESET_SET,
 12000
};

 char enter[] = "Enter Duty Cycle\r\n";
 char enter_new[] = "Enter New Duty Cycle\r\n";
 char digits[3];
 uint8_t RXByteCounter = 0;

void transmit(char *str);
int convert(uint32_t ByteCounter, char *data);

void main(void){

 WDT_A_holdTimer();

 GPIO_setAsPeripheralModuleFunctionInputPin(GPIO_PORT_P1, GPIO_PIN2 |
     GPIO_PIN3, GPIO_PRIMARY_MODULE_FUNCTION);
 GPIO_setAsPeripheralModuleFunctionOutputPin(GPIO_PORT_P2, GPIO_PIN4,
     GPIO_PRIMARY_MODULE_FUNCTION);

 CS_setDCOCenteredFrequency(CS_DCO_FREQUENCY_12);

 UART_initModule(EUSCI_A0_BASE, &uartConfig);
 UART_enableModule(EUSCI_A0_BASE);
 UART_enableInterrupt(EUSCI_A0_BASE, EUSCI_A_UART_RECEIVE_INTERRUPT);

 Interrupt_enableInterrupt(INT_EUSCIA0);
```

```
  Interrupt_enableSleepOnIsrExit();
  Interrupt_enableMaster();

  transmit(enter);

  PCM_setPowerState(PCM_LPM0_LDO_VCORE0);
}
void EUSCIA0_IRQHandler(void){
 char RXData;
 uint32_t status = UART_getEnabledInterruptStatus(EUSCI_A0_BASE);
 UART_clearInterruptFlag(EUSCI_A0_BASE, status);

 if(status & EUSCI_A_UART_RECEIVE_INTERRUPT_FLAG){
  RXData = UART_receiveData(EUSCI_A0_BASE);
  if(RXData != 0x0D && RXByteCounter <= 3) digits[RXByteCounter++] =
      RXData;
  else{
   pwmConfig.dutyCycle = 120 * convert(RXByteCounter, digits);
   RXByteCounter = 0;
   Timer_A_generatePWM(TIMER_A0_BASE, &pwmConfig);
   transmit(enter_new);}}
}

void transmit(char *str){
 while(*str != 0){
  UART_transmitData(EUSCI_A0_BASE, *str++);}
}

int convert(uint32_t ByteCounter, char *data){
 char hundreds = '0', tens = '0', ones = '0';
 if(ByteCounter == 1) ones = data[0];
 else if(ByteCounter == 2){
  ones = data[1];
  tens = data[0];}
 else{
  ones = data[2];
  tens = data[1];
  hundreds = data[0];}
 return (int)(((hundreds - 0x30)*100) + ((tens - 0x30)*10) + (ones -
     0x30));
}
```

We next repeat the examples in Listings 9.2 and 9.3 using DriverLib functions. We provide the corresponding C codes in Listings 9.5 and 9.6. Here in the first C code in Listing 9.5, we enable the port interrupts. Afterwards, we enable the EUSCI_A0 UART transmit interrupt in the port ISR. Finally, we clear the enable interrupt flag and transmit the values using the function UART_transmitData and disable the EUSCI_A0 UART transmit interrupt using the function UART_disableInterrupt in the EUSCI_A0 ISR. In the C code given in Listing 9.6, we toggle the red, green, and blue colors of LED2 according to the received data in the EUSCI_A0 ISR.

Listing 9.5 The Transmitter Part of the UART Communication Between Two MSP432 LaunchPads, Using DriverLib Functions

```c
#include <ti/devices/msp432p4xx/driverlib/driverlib.h>

// http://software-dl.ti.com/msp430/msp430_public_sw/mcu/msp430/
    MSP430BaudRateConverter/index.html
const eUSCI_UART_Config uartConfig =
{
 EUSCI_A_UART_CLOCKSOURCE_SMCLK, // SMCLK Clock Source
 13, // BRDIV = 13
 0, // UCxBRF = 0
 37, // UCxBRS = 37
 EUSCI_A_UART_NO_PARITY, // No Parity
 EUSCI_A_UART_LSB_FIRST, // MSB First
 EUSCI_A_UART_ONE_STOP_BIT, // One stop bit
 EUSCI_A_UART_MODE, // UART mode
 EUSCI_A_UART_OVERSAMPLING_BAUDRATE_GENERATION // Oversampling
};

 char TXData[10] = "RGBBGRBGGR";
 int i = 0;

void main(void){

 WDT_A_holdTimer();

 GPIO_setAsPeripheralModuleFunctionInputPin(GPIO_PORT_P3, GPIO_PIN2 |
     GPIO_PIN3, GPIO_PRIMARY_MODULE_FUNCTION);
 GPIO_setAsInputPinWithPullUpResistor(GPIO_PORT_P1, GPIO_PIN1);
 GPIO_clearInterruptFlag(GPIO_PORT_P1, GPIO_PIN1);
 GPIO_enableInterrupt(GPIO_PORT_P1, GPIO_PIN1);

 CS_setDCOCenteredFrequency(CS_DCO_FREQUENCY_12);

 UART_initModule(EUSCI_A2_BASE, &uartConfig);
 UART_enableModule(EUSCI_A2_BASE);

 Interrupt_enableInterrupt(INT_PORT1);
 Interrupt_enableInterrupt(INT_EUSCIA2);
 Interrupt_enableSleepOnIsrExit();
 Interrupt_enableMaster();

 PCM_setPowerState(PCM_LPM0_LDO_VCORE0);
}

void EUSCIA2_IRQHandler(void){
 uint32_t status = UART_getEnabledInterruptStatus(EUSCI_A2_BASE);
 UART_clearInterruptFlag(EUSCI_A2_BASE, status);

 if(status & EUSCI_A_UART_TRANSMIT_INTERRUPT_FLAG){
   UART_transmitData(EUSCI_A2_BASE, TXData[i++%10]);
   UART_disableInterrupt(EUSCI_A2_BASE, EUSCI_A_UART_TRANSMIT_INTERRUPT);}
}

void PORT1_IRQHandler(void){
 uint32_t status = GPIO_getEnabledInterruptStatus(GPIO_PORT_P1);
 GPIO_clearInterruptFlag(GPIO_PORT_P1, status);

 if(status & GPIO_PIN1)
   UART_enableInterrupt(EUSCI_A2_BASE, EUSCI_A_UART_TRANSMIT_INTERRUPT);
}
```

Digital Communication

Listing 9.6 The Receiver Part of the UART Communication Between Two MSP432 LaunchPads, Using DriverLib Functions

```c
#include <ti/devices/msp432p4xx/driverlib/driverlib.h>

// http://software-dl.ti.com/msp430/msp430_public_sw/mcu/msp430/
    MSP430BaudRateConverter/index.html
const eUSCI_UART_Config uartConfig =
{
 EUSCI_A_UART_CLOCKSOURCE_SMCLK, // SMCLK Clock Source
 13, // BRDIV = 13
 0, // UCxBRF = 0
 37, // UCxBRS = 37
 EUSCI_A_UART_NO_PARITY, // No Parity
 EUSCI_A_UART_LSB_FIRST, // MSB First
 EUSCI_A_UART_ONE_STOP_BIT, // One stop bit
 EUSCI_A_UART_MODE, // UART mode
 EUSCI_A_UART_OVERSAMPLING_BAUDRATE_GENERATION // Oversampling
};

void main(void){

 WDT_A_holdTimer();

 GPIO_setAsPeripheralModuleFunctionInputPin(GPIO_PORT_P3, GPIO_PIN2 |
     GPIO_PIN3, GPIO_PRIMARY_MODULE_FUNCTION);
 GPIO_setAsOutputPin(GPIO_PORT_P2, GPIO_PIN0 | GPIO_PIN1 | GPIO_PIN2);
 GPIO_setOutputLowOnPin(GPIO_PORT_P2, GPIO_PIN0 | GPIO_PIN1 |
     GPIO_PIN2);

 CS_setDCOCenteredFrequency(CS_DCO_FREQUENCY_12);

 UART_initModule(EUSCI_A2_BASE, &uartConfig);
 UART_enableModule(EUSCI_A2_BASE);
 UART_enableInterrupt(EUSCI_A2_BASE, EUSCI_A_UART_RECEIVE_INTERRUPT);

 Interrupt_enableInterrupt(INT_EUSCIA2);
 Interrupt_enableSleepOnIsrExit();
 Interrupt_enableMaster();

 PCM_setPowerState(PCM_LPM0_LDO_VCORE0);
}

void EUSCIA2_IRQHandler(void){
 uint32_t status = UART_getEnabledInterruptStatus(EUSCI_A2_BASE);
 UART_clearInterruptFlag(EUSCI_A2_BASE, status);

 if(status & EUSCI_A_UART_RECEIVE_INTERRUPT_FLAG){
  if(UART_receiveData(EUSCI_A2_BASE) == 'R')
   GPIO_toggleOutputOnPin(GPIO_PORT_P2, GPIO_PIN0);
  else if(UART_receiveData(EUSCI_A2_BASE) == 'G')
   GPIO_toggleOutputOnPin(GPIO_PORT_P2, GPIO_PIN1);
  else if(UART_receiveData(EUSCI_A2_BASE) == 'B')
   GPIO_toggleOutputOnPin(GPIO_PORT_P2, GPIO_PIN2);}
}
```

Listing 9.7 The UART PWM Application, Using Energia Functions

```
char enter[] = "Enter Duty Cycle\r\n";
char enter_new[] = "Enter New Duty Cycle\r\n";
char digits[3];
uint8_t RXByteCounter = 0;

void setup(){
 Serial.begin(9600);
 Serial.print(enter);
}

void loop(){
 while (Serial.available()) {
  char inChar = (char)Serial.read();

  if(inChar != 0x0A && RXByteCounter <= 3)
   digits[RXByteCounter++] = inChar;
  else{
   int duty = 255 * convert(RXByteCounter, digits);
   analogWrite(RED_LED, duty);
   RXByteCounter = 0;
   Serial.print(enter_new);}}
}

float convert(uint32_t ByteCounter, char *data){
 char hundreds = '0', tens = '0', ones = '0';
 if(ByteCounter == 1) ones = data[0];
 else if(ByteCounter == 2){
  ones = data[1];
  tens = data[0];}
 else{
  ones = data[2];
  tens = data[1];
  hundreds = data[0];}
 return (float)(((hundreds - 0x30)*100) + ((tens - 0x30)*10) + (ones - 
    0x30)) / 100.00;
}
```

9.2.4.3 Energia Examples

We repeat the example in Listing 9.1 using Energia. We provide the corresponding code in Listing 9.7. Here, we first configure the UART module. We read the incoming data in the loop function. We set the PWM duty cycle using the `analogWrite` function.

We next repeat the examples in Listings 9.2 and 9.3 using Energia functions. We provide the corresponding codes in Listings 9.8 and 9.9. In the first code given in Listing 9.8, we configure the UART module and port interrupt. In the port ISR, we transmit the values using the function `print`. In the second code given in Listing 9.9, we toggle the red, green, and blue colors of LED2 according to the received data in the loop function.

9.3 Serial Peripheral Interface

Serial peripheral interface (SPI) is the synchronous communication mode. It can be used between multiple master devices and one slave device. It can also be used for one master

Listing 9.8 The Transmitter Part of the UART Communication Between Two MSP432 LaunchPads, Using Energia Functions

```
char TXData[11] = "RGBBGRBGGR";
int i = 0;

void setup(){
 Serial.begin(9600);
 pinMode(PUSH1, INPUT_PULLUP);
 attachInterrupt(PUSH1, pushButton, FALLING);
}

void loop(){}

void pushButton(){
 delay(100);
 Serial.print(TXData[i++%10]);
}
```

Listing 9.9 The Receiver Part of the UART Communication Between Two MSP432 LaunchPads, Using Energia Functions

```
boolean RED_LED_state = LOW;
boolean GREEN_LED_state = LOW;
boolean BLUE_LED_state = LOW;

void setup(){
 Serial.begin(9600);
}

void loop(){
 while (Serial.available()){
  char data = (char)Serial.read();

  if(data == 'R'){
   RED_LED_state = !RED_LED_state;
   digitalWrite(RED_LED, RED_LED_state);}
  else if(data == 'G'){
   GREEN_LED_state = !GREEN_LED_state;
   digitalWrite(GREEN_LED, GREEN_LED_state);}
  else if(data == 'B'){
   BLUE_LED_state = !BLUE_LED_state;
   digitalWrite(BLUE_LED, BLUE_LED_state);}}
}
```

and one or more slave devices. As in the UART mode, we will only focus on the SPI mode between one master and slave device in this chapter. Block diagram of the SPI mode is as in Fig. 9.4. SPI is the only communication mode available in both eUSCI_Ax and eUSCI_Bx modules. Therefore, the character "x" is used in register or variable names to indicate that the same register can be used for eUSCI_Ax or eUSCI_Bx.

As can be seen in Fig. 9.4, the SPI mode has four pins for communication. These pins are slave in-master out (UCxSIMO), master in-slave out (UCxSOMI), SPI clock (UCxCLK), and slave-transmit enable UCxSTE. UCxSIMO and UCxSOMI pins are used for data transmission. UCxSIMO is the data output line and UCxSOMI is the data input line

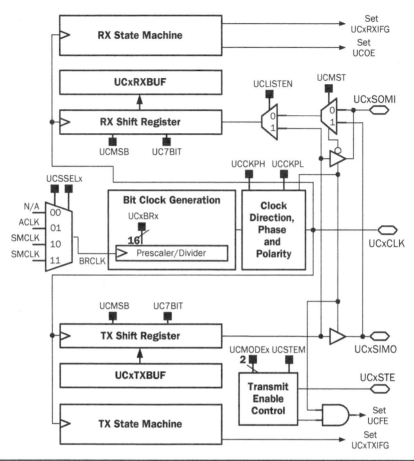

FIGURE 9.4 Block diagram of the SPI mode.

for the master device. UCxSIMO is the data input line and UCxSOMI is the data output line for the slave device. UCxCLK is the SPI clock generated by the master device. It ensures synchronization between the master and slave devices. UCxSTE is used to enable the chosen master in the multiple master mode or chosen slave in the multiple slave mode. When this pin is used, the SPI mode is called four-pin. In one master and slave mode, some slave devices need this pin to start or end the SPI communication. Therefore, four-pin SPI is a necessity for them. If the SPI communication is established between one master and slave (and UCxSTE is not needed), then UCxSIMO, UCxSOMI, and UCxCLK pins will be enough. The UCxSTE pin can be connected to the ground in this setting. This SPI mode is called three-pin.

9.3.1 SPI Registers

There are eight 16-bit registers available to control the SPI for both eUSCI_Ax and eUSCI_Bx modules. They are UCAxCTLW0, UCAxBRW, UCAxSTATW, UCAxRXBUF, UCAxTXBUF, UCAxIE, UCAxIFG, and UCAxIV registers for eUSCI_Ax module. For the eUSCI_Bx module, these are UCBxCTLW0, UCBxBRW, UCBxSTATW, UCBxRXBUF, UCBxTXBUF, UCBxIE, UCBxIFG, and UCBxIV registers. We only explain SPI registers for the eUSCI_Ax module since they are exactly same for the eUSCI_Bx module

Bits	15	14	13	12	11
	UCCKPH	UCCKPL	UCMSB	UC7BIT	UCMST
Bits	10-9	8	7-6	1	0
	UCMODEx	UCSYNC	UCSSELx	UCSTEM	UCSWRST

TABLE 9.11 eUSCI_Ax Control Word Register 0 (UCAxCTLW0)

Mode	UCCKPL	UCCKPH
0	0	1
1	0	0
2	1	1
3	1	0

TABLE 9.12 SPI Clock Modes

(only with different names). Structures EUSCI_Ax and EUSCI_Bx are used to control the mentioned registers. In these, the registers of interest are called CTLW0, BRW, STATW, RXBUF, TXBUF, IE, IFG, and IV.

The SPI mode is configured by the eUSCI_Ax control word register 0 (UCAxCTLW0). To note here, the UCAxCTLW0 register is also used in the UART mode. Here, it is used for SPI with different entries. The reader should be aware of this overlap. The entries of the UCAxCTLW0 register are tabulated in Table 9.11.

In Table 9.11, UCCKPH and UCCKPL bits are used together to adjust the SPI clock modes. The UCCKPL bit is used to set the clock polarity. When this bit is set with the constant EUSCI_A_CTLW0_CKPL, the clock is kept high in the idle state. When it is reset, the clock is kept low in the idle state. The UCCKPL bit does not affect the transmission format. The UCCKPH bit, on the other hand, has a direct effect on the transmission format. When this bit is set with the constant EUSCI_A_CTLW0_CKPH, the data is read on the first clock edge and sent on the next edge. When this bit is reset, the data is sent on the first clock edge and read on the next edge. UCCKPL and UCCKPH bits must be the same for both master and slave devices to set up an SPI communication between them.

Clock modes for the SPI are tabulated in Table 9.12. Here, modes zero and three are the most commonly used ones. In these, the data is read on the rising edge and sent on the falling edge of the clock. Mode zero needs the UCxSTE pin. Therefore, it is preferred in the four-pin SPI mode. Unlike mode zero, mode three does not need the UCxSTE pin. Therefore, it is used in the three-pin SPI mode.

The UCMSB bit in Table 9.11 is used to choose the start bit for data transfer. If this bit is set with the constant EUSCI_A_CTLW0_MSB, then the transmission starts from the MSB. Otherwise, the transmission starts from the LSB. The UC7BIT bit is used to select the data length. When this bit is reset, the data length is taken as eight bits. When it is set with the constant EUSCI_A_CTLW0_SEVENBIT, the data length is taken as seven bits. The UCMST bit is used to decide on the usage type of the device. When this bit is set with the constant EUSCI_A_CTLW0_MST, the device is used as master. Otherwise, it is used as slave. UCMODEx bits are used to select the synchronization mode. Constants for these bits are EUSCI_A_CTLW0_MODE_0 (three-pin SPI mode), EUSCI_A_CTLW0_MODE_1 (four-pin SPI mode with UCxSTE active high), EUSCI_A_CTLW0_MODE_2 (four-pin SPI mode with UCxSTE active low), and EUSCI_A_CTLW0_MODE_3 (I^2C mode).

Bits	7	0
	UCLISTEN	UCBUSY

TABLE 9.13 eUSCI_Ax Status Register (UCAxSTATW)

Finally, the UCSYNC bit is used to select the communication mode. When this bit is reset, asynchronous mode is selected. When it is set with the constant EUSCI_A_CTLW0_SYNC, synchronous mode is selected. Therefore, this bit must be set for the SPI mode. UCSELx bits are used to select the SPI clock. Constants for these bits are EUSCI_A_CTLW0_UCSSEL_1 (for ACLK) and EUSCI_A_CTLW0_UCSSEL_2 (for SMCLK). After the clock is selected, it can be divided by the 16-bit coefficient UCBRx as explained in Sec. 9.1.2. The UCSTEM bit is used to select the usage of STE pin for the master device. When this bit is set with the constant EUSCI_A_CTLW0_STEM, STE pin is used to generate the enable signal for a four-pin slave. Otherwise, it is used to prevent conflicts with other master devices. This bit is ignored when the three-pin SPI mode is used or the device is set as slave. The UCSWRST bit is used to reset the USCI module. When this bit is set with the constant EUSCI_A_CTLW0_SWRST, the USCI module is reset. Otherwise, the USCI module will be ready for operation.

The SPI mode has a status register called UCAxSTATW that is specifically used to observe the changes in the system. The entries of this register are given in Table 9.13. In this table, the UCLISTEN bit is used to generate an internal loop between the transmitter and receiver on the same device. When this bit is set with the constant EUSCI_A_STATW_LISTEN, the loopback is enabled. When it is reset, the loopback is disabled. This property can be used to troubleshoot the communication codes on a single device. The UCBUSY bit shows whether the eUSCI module is busy or not. This bit is set when the transmit or receive operation is performed. It is reset, when the system is inactive. The status of the bit can be checked by using the constant EUSCI_A_STATW_BUSY. For further information on the remaining bits of this register, please see [4].

9.3.1.1 SPI Transmit/Receive Operations

Transmit and receive operations must be carried out simultaneously in the SPI mode. Therefore, the data must be received from the slave device or transmitted from the master device even if it is completely redundant. Next, we provide transmit/receive operations for the master and slave modes separately.

One transmit-receive cycle for the SPI **master mode** is as follows. The eUSCI module is enabled. Transfer starts when the data is written to UCAxTXBUF. Then, the data is transferred to the transfer shift register from UCAxTXBUF. UCTXIFG is set to indicate that UCAxTXBUF is ready to accept new data. The data in the transfer shift register is sent to the UCxSIMO pin starting with MSB or LSB order (based on the UCMSB bit setting). Meanwhile, the received data is kept waiting at the UCxSOMI pin until the next clock edge. The data in the UCxSOMI pin is moved to the receive shift register with the next clock edge. Then, the data is transferred to UCAxRXBUF from the receive shift register. This operation is repeated until the seven or eight bits (depending on the setting of the UC7BIT) are transferred. As the transfer is complete, the UCRXIFG bit is set to indicate that transmit-receive cycle is complete.

UCIVx Content	Interrupt Source	Interrupt Flag	Interrupt Priority
0x00	No interrupt pending	–	–
0x02	Receive buffer full	UCRXIFG	Highest
0x04	Transmit buffer empty	UCTXIFG	Lowest

TABLE 9.14 UCAxIV Register

When the UCSTEM bit is set in four-pin master mode, UCxSTE pin acts as digital output. In this mode, the slave enable signal for a single slave is automatically generated on the UCxSTE pin after the data is written to the UCAxTXBUF. Then, the data is transferred to the transfer shift register from UCAxTXBUF and same steps are repeated as above.

One transmit-receive cycle for the **slave mode** is as follows. The UCxCLK supplied by the master device is used to start the data transfer. Before this clock is enabled, the data should be transferred to the transmit shift register from UCAxTXBUF. UCTXIFG is also set to indicate that UCAxTXBUF is ready to accept new data. The data in the transmit shift register is sent to the UCxSOMI pin starting with MSB or LSB order (based on the UCMSB bit setting). This data waits until the clock is activated. The data kept in the UCxSOMI pin is sent to output as the clock is activated. Meanwhile, the received data is kept waiting at the UCxSIMO pin until the next clock edge. The data in the UCxSIMO pin is moved to the receive shift register with the next clock edge. Then, the data is transferred to UCAxRXBUF from the receive shift register. This operation is repeated until the seven or eight bits (depending on the setting of the UC7BIT) are transferred. As the transfer is complete, the UCRXIFG bit is set to indicate that transmit-receive cycle is complete.

In four-pin slave mode, the operations explained above is also controlled by the UCxSTE pin. This pin is a digital input for the slave and controlled by the master device. When slave enable signal is obtained from the UCxSTE pin, the slave device is in active state and all transmit and receive operations explained above are performed normally. However, when the slave device is set inactive, its UCxSOMI pin is reconfigured as input. Any receive operation in progress in the UCxSIMO pin is stopped. Until the slave device is set active by reconfiguring the UCxSTE bit, ongoing shift operations are also stopped.

9.3.1.2 SPI Interrupts

SPI has three registers to control transmit and receive interrupts. These are eUSCI_Ax interrupt enable register (UCAxIE), eUSCI_Ax interrupt flag register (UCAxIFG), and eUSCI_Ax interrupt vector register (UCAxIV).

In the UCAxIE register, UCTXIE and UCRXIE bits are used to enable transmit and receive interrupts, respectively. When these bits are set with constants EUSCI_A_IE_TXIE and EUSCI_A_IE_RXIE transmit and receive interrupts are enabled, respectively. Otherwise, interrupts are disabled. Related to these, UCTXIFG and UCRXIFG bits of the UCAxIFG register are the transmit and receive interrupt flags, respectively. Status of these bits can be observed by using the constants EUSCI_A_IFG_TXIFG and EUSCI_A_IFG_RXIFG for the transmit and receive interrupts, respectively.

The UCAxIV register is the interrupt vector used for transmit and receive interrupts. The contents of this register are tabulated in Table 9.14.

The interrupt operations for the transmitter and the receiver are similar in the SPI mode. More specifically, the interrupt-based communication operation works as follows. Initially, the UCTXIE and UCRXIE bits should be set in the master and slave devices to enable transmit and receive interrupts. In the transmitter, an interrupt is requested when UCAxTXBUF is ready for another byte. Then, UCTXIFG is set. This flag is automatically cleared when a new byte is written to UCAxTXBUF. In the receiver, an interrupt is requested when a byte is loaded to UCAxRXBUF. Then, UCRXIFG is set. This flag is automatically cleared when data in UCAxRXBUF is read.

9.3.2 eUSCI Module in SPI Mode via DriverLib Functions

The eUSCI module in SPI mode can be controlled via DriverLib functions below. Here, moduleInstance is the eUSCI base to be initialized and controlled. It can be one of EUSCI_A0_BASE, EUSCI_A1_BASE, EUSCI_A2_BASE, EUSCI_A3_BASE, EUSCI_B0_BASE, EUSCI_B1_BASE, EUSCI_B2_BASE, and EUSCI_B3_BASE. The complete function list and macro definitions can be found in [11].

```
bool SPI_initMaster(uint32_t moduleInstance, const
    eUSCI_SPI_MasterConfig *config)
/*
Initializes the eUSCI module in SPI master mode by modifying the
    UCAxCTLW0 register.
config: the configuration structure. It contains clock, clock divider,
    desired SPI clock, shift register direction, clock phase, clock
    polarity and SPI mode values. Please see the datasheet for more
    information.
*/

bool SPI_initSlave(uint32_t moduleInstance, const eUSCI_SPI_SlaveConfig
    *config)
/*
Initializes the eUSCI module in SPI slave mode by modifying the
    UCAxCTLW0 register.
config: the configuration structure. It contains shift register
    direction, clock phase, clock polarity, and SPI mode values. Please
     see the datasheet for more information.
*/

void SPI_transmitData(uint32_t moduleInstance, uint_fast8_t
    transmitData)
/*
Sends a byte by modifying the UCAxTXBUF register.
transmitData: data to be sent.
*/

uint8_t SPI_receiveData(uint32_t moduleInstance)
/*
Returns the byte sent by SPI by reading the UCAxRXBUF register.
*/

void SPI_enableModule(uint32_t moduleInstance)
/*
Enables the eUSCI module in SPI mode by modifying the UCSWRST bit of
    the UCAxCTLW0 register.
*/
```

```
void SPI_disableModule(uint32_t moduleInstance)
/*
Disables the eUSCI module in SPI mode by modifying the UCSWRST bit of
    the UCAxCTLW0 register.
*/

void SPI_enableInterrupt(uint32_t moduleInstance, uint_fast8_t mask)
/*
Enables SPI interrupts by modifying the UCAxIFG and UCAxIE registers.
mask: the interrupt type to be enabled. It can be logical OR form of
    EUSCI_SPI_RECEIVE_INTERRUPT and EUSCI_SPI_TRANSMIT_INTERRUPT
    values.
*/

void SPI_disableInterrupt(uint32_t moduleInstance, uint_fast8_t mask)
/*
Disables SPI interrupts by modifying the UCAxIFG and UCAxIE registers.
mask: the interrupt type to be disabled. It can be logical OR form of
    EUSCI_SPI_RECEIVE_INTERRUPT and EUSCI_SPI_TRANSMIT_INTERRUPT
    values.
*/

uint_fast8_t SPI_getEnabledInterruptStatus (uint32_t moduleInstance)
/*
Returns the status of enabled interrupt flags by reading UCAxIFG and
    UCAxIE registers.
*/

void SPI_clearInterruptFlag (uint32_t moduleInstance, uint_fast8_t
   mask)
 /*
Clears SPI interrupts by modifying the UCAxIFG register.
mask: the interrupt type to be cleared. It can be logical OR form of
    EUSCI_SPI_RECEIVE_INTERRUPT and EUSCI_SPI_TRANSMIT_INTERRUPT
    values.
*/
```

9.3.3 Coding Practices for the SPI Mode

In this section, we provide sample C codes on the SPI communication mode. In these, the button S1 connected to pin P1.1 on the MSP432 LaunchPad is used as input. The red, green, and blue colors of LED2 connected to pins P2.0, P2.1, and P2.2 on the MSP432 LaunchPad are used as output.

9.3.3.1 Register-Level Examples

Three-pin SPI mode is used to establish a digital communication between the two MSP432 LaunchPads in the first example. The connection diagram for this example is as in Fig. 9.5. The C code for the master device (used as transmitter) is given in Listing 9.10. The C code for the slave device (used as receiver) is given in Listing 9.11. In this example, when the button S1 of the master device is pressed, the master sends the next byte from TXData array to control the red, green, and blue colors of LED2 on the slave device.

228 Chapter Nine

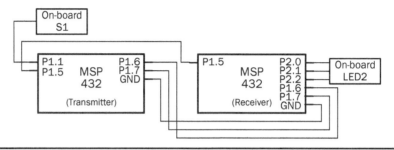

FIGURE 9.5 The connection diagram for the SPI application in three-pin mode.

Listing 9.10 The Master Transmitter Code for the SPI Example in Three-Pin Mode, Using Registers

```
#include "msp.h"

#define S1 BIT1
#define LED1 BIT0

char TXData[10] = "RGBBGRBGGR";
char RXData;
int i = 0;

void main(void){

    WDT_A->CTL = WDT_A_CTL_PW | WDT_A_CTL_HOLD;

    P1->DIR = LED1;
    P1->REN = S1;
    P1->OUT = S1;
    P1->IE = S1;
    P1->IES = S1;
    P1->IFG = 0x00;
    P1->SEL0 |= BIT5 | BIT6 | BIT7; // Set P1.5, P1.6 and P1.7 as UCB0CLK,
        UCB0SIMO and UCB0SOMI

    EUSCI_B0->CTLW0 |= EUSCI_B_CTLW0_SWRST; // Hold EUSCI_B0 module in
        reset state
    EUSCI_B0->CTLW0 |= EUSCI_B_CTLW0_MST|EUSCI_B_CTLW0_SYNC|
        EUSCI_B_CTLW0_CKPL|EUSCI_B_CTLW0_MSB;
    EUSCI_B0->CTLW0 |= EUSCI_B_CTLW0_UCSSEL_2; // Select SMCLK as EUSCI_B0
        clock
    EUSCI_B0->BRW = 0x02; // Set BITCLK = BRCLK / (UCBRx+1) = 3 MHz / 3 =
        1 MHz
    EUSCI_B0->CTLW0 &= ~EUSCI_B_CTLW0_SWRST; // Clear SWRST to resume
        operation
    EUSCI_B0->IFG &= ~EUSCI_B_IFG_RXIFG; // Clear EUSCI_B0 RX interrupt
        flag
    EUSCI_B0->IE |= EUSCI_B_IE_RXIE; // Enable EUSCI_B0 RX interrupt

    NVIC->ISER[1] = 0x00000008; // Port P1 interrupt is enabled in NVIC
    NVIC->ISER[0] = 0x00100000; // EUSCI_B0 interrupt is enabled in NVIC
    __enable_irq(); // All interrupts are enabled
    SCB->SCR |= SCB_SCR_SLEEPONEXIT_Msk; // Sleep on exit
    __sleep(); // enter LPM0
```

```c
}
void EUSCIB0_IRQHandler(void){
 uint32_t status = EUSCI_B0->IFG; // Get EUSCI_B0 interrupt flag

 if(status & EUSCI_B_IFG_RXIFG){ // Check if receive interrupt occurs
  RXData = EUSCI_B0->RXBUF; // Load current receive buffer value to
      RXData
  P1->OUT &= ~LED1;} // Turn off LED1
 else if(status & EUSCI_B_IFG_TXIFG){ // Check if transmit interrupt
     occurs
  EUSCI_B0->TXBUF = TXData[i++%10]; // Load current TXData value to
      transmit buffer
  EUSCI_B0->IE &= ~EUSCI_B_IE_TXIE;} // Disable EUSCI_B0 TX interrupt
}
void PORT1_IRQHandler(void){
 uint32_t status = P1->IFG;
 P1->IFG &= ~S1;

 if(status & S1){
  P1->OUT |= LED1;
  EUSCI_B0->IE |= EUSCI_B_IE_TXIE;} // Enable EUSCI_A0 TX interrupt
}
```

Also, the slave device transmits the received byte back to master to create a feedback for error checking.

9.3.3.2 DriverLib Examples

We next repeat the example in the previous section using DriverLib functions. We provide the corresponding C codes in Listings 9.12 and 9.13.

In Listing 9.12, we first set the peripheral pins and port interrupt. Afterwards, we initialize and enable the EUSCI_B0 module in SPI master mode by the functions SPI_initMaster and SPI_enableModule. Here, we set the configuration structure such that:

```
selectClockSource = SMCLK
clockSourceFrequency = 3MHZ
desiredSpiClock = 1MHZ
```

The EUSCI_B0 module generates 1-MHz SPI clock based on these values. The configuration structure sets the MSB bit to be sent first in three-pin mode, clock inactivity polarity is set high, and the data is sent on the first clock edge. Then, we enable the EUSCI_B0 SPI receive interrupt using the function SPI_enableInterrupt. In the port ISR, we enable the EUSCI_B0 SPI transmit interrupt and turn on LED1 to indicate that the module is ready to send data. Then, we clear the enabled interrupt flag, read received values using the function SPI_receiveData, transmit the values using the function SPI_transmitData, and disable the EUSCI_B0 SPI transmit interrupt using the function SPI_disableInterrupt in the EUSCI_B0 ISR. As we read the received value, LED1 turns off.

In Listing 9.13, we set the peripheral pins. Afterwards, we initialize the EUSCI_B0 module in SPI slave mode by the function SPI_initSlave. Here, we set the configuration structure such that the MSB bit is received first, three-pin mode is used, clock

Listing 9.11 The Slave Receiver Code for the SPI Example in Three-Pin Mode, Using Registers

```c
#include "msp.h"

#define LED2_RED BIT0
#define LED2_GREEN BIT1
#define LED2_BLUE BIT2

 char RXData;

void main(void){

 WDT_A->CTL = WDT_A_CTL_PW | WDT_A_CTL_HOLD;

 P2->DIR = LED2_RED|LED2_GREEN|LED2_BLUE;
 P2->OUT = 0x00;
 P1->SEL0 |= BIT5 | BIT6 | BIT7; // Set P1.5, P1.6 and P1.7 as UCB0CLK,
    UCB0SIMO and UCB0SOMI

 EUSCI_B0->CTLW0 |= EUSCI_B_CTLW0_SWRST; // Hold EUSCI_B0 module in
    reset state
 EUSCI_B0->CTLW0 |= EUSCI_B_CTLW0_SYNC|EUSCI_B_CTLW0_CKPL|
    EUSCI_B_CTLW0_MSB;
 EUSCI_B0->CTLW0 &= ~EUSCI_B_CTLW0_SWRST; // Clear SWRST to resume
    operation

 EUSCI_B0->IFG &= ~EUSCI_B_IFG_RXIFG; // Clear EUSCI_B0 RX interrupt
    flag
 EUSCI_B0->IE |= EUSCI_B_IE_RXIE; // Enable EUSCI_B0 RX interrupt

 NVIC->ISER[0] = 0x00100000; // EUSCI_B0 interrupt is enabled in NVIC
 __enable_irq(); // All interrupts are enabled

 SCB->SCR |= SCB_SCR_SLEEPONEXIT_Msk; // Sleep on exit
 __sleep(); // enter LPM0
}

void EUSCIB0_IRQHandler(void){
 uint32_t status = EUSCI_B0->IFG; // Get EUSCI_B0 interrupt flag

 if(status & EUSCI_A_IFG_RXIFG){ // Check if receive interrupt occurs
  RXData = EUSCI_B0->RXBUF;  // Load current RXData value to transmit
    buffer
  if( RXData == 'R') P2->OUT ^= LED2_RED; // Toggle P2.0 if 'R' is
    received
  else if( RXData == 'G') P2->OUT ^= LED2_GREEN; // Toggle P2.1 if 'G'
    is received
  else if( RXData == 'B') P2->OUT ^= LED2_BLUE; // Toggle P2.2 if 'B'
    is received
 EUSCI_B0->IE |= EUSCI_B_IE_TXIE;} // Enable EUSCI_B0 TX interrupt
  else if(status & EUSCI_B_IFG_TXIFG){ // Check if transmit interrupt
    occurs
  EUSCI_B0->TXBUF = RXData; // Load current RXData value to transmit
    buffer
  EUSCI_B0->IE &= ~EUSCI_B_IE_TXIE;} // Disable EUSCI_B0 TX interrupt
}
```

Digital Communication   231

Listing 9.12 The Master Transmitter Code for the SPI Application in Three-Pin Mode, Using DriverLib Functions

```c
#include <ti/devices/msp432p4xx/driverlib/driverlib.h>

const eUSCI_SPI_MasterConfig spiMasterConfig =
{
 EUSCI_B_SPI_CLOCKSOURCE_SMCLK,   // SMCLK Clock Source
 3000000,                          // SMCLK = DCO = 3MHZ
 1000000,                          // SPICLK = 1MHZ
 EUSCI_B_SPI_MSB_FIRST,            // MSB First
 EUSCI_B_SPI_PHASE_DATA_CHANGED_ONFIRST_CAPTURED_ON_NEXT,   // Phase
 EUSCI_B_SPI_CLOCKPOLARITY_INACTIVITY_HIGH, // High polarity
 EUSCI_B_SPI_3PIN                  // 3Wire SPI Mode
};

 char TXData[10] = "RGBBGRBGGR";
 char RXData;
 int i = 0;

void main(void){

 WDT_A_holdTimer();

 GPIO_setAsPeripheralModuleFunctionInputPin(GPIO_PORT_P1, GPIO_PIN5 |
    GPIO_PIN6 | GPIO_PIN7, GPIO_PRIMARY_MODULE_FUNCTION);
 GPIO_setAsOutputPin(GPIO_PORT_P1, GPIO_PIN0);
 GPIO_setOutputLowOnPin(GPIO_PORT_P1, GPIO_PIN0);
 GPIO_setAsInputPinWithPullUpResistor(GPIO_PORT_P1, GPIO_PIN1);
 GPIO_clearInterruptFlag(GPIO_PORT_P1, GPIO_PIN1);
 GPIO_enableInterrupt(GPIO_PORT_P1, GPIO_PIN1);

 SPI_initMaster(EUSCI_B0_BASE, &spiMasterConfig);
 SPI_enableModule(EUSCI_B0_BASE);
 SPI_enableInterrupt(EUSCI_B0_BASE, EUSCI_B_SPI_RECEIVE_INTERRUPT);

 Interrupt_enableInterrupt(INT_PORT1);
 Interrupt_enableInterrupt(INT_EUSCIB0);
 Interrupt_enableSleepOnIsrExit();
 Interrupt_enableMaster();

 PCM_setPowerState(PCM_LPM0_LDO_VCORE1);
}

void EUSCIB0_IRQHandler(void){
 uint32_t status = SPI_getEnabledInterruptStatus(EUSCI_B0_BASE);
 SPI_clearInterruptFlag(EUSCI_B0_BASE, status);

 if(status & EUSCI_B_SPI_RECEIVE_INTERRUPT){
  RXData = SPI_receiveData(EUSCI_B0_BASE);
  GPIO_setOutputLowOnPin(GPIO_PORT_P1, GPIO_PIN0);}
 else if(status & EUSCI_B_SPI_TRANSMIT_INTERRUPT){
  SPI_transmitData(EUSCI_B0_BASE, TXData[i++%10]);
  SPI_disableInterrupt(EUSCI_B0_BASE, EUSCI_B_SPI_TRANSMIT_INTERRUPT);}
}

void PORT1_IRQHandler(void){
 uint32_t status = GPIO_getEnabledInterruptStatus(GPIO_PORT_P1);
 GPIO_clearInterruptFlag(GPIO_PORT_P1, status);

 if(status & GPIO_PIN1){
  GPIO_setOutputHighOnPin(GPIO_PORT_P1, GPIO_PIN0);
  SPI_enableInterrupt(EUSCI_B0_BASE, EUSCI_B_SPI_TRANSMIT_INTERRUPT);}
}
```

Listing 9.13 The Slave Receiver Code for the SPI Application in Three-Pin Mode, Using DriverLib Functions

```c
#include <ti/devices/msp432p4xx/driverlib/driverlib.h>

const eUSCI_SPI_SlaveConfig spiSlaveConfig =
{
 EUSCI_B_SPI_MSB_FIRST,                                    // MSB First
 EUSCI_B_SPI_PHASE_DATA_CHANGED_ONFIRST_CAPTURED_ON_NEXT,  // Phase
 EUSCI_B_SPI_CLOCKPOLARITY_INACTIVITY_HIGH,                // Normal Polarity
 EUSCI_B_SPI_3PIN                                          // 3wire mode
};

 char RXData;
 int i = 0;

void main(void){

 WDT_A_holdTimer();

 GPIO_setAsPeripheralModuleFunctionInputPin(GPIO_PORT_P1, GPIO_PIN5 |
     GPIO_PIN6 | GPIO_PIN7, GPIO_PRIMARY_MODULE_FUNCTION);
 GPIO_setAsOutputPin(GPIO_PORT_P2, GPIO_PIN0 | GPIO_PIN1 | GPIO_PIN2);
 GPIO_setOutputLowOnPin(GPIO_PORT_P2, GPIO_PIN0 | GPIO_PIN1 | GPIO_PIN2
     );

 SPI_initSlave(EUSCI_B0_BASE, &spiSlaveConfig);
 SPI_enableModule(EUSCI_B0_BASE);
 SPI_enableInterrupt(EUSCI_B0_BASE, EUSCI_B_SPI_RECEIVE_INTERRUPT);

 Interrupt_enableInterrupt(INT_EUSCIB0);
 Interrupt_enableSleepOnIsrExit();
 Interrupt_enableMaster();

 PCM_setPowerState(PCM_LPM0_LDO_VCORE1);
}

void EUSCIB0_IRQHandler(void){
 uint32_t status = SPI_getEnabledInterruptStatus(EUSCI_B0_BASE);
 SPI_clearInterruptFlag(EUSCI_B0_BASE, status);

 if(status & EUSCI_B_SPI_RECEIVE_INTERRUPT){
  RXData = SPI_receiveData(EUSCI_B0_BASE);
  if(RXData == 'R') GPIO_toggleOutputOnPin(GPIO_PORT_P2, GPIO_PIN0);
  else if(RXData == 'G') GPIO_toggleOutputOnPin(GPIO_PORT_P2,
     GPIO_PIN1);
  else if(RXData == 'B') GPIO_toggleOutputOnPin(GPIO_PORT_P2,
     GPIO_PIN2);
  SPI_enableInterrupt(EUSCI_B0_BASE, EUSCI_B_SPI_TRANSMIT_INTERRUPT);}
  else if(status & EUSCI_B_SPI_TRANSMIT_INTERRUPT){
   SPI_transmitData(EUSCI_B0_BASE, RXData);
   SPI_disableInterrupt(EUSCI_B0_BASE, EUSCI_B_SPI_TRANSMIT_INTERRUPT);}
}
```

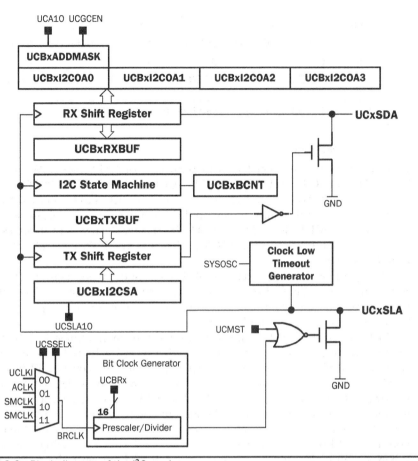

FIGURE 9.6 Block diagram of the I²C mode.

inactivity polarity is high, and data is captured on the first clock edge. We enable the EUSCI_B0 SPI receive interrupt. In the EUSCI_B0 ISR, we clear the enabled interrupt flag, read received values using the function `SPI_receiveData` and toggle the red, green, and blue colors of LED2 according to the received data.

9.4 Inter-Integrated Circuit

Inter-integrated circuit (I²C) is the second synchronous communication mode supported by MSP432. It can be used between multiple master and slave devices. The master and slave devices are represented by address values in I²C. Besides, a simple protocol establishes an effective communication between multiple master and slave devices. The block diagram of the I²C mode is as in Fig. 9.6.

As can be seen in Fig. 9.6, the I²C mode has two bidirectional pins for communication. These are the serial data (SDA) pin and serial clock (SCL) pin. These pins must be connected to the positive supply voltage (V_{CC}) via external pull-up resistors having values around 10 kΩ. Unlike SPI, here the transmit and receive operations are done on a single line. This saves pins, but slows down the communication speed. More information on I²C can be found in [19].

Bits	15	14	13	11	10-9	8
	UCA10	UCSLA10	UCMM	UCMST	UCMODEx	UCSYNC
Bits	7-6	4	3	2	1	0
	UCSSELx	UCTR	UCTXNACK	UCTXSTP	UCTXSTT	UCSWRST

TABLE 9.15 eUSCI_Bx Control Word Register 0 (UCBxCTLW0)

9.4.1 I²C Registers

There are eighteen 16-bit registers available to control the I²C mode. We will explain only UCBxCTLW0, UCBxCTLW1, UCBxBRW, UCBxSTATW, UCBxTBCNT, UCBxRXBUF, UCBxTXBUF, UCBxI2COA0, UCBxI2CSA, UCAxIE, UCAxIFG, and UCAxIV registers in this book. For further information on the remaining registers, please see [4]. A structure named EUSCI_Bx is used to control I²C registers. The registers of interest are named CTLW0, CTLW1, BRW, STATW, TBCNT, RXBUF, TXBUF, I2COA0, I2CSA, IE, IFG, and IV in this structure.

The I²C mode is configured by the eUSCI_Bx control word register 0 (UCBxCTLW0). To note here, the UCBxCTLW0 register is also used in the SPI mode. Here, it is used for I²C with different entries. The reader should be aware of this overlap. The entries of the UCBxCTLW0 register are tabulated in Table 9.15.

In Table 9.15, the UCA10 bit is used to select the own-address length of the device. When this bit is set with the constant EUSCI_B_CTLW0_A10, 10-bit address is used. Otherwise, seven-bit address is used. The UCSLA10 bit (with constant EUSCI_B_CTLW0_SLA10) is used to set the slave address length similar to the UCA10 bit settings. The UCMM bit (with constant EUSCI_B_CTLW0_MM) is used to choose the master device number. This bit should be reset if there is only one master in the system. Otherwise, it should be set to indicate that more than one master device will be used in communication. The UCMST bit is used to decide on whether the device is acting as master or slave. When this bit is set with the constant EUSCI_B_CTLW0_MST, the device will be used as master. Otherwise, the device will be used as slave. The UCMODEx bits are used to select the synchronous communication mode. For I²C, they should be set to the constant EUSCI_B_CTLW0_MODE_3. Finally, the UCSYNC bit is used to choose the communication mode. When this bit is reset, asynchronous mode is chosen. When it is set with the constant EUSCI_B_CTLW0_SYNC, synchronous mode is chosen. Therefore, this bit must be set for the I²C mode. UCSSELx bits are used to select the I²C clock. Constants for these bits are EUSCI_B_CTLW0_UCSSEL_0 (for UCLKI), EUSCI_B_CTLW0_UCSSEL_1 (for ACLK) and EUSCI_B_CTLW0_UCSSEL_2 (for SMCLK). After the clock is selected, it can be divided by the 16-bit coefficient UCBRx as in SPI mode. The UCTR bit is used to select whether the device is a transmitter or receiver. When this bit is set with the constant EUSCI_B_CTLW0_TR, the device becomes a transmitter. When it is reset, the device becomes a receiver. The UCTXNACK bit is used to adjust the not acknowledge (NACK) bit setting. When this bit is set with the constant EUSCI_B_CTLW0_TXNACK, the NACK bit is generated. When this bit is reset, sending an ACK bit occurs normally. The UCTXNACK bit is reset automatically after the NACK bit is sent. More information on these settings can be found in [19]. UCTXSTT and UCTXSTP bits are used to transmit start and stop conditions, respectively. When these bits are set with constants EUSCI_B_CTLW0_TXSTT and EUSCI_B_CTLW0_TXSTP, start or stop conditions are generated. These can be produced only by the master device. Therefore, they are not used

Bits	3-2	1-0
	UCASTPx	UCGLITx

TABLE 9.16 eUSCI_Bx Control Word Register 1 (UCBxCTLW1)

Bits	15-8	4
	UCBCNTx	UCBBUSY

TABLE 9.17 eUSCI_Bx Status Register (UCBxSTATW)

in slave mode. When these bits are set in the master receiver mode to generate a repeated start or stop condition, they are followed by a NACK bit. The UCTXSTT bit is reset automatically after the start condition and the address data is sent. The UCTXSTP bit is reset automatically after the stop condition is generated. The UCSWRST bit is used to reset the USCI module. When this bit is set with the constant EUSCI_B_CTLW0_SWRST, the eUSCI module is reset. When the UCSWRST bit is reset, the eUSCI module will be ready for operation.

The eUSCI_Bx control word register 1 (UCBxCTLW1) can be used for additional configurations. The entries of this register are tabulated in Table 9.16. In this table, the UCASTPx bits are used to generate automatic stop condition. Constants for these bits are EUSCI_B_CTLW1_ASTP_0, EUSCI_B_CTLW1_ASTP_1, and EUSCI_B_CTLW1_ASTP_2. When the first constant is used, the stop condition is generated by setting the UCTXSTP bit. It is not automatically generated in this mode. When the second constant is used, UCBCNTIFG is set when the byte counter reaches the threshold value defined in the UCBxTBCNT register. When the third constant is used, stop condition is generated automatically after the byte counter reaches the threshold value defined in the UCBxTBCNT register. UCBCNTIFG is also set when the byte counter reaches the threshold value. The UCGLITx bits are used to select the glitch filter length for both SDA and SCL lines. Constants for these bits are EUSCI_B_CTLW1_GLIT_0 (pulses with maximum 50-nanosecond duration are filtered), EUSCI_B_CTLW1_GLIT_1 (pulses with maximum 25-nanosecond duration are filtered), EUSCI_B_CTLW1_GLIT_2 (pulses with maximum 12.5-nanosecond duration are filtered), and EUSCI_B_CTLW1_GLIT_3 (pulses with maximum 6.25-nanosecond duration are filtered). Glitch filter with 50-nanosecond length is the best choice for I^2C standards. For further information on the remaining bits of this register, please see [4].

The I^2C mode has a status register called UCBxSTATW that is specifically used to observe the changes in the system. The entries of this register are given in Table 9.17. In this table, the UCBCNTx bits are used to hold the byte counter value. After each start or restart condition, the number of bytes received or transmitted is kept in these bits. The UCBBUSY bit is used to indicate that whether the I^2C bus (clk and data) is busy or not. When this bit is reset, it means the bus is inactive. When it is set, it means that the bus is active. The status of this bit can be checked by the constant EUSCI_B_STATW_BBUSY. For further information on the remaining bits of this register, please see [4].

The I^2C mode has a register to keep the threshold value to generate automatic stop condition called eUSCI_Bx byte counter threshold register (UCBxTBCNT). The number written in this register is compared with the UCBCNTx byte counter. The automatic stop condition is generated after the UCBCNTx byte counter reaches the value in

UCBxTBCNT if UCASTPx bits are set to the value 2 with the constant EUSCI_B_CTLW1_ASTP_2.

The I²C mode has two additional address registers. These are eUSCI_Bx I²C own address register 0 (UCBxI2COA0) and eUSCI_Bx I²C slave address register (UCBxI2CSA). UCBxI2COA0 keeps the device's own address. It is used when the device is set as slave. The fifteenth bit of this register (UCGCEN) is used to respond to a general call. When this bit is set with the constant EUSCI_B_I2COA0_GCEN, the device responds to a general call. When it is reset, the device does not respond to any general calls. Also, the tenth bit of this register (UCOAEN) is used to enable or disable entered own address. When this bit is set with the constant EUSCI_B_I2COA0_OAEN, the own address is enabled. Otherwise, it is disabled. UCB0I2CSA keeps the address of the slave device to be connected by the master. Therefore, it is used when the device is in master mode. When the master device wants to communicate with another slave device, this register must be filled with the address of the new slave device.

9.4.1.1 I²C Transmit/Receive Operations

Data transfer in the I²C mode is carried out on a byte basis. Every bit of the byte is transferred during one SCL pulse. Communication starts when the master device sends the start condition to the slave device. This is done by generating a high to low transition on the SDA, while SCL is high. Then, the slave address is transmitted by the master device in the next one or two bytes according to the addressing mode.

In the seven-bit addressing mode, the address information is sent in one byte. In this byte, first seven bits represent the slave address. The eight bit is $R/\overline{W}$. When the $R/\overline{W}$ bit is zero, it means that the master will transmit data to the slave. When the $R/\overline{W}$ bit is one, it means that the master will receive data from the slave. After this byte is transmitted by the master, the slave sends an acknowledge (ACK) bit to the master to indicate that the address information is received. Actually, this ACK bit is sent by the receiver (master or slave) after each received byte throughout communication to show that the transmitted byte is received.

In the 10-bit addressing mode, the address information is sent in two bytes. The first byte is formed by the constant binary number 11110, first two bits of the slave address, and the $R/\overline{W}$ bit. The second byte contains the remaining eight bits of the slave address. After receiving each byte, the slave sends an ACK bit. After the address information is acknowledged by the slave device, data is transmitted or received byte by byte according to the $R/\overline{W}$ bit. As in the seven-bit addressing mode, the ACK bit is sent by the receiver (master or slave) after receiving each byte.

The UCBBUSY bit is set to indicate that the bus is busy during the communication period. As data transfer is complete, the communication halts. This is done by the master device via sending the stop condition (a low to high transition on the SDA while the SCL is high).

Sometimes, the direction of data transfer has to be changed during I²C communication. This can be achieved by sending a start condition followed by the address information and new $R/\overline{W}$ bit after an ACK bit anywhere in data transfer. This way, the direction can be changed without stoping data transfer since no stop condition is generated. This is called restart condition.

There are four transmit/receive operation options for the I²C mode. These are slave transmitter, slave receiver, master transmitter, and master receiver. We explore each in detail next.

In the **slave transmitter mode**, the device is set as slave by setting UCSYNC and resetting UCMST bits. The slave must be set as receiver by resetting the UCTR bit. This is done to get the address information from the master. Then, the slave address is written to the UCBxI2COA0 register. This address can be either seven or 10 bits long based on the UCA10 bit value. After the start condition is detected by the slave, its own address is compared with the received one coming from the master (from the UCBxI2CSA register). If both addresses match with a set $R/\overline{W}$ bit, UCSTTIFG is set and the slave is configured as the transmitter by setting the UCTR bit automatically. The UCTXIFG0 bit is also set automatically in this step. Then, the first data bit is written to the UCBxTXBUF and an ACK bit is sent by the slave to indicate that the address information is acknowledged. Afterwards, UCTXIFG0 is reset and the data byte is transmitted. UCTXIFG0 is set again as soon as the data in UCBxTXBUF is transferred to the transmit shift register. After the data is transmitted to the master, there are three options for the system. First, the master can send an ACK bit, new data byte is transmitted, and the transfer proceeds. Second, the master can send a NACK bit followed by the stop condition to end data transfer. UCSTPIFG is set after the stop condition is received by the slave. Third, the master can send a NACK bit followed by the restart condition to restart the data transfer. Then, the data transfer cycle returns to the step where the start condition and address information is received by the slave.

The **slave receiver mode** has the same configuration steps with the previous one. Only in the slave receiver mode, the master is configured as transmitter and received $R/\overline{W}$ bit is reset. If both addresses match with a reset $R/\overline{W}$ bit, UCSTTIFG is set, UCSTPIFG is reset, and the slave is configured as receiver by resetting the UCTR bit automatically. UCRXIFG0 is set automatically after the first data byte is received. Then, the received data is read from UCBxRXBUF and an ACK bit is sent by the slave. There are four options for the system after the ACK bit is sent. First, the master can transfer a new data byte and the transfer proceeds. Second, the master can send a stop condition. Here, UCSTPIFG is set after the stop condition. Third, the master can send a restart condition. Then, the data transfer cycle returns to the step where the start condition and address information is received by the slave. Fourth, the slave device can send a NACK bit instead of an ACK bit to the master if the UCTXNACK bit is set during the last data cycle. The master device must respond to this by generating a stop or restart condition. If a NACK is transmitted before the last data in UCBxRXBUF is read, new data is written to the UCBxRXBUF and the last data is lost. To prevent this, the data in UCBxRXBUF must be read before UCTXNACK is set. After the NACK is transmitted, UCTXNACK bit is reset automatically. UCSTPIFG is set after the stop condition. If a restart condition occurs, data transfer cycle returns to the step where the start condition and address information is received by the slave. The fourth option can be used if the slave wants to restart or stop the communication.

In the **master transmitter mode**, first the device is set as master by setting UCSYNC and UCMST bits. Then, the target slave address is written to the UCBxI2CSA register in accordance with the UCSLA10 bit (7- or 10-bit addressing modes). Also the UCTR bit must be set to indicate that the master is used as the transmitter. The master generates a start condition to initiate the communication if the UCTXSTT bit is set by software. When this start condition is generated, UCTXIFG0 is set to show that UCBxTXBUF is ready for new data. Then, the slave address is transmitted with the $R/\overline{W}$ bit being reset. An ACK bit is expected from the slave as the first data byte is written to UCBxTXBUF. The UCTXSTT bit is reset automatically after the ACK bit is received and the UCTXSTT flag is reset automatically as soon as the complete address is sent. Also, UCTXIFG0 is set again as soon as data in UCBxTXBUF is transferred to the transmit shift register. Then, data byte is transmitted from the master to the slave. After this transmission, there are four options.

First, the slave can send an ACK bit, new data byte is transmitted, and transfer proceeds. Second, the master can generate a stop condition after the last ACK bit is received from the slave if UCTXSTP is set. When the data is transferred from UCBxTXBUF to the transmit shift register, UCTXIFG0 is set to show that data transmission has started and the UCTXSTP bit may be set. UCTXSTP is reset automatically after the stop condition is generated. When EUSCI_B_CTLW1_ASTP_2 constant is used for UCASTPx bits, the byte counter is used for generating stop condition and the user does not need to set the UCTXSTP bit. Third, the master can generate a restart condition after the last ACK bit is received from the slave if UCTXSTT is set. Then, the data transfer cycle returns to the step where the start condition and address information is received by the slave. If desired, UCTR and UCBxI2CSA can be changed here. Fourth, the slave can send a NACK bit. This sets the UCNACKIFG bit. The master must respond to this by generating a stop or restart condition. The data in UCBxTXBUF is discarded here. If this data needs to be transmitted after a restart condition, it must be rewritten to UCBxTXBUF. The fourth option can be used if the slave wants to restart or stop the communication.

The following scenario may occur in the first address transmit operation by the master. If the address information cannot be acknowledged by the slave, it sends a NACK bit to the master and UCNACKIFG is set. The master device must respond to this by generating a stop or restart condition. This is also the case for the master receiver mode to be explained next.

The **master receiver mode** has the same configuration steps with the previous one. Only the UCTR bit must be reset to indicate that the master is used as a receiver. Here, the master generates a start condition to initiate the communication if the UCTXSTT bit is set by software. Then, the slave address is transmitted with $R/\overline{W} = 1$. As the ACK bit is received from the slave (for the address information), the UCTXSTT bit is reset automatically and the first data byte can be received. The UCTXSTT flag is also reset automatically as soon as the complete address is sent. After this byte is received, the ACK bit is sent by the master and UCRXIFG0 is set to indicate that the data is loaded to UCBxRXBUF. After the UCRXIFG0 bit is set, there are three options for the system. First, the master can send an ACK bit, the new data byte is received, and the transfer proceeds. Second, the master can generate a stop condition by setting UCTXSTP and sending a NACK bit. Here, UCTXSTP is reset automatically after the stop condition is generated. Third, the master can generate a restart condition by setting UCTXSTT and sending a NACK bit. Here, the data transfer cycle returns to the step where the start condition and address information is received by the slave. If desired, UCTR and UCBxI2CSA can be changed here.

9.4.1.2 I²C Interrupts

I²C has three registers to control transmit and receive interrupts. These are eUSCI_Bx interrupt enable register (UCBxIE), eUSCI_Bx interrupt flag register (UCBxIFG), and eUSCI_Bx interrupt vector register (UCBxIV).

The eUSCI_Bx interrupt enable register (UCBxIE) is used to enable interrupts for different I²C conditions. The entries of this register are given in Table 9.18. The UCBCNTIE bit (with constant EUSCI_B_IE_BCNTIE) is used to enable byte counter interrupt.

Bits	6	5	3	2	1	0
	UCBCNTIE	UCNACKIE	UCSTPIE	UCSTTIE	UCTXIE0	UCRXIE0

TABLE 9.18 eUSCI_Bx Interrupt Enable Register (UCBxIE)

Bits	6	5	3	2	1	0
	UCBCNTIFG	UCNACKIFG	UCSTPIFG	UCSTTIFG	UCTXIFG0	UCRXIFG0

TABLE 9.19 eUSCI_Bx Interrupt Flag Register (UCBxIFG)

UCNACKIE bit (with constant EUSCI_B_IE_NACKIE) is used to enable not acknowledge interrupt. UCSTPIE and UCSTTIE bits (with constants EUSCI_B_IE_STPIE and EUSCI_B_IE_STTIE) are used to enable interrupts for stop and start conditions, respectively. Finally, UCTXIE0 and UCRXIE0 bits (with constants EUSCI_B_IE_TXIE0 and EUSCI_B_IE_RXIE0) are used to enable transmit and receive interrupts, respectively. For further information on the remaining bits of this register, please see [4].

The eUSCI_Bx interrupt flag register (UCBxIFG) is used to keep flags for different I^2C interrupts. The entries of this register are given in Table 9.19. UCBCNTIFG bit (with constant EUSCI_B_IFG_BCNTIFG) is the interrupt flag for byte counter interrupt. UCNACKIFG bit (with constant EUSCI_B_IFG_NACKIFG) is the interrupt flag for not acknowledge interrupt. UCSTPIFG and UCSTTIFGE bits (with constants EUSCI_B_IFG_STPIFG and EUSCI_B_IFG_STTIFG) are the flags for stop and start condition interrupts, respectively. Finally, UCTXIFG0 and UCRXIFG0 bits (with constants EUSCI_B_IFG_TXIFG0 and EUSCI_B_IFG_RXIFG0) are the flags for transmit and receive interrupts, respectively. For further information on the remaining bits of this register, please see [4].

The UCBxIV register is the interrupt vector used for I^2C interrupts. The content of this register is given in Table 9.20.

The interrupt-based communication operation in the I^2C is the same as in SPI mode. The only difference is the UCTXIFG0 bit. This bit should be set for different conditions for master and slave modes. In the master mode, UCTXIFG0 is set when UCTXSTT bit was set by the user. In the slave mode, UCTXIFG0 is set when own address or start bit was received.

9.4.2 eUSCI Module in I^2C Mode via DriverLib Functions

The eUSCI module in I^2C mode can be controlled via DriverLib functions given below. Here, moduleInstance is the eUSCI base to be initialized and controlled. It can be one of EUSCI_B0_BASE, EUSCI_B1_BASE, EUSCI_B2_BASE, and EUSCI_B3_BASE values. The complete function list and macro definitions can be found in [11].

```
void I2C_initMaster (uint32_t moduleInstance, const
   eUSCI_I2C_MasterConfig *config)
/*
Initializes the eUSCI module in I2C master mode by modifying the
   UCBxCTL0, UCBxCTL1, UCBxBR0, and UCBxBR1 registers.
config: the configuration structure. It contains clock, desired I2C
   clock, I2C data rate, auto stop values. Please see the datasheet
   for more information.
*/
```

UCIVx Content	Interrupt Source	Interrupt Flag	Interrupt Priority
0x00	No interrupt pending	–	–
0x02	Arbitration lost	UCALIFG	Highest
0x04	Not acknowledgment	UCNACKIFG	
0x06	Start condition received	UCSTTIFG	
0x08	Stop condition received	UCSTPIFG	
0x0A	Slave 3 data received	UCRXIFG3	
0x0C	Slave 3 transmit buffer empty	UCTXIFG3	
0x0E	Slave 2 data received	UCRXIFG2	
0x10	Slave 2 transmit buffer empty	UCTXIFG2	
0x12	Slave 1 data received	UCRXIFG1	
0x14	Slave 1 transmit buffer empty	UCTXIFG1	
0x16	Data received	UCRXIFG0	
0x18	Transmit buffer empty	UCTXIFG0	
0x1A	Byte counter zero	UCBCNTIFG	
0x1C	Clock low timeout	UCCLTOIFG	
0x1E	Ninth bit position	UCBIT9IFG	Lowest

TABLE 9.20 UCBxIV Register

```
void I2C_initSlave (uint32_t moduleInstance, uint_fast16_t
    slaveAddress, uint_fast8_t slaveAddressOffset, uint32_t
    slaveOwnAddressEnable)
/*
Initializes the eUSCI module in I2C slave mode by modifying the
    UCBxCTL0, UCBxCTL1, and UCBxI2COAx registers.
slaveAddress: the slave address to be used in I2C communication. It can
    be a 7 or 10-bit value.
slaveAddressOffset: the address offset. It defines the x of UCBxI2COAx
    register to be used. It can be one of
    EUSCI_B_I2C_OWN_ADDRESS_OFFSET0, EUSCI_B_I2C_OWN_ADDRESS_OFFSET1,
    EUSCI_B_I2C_OWN_ADDRESS_OFFSET2, or EUSCI_B_I2C_OWN_ADDRESS_OFFSET3
    values.
slaveOwnAddressEnable: enables or disables the selected address.
*/

void I2C_enableModule (uint32_t moduleInstance)
/*
Enables the eUSCI module in I2C mode by modifying the UCSWRST bit of
    the UCBxCTL1 register.
*/

void I2C_disableModule (uint32_t moduleInstance)
/*
Disables the eUSCI module in I2C mode by modifying the UCSWRST bit of
    the UCBxCTL1 register.
*/
```

```
void I2C_setSlaveAddress (uint32_t moduleInstance, uint_fast16_t
    slaveAddress)
/*
Sets the slave address that data to be sent in I2C master mode by
    modifying the UCBxI2CSA register.
slaveAddress: the 7 or 10-bit address to be set.
*/

void I2C_setMode (uint32_t moduleInstance, uint_fast8_t mode)
/*
Sets the I2C transmit or receive mode by modifying the UCTR bit of the
    UCBxCTL1 register.
mode: the mode to be set. It can be one of EUSCI_B_I2C_TRANSMIT_MODE or
    EUSCI_B_I2C_RECEIVE_MODE values.
*/

void I2C_slavePutData (uint32_t moduleInstance, uint8_t transmitData)
/*
Sends a byte from a slave device by modifying the UCBxTXBUF register.
transmitData: 8-bit data to be sent.
*/

uint8_t I2C_slaveGetData (uint32_t moduleInstance)
/*
Returns the byte sent to slave device by reading the UCBxRXBUF
    register.
*/

uint8_t I2C_isBusBusy (uint32_t moduleInstance)
/*
Returns the I2C status by reading the UCBBUSY bit in the UCBxSTAT
    register.
*/

void I2C_masterSendSingleByte (uint32_t moduleInstance, uint8_t txData)
/*
Sends start condition, a single byte and stop condition from a master
    device by modifying the UCBxIE, UCBxCTL1, UCBxIFG, UCBxTXBUF, and
    UCBxIE registers.
txData: 8-bit data to be sent.
*/

void I2C_masterSendMultiByteStart (uint32_t moduleInstance, uint8_t
    txData)
/*
Sends start condition and the first byte from a master device by
    modifying the UCBxIE, UCBxCTL1, UCBxIFG, UCBxTXBUF and UCBxIE
    registers.
txData: the first 8-bit data to be sent.
*/

void I2C_masterSendMultiByteNext (uint32_t moduleInstance, uint8_t
    txData)
/*
```

242 Chapter Nine

```
Continues to send multiple bytes from a master device by modifying the
    UCBxTXBUF register.
txData: the single 8-bit data to be sent next.
*/

void I2C_masterSendMultiByteStop (uint32_t moduleInstance)
/*
Sends a stop condition from the master device by modifying the UCTXSTP
    bit of the UCBxCTL1 register.
*/

void I2C_masterReceiveStart (uint32_t moduleInstance)
/*
Starts the reception of data in I2C master mode by modifying the
    UCTXSTT bit of the UCBxCTL1 register.
*/

uint8_t I2C_masterReceiveMultiByteNext (uint32_t moduleInstance)
/*
Returns the received data from a multi byte transmission by reading the
    UCBxRXBUF register.
*/

void I2C_masterReceiveMultiByteStop (uint32_t moduleInstance)
/*
Sends stop condition at the end of the multi byte reception at master
    side by modifying the UCTXSTP bit of the UCBxCTL1 register.
*/

uint8_t I2C_masterReceiveSingleByte (uint32_t moduleInstance)
/*
Sends start and stop condition and receives a single byte from the
    master device by modifying the UCBxIE, UCBxCTL1, UCBxIFG, UCBxTXBUF
    , and UCBxIE registers.
*/

void I2C_enableInterrupt (uint32_t moduleInstance, uint_fast16_t mask)
/*
Enables I2C interrupts by modifying the UCAxIFG and UCAxIE registers.
mask: the interrupt type to be enabled. It can be logical OR form of
    EUSCI_SPI_RECEIVE_INTERRUPT and EUSCI_SPI_TRANSMIT_INTERRUPT
    values.
*/

void I2C_disableInterrupt (uint32_t moduleInstance, uint_fast16_t mask)
/*
Disables I2C interrupts by modifying the UCAxIFG and UCAxIE registers.
mask: the interrupt type to be disabled. It can be logical OR form of
    EUSCI_SPI_RECEIVE_INTERRUPT and EUSCI_SPI_TRANSMIT_INTERRUPT
    values.
*/

void I2C_clearInterruptFlag (uint32_t moduleInstance, uint_fast16_t
    mask)
/*
```

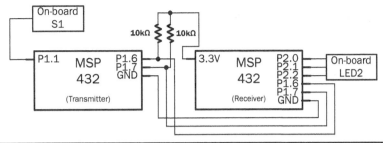

FIGURE 9.7 The connection diagram for the master transmitter and slave receiver I²C example.

```
Clears I2C interrupts by modifying the UCAxIFG register.
mask: the interrupt type to be cleared. It can be logical OR form of
    EUSCI_SPI_RECEIVE_INTERRUPT and EUSCI_SPI_TRANSMIT_INTERRUPT
    values.
*/

uint_fast16_t I2C_getEnabledInterruptStatus (uint32_t moduleInstance)
/*
Returns the status of enabled interrupt flags by reading UCAxIFG and
    UCAxIE registers.
*/
```

9.4.3 Coding Practices for the I²C Mode

We provide sample C codes on the usage of I²C communication mode in this section. In these, the button S1 connected to pin P1.1 on the MSP432 LaunchPad is used as input. The red, green, and blue colors of LED2 connected to pins P2.0, P2.1, and P2.2 on the MSP432 LaunchPad are used as output.

9.4.3.1 Register-Level Examples

In the first example, the I²C mode is used to establish a digital communication between two MSP432 LaunchPads. The connection diagram for this example is given in Fig. 9.7. The C code for the master transmitter device is given in Listing 9.14. The C code for the slave receiver device is given in Listing 9.15. In this example, when the button S1 of the master device is pressed, it sends the next character from the TXData array to control the red, green, and blue colors of LED2 on the slave device.

The I²C mode is again used to establish a digital communication between the two MSP432 LaunchPads in Listings 9.16 and 9.17. The connection diagram for this example is given in Fig. 9.8. This time the slave becomes the transmitter and the master becomes the receiver. The C code for the slave transmitter device is given in Listing 9.16. The C code for the master receiver device is given in Listing 9.17. In this example, when the button S1 of the master device is pressed, slave sends the next character from the TXData array to control the red, green, and blue colors of LED2 on the master device.

9.4.3.2 DriverLib Examples

We next repeat the examples in Listings 9.14 and 9.15 using DriverLib functions. We provide the corresponding C codes in Listings 9.18 and 9.19, respectively.

In Listing 9.18, we first set the peripheral pins and port interrupt. Afterwards, we initialize and enable the EUSCI_B0 module in I²C master mode by the functions I2C_initMaster and I2C_enableModule. Then, we set the mode as transmitter by the

244 Chapter Nine

Listing 9.14 The Master Transmitter I²C Example, Using Registers

```c
#include "msp.h"

#define S1 BIT1
#define SLAVE_ADDRESS 0x48

 char TXData[10] = "RGBBGRBGGR";
 int i = 0;

void main(void){

  WDT_A->CTL = WDT_A_CTL_PW | WDT_A_CTL_HOLD;

  P1->DIR &= ~S1;
  P1->REN = S1;
  P1->OUT = S1;
  P1->IE = S1;
  P1->IES = S1;
  P1->IFG = 0x00;
  P1->SEL0 |= BIT6 | BIT7; // P1.6 and P1.7 as UCB0SDA and UCB0SCL

  EUSCI_B0->CTLW0 |= EUSCI_B_CTLW0_SWRST; // Hold EUSCI_B0 module in
      reset state
  EUSCI_B0->CTLW0 |= EUSCI_B_CTLW0_MODE_3|EUSCI_B_CTLW0_MST|
      EUSCI_B_CTLW0_SYNC;
  EUSCI_B0->CTLW0 |= EUSCI_B_CTLW0_UCSSEL_2; // Select SMCLK as EUSCI_B0
      clock
  EUSCI_B0->BRW = 0x001E; // Set BITCLK = BRCLK / (UCBRx+1) = 3 MHz / 30
      = 100 kHz
  EUSCI_B0->I2CSA = SLAVE_ADDRESS;
  EUSCI_B0->CTLW0 &= ~EUSCI_B_CTLW0_SWRST; // Clear SWRST to resume
      operation

  NVIC->ISER[1] = 0x00000008; // Port P1 interrupt is enabled in NVIC
  NVIC->ISER[0] = 0x00100000; // EUSCI_B0 interrupt is enabled in NVIC
  __enable_irq(); // All interrupts are enabled

  while (EUSCI_B0->CTLW0 & EUSCI_B_CTLW0_TXSTP);
  EUSCI_B0->CTLW0 |= EUSCI_B_CTLW0_TR | EUSCI_B_CTLW0_TXSTT;
  SCB->SCR |= SCB_SCR_SLEEPONEXIT_Msk; // Sleep on exit
  __sleep(); // enter LPM0
}

void EUSCIB0_IRQHandler(void){
 uint32_t status = EUSCI_B0->IFG; // Get EUSCI_B0 interrupt flag
 EUSCI_B0->IFG &=~ EUSCI_B_IFG_TXIFG0; // Clear EUSCI_B0 TX interrupt
     flag
 if(status & EUSCI_B_IFG_TXIFG0){ // Check if transmit interrupt occurs
  EUSCI_B0->TXBUF = TXData[i++%10]; // Load current TXData value to
      transmit buffer
  EUSCI_B0->IE &= ~EUSCI_B_IE_TXIE0;} // Disable EUSCI_B0 TX interrupt
}

void PORT1_IRQHandler(void){
 uint32_t status = P1->IFG;
 P1->IFG &= ~S1;
 if(status & S1) EUSCI_B0->IE |= EUSCI_B_IE_TXIE0; // Enable EUSCI_A0
     TX interrupt
}
```

Listing 9.15 The Slave Receiver I²C Example, Using Registers

```c
#include "msp.h"

#define LED2_RED BIT0
#define LED2_GREEN BIT1
#define LED2_BLUE BIT2

#define SLAVE_ADDRESS 0x48

char RXData;

void main(void){

 WDT_A->CTL = WDT_A_CTL_PW | WDT_A_CTL_HOLD;

 P2->DIR = LED2_RED|LED2_GREEN|LED2_BLUE;
 P2->OUT = 0x00;
 P1->SEL0 |= BIT6 | BIT7; // P1.6 and P1.7 as UCB0SDA and UCB0SCL

 EUSCI_B0->CTLW0 |= EUSCI_B_CTLW0_SWRST; // Hold EUSCI_B0 module in
     reset state
 EUSCI_B0->CTLW0 |= EUSCI_B_CTLW0_MODE_3|EUSCI_B_CTLW0_SYNC;
 EUSCI_B0->I2COA0 = SLAVE_ADDRESS | EUSCI_B_I2COA0_OAEN;
 EUSCI_B0->CTLW0 &= ~EUSCI_B_CTLW0_SWRST; // Clear SWRST to resume
     operation
 EUSCI_B0->IFG &= ~EUSCI_B_IFG_RXIFG0; // Clear EUSCI_B0 RX interrupt
     flag
 EUSCI_B0->IE |= EUSCI_B_IE_RXIE0; // Enable EUSCI_B0 RX interrupt

 NVIC->ISER[0] = 0x00100000; // EUSCI_B0 interrupt is enabled in NVIC
 __enable_irq(); // All interrupts are enabled

 SCB->SCR |= SCB_SCR_SLEEPONEXIT_Msk; // Sleep on exit
 __sleep(); // enter LPM0
}

void EUSCIB0_IRQHandler(void){
 uint32_t status = EUSCI_B0->IFG; // Get EUSCI_B0 interrupt flag
 EUSCI_B0->IFG &=~ EUSCI_B_IFG_RXIFG0; // Clear EUSCI_B0 RX interrupt
     flag
 if(status & EUSCI_B_IFG_RXIFG0){ // Check if receive interrupt occurs
  RXData = EUSCI_B0->RXBUF;   // Load current RXData value to transmit
      buffer
  if(RXData == 'R') P2->OUT ^= LED2_RED; // Toggle P2.0 if 'R' is
      received
   else if(RXData == 'G') P2->OUT ^= LED2_GREEN; // Toggle P2.1 if 'G'
       is received
   else if(RXData == 'B') P2->OUT ^= LED2_BLUE;} // Toggle P2.2 if 'B'
       is received
}
```

Chapter Nine

Listing 9.16 The Slave Transmitter I²C Example, Using Registers

```c
#include "msp.h"

#define SLAVE_ADDRESS 0x48

 char TXData[10] = "RGBBGRBGGR";
 int i = 0;

void main(void){

 WDT_A->CTL = WDT_A_CTL_PW | WDT_A_CTL_HOLD;

 P1->SEL0 |= BIT6 | BIT7; // P1.6 and P1.7 as UCB0SDA and UCB0SCL

 EUSCI_B0->CTLW0 |= EUSCI_B_CTLW0_SWRST; // Hold EUSCI_B0 module in
     reset state
 EUSCI_B0->CTLW0 |= EUSCI_B_CTLW0_MODE_3|EUSCI_B_CTLW0_SYNC;
 EUSCI_B0->I2COA0 = SLAVE_ADDRESS | EUSCI_B_I2COA0_OAEN;
 EUSCI_B0->CTLW0 &= ~EUSCI_B_CTLW0_SWRST; // Clear SWRST to resume
     operation
 EUSCI_B0->IE |= EUSCI_B_IE_STTIE; // Enable EUSCI_B0 Start interrupt

 NVIC->ISER[0] = 0x00100000; // EUSCI_B0 interrupt is enabled in NVIC
 __enable_irq(); // All interrupts are enabled

 SCB->SCR |= SCB_SCR_SLEEPONEXIT_Msk; // Sleep on exit
 __sleep(); // enter LPM0
}

void EUSCIB0_IRQHandler(void){
 uint32_t status = EUSCI_B0->IFG; // Get EUSCI_B0 interrupt flag
 if(status & EUSCI_B_IFG_STTIFG){ // Check if start interrupt occurs
  EUSCI_B0->IE |= EUSCI_B_IE_TXIE0; // Enable EUSCI_B0 TX0 interrupt
  EUSCI_B0->TXBUF = TXData[i++%10];} // Load current TXData value to
      transmit buffer
 if(status & EUSCI_B_IFG_TXIFG0){ // Check if transmit interrupt occurs
  EUSCI_B0->IE &= ~EUSCI_B_IE_TXIE0; // Disable EUSCI_B0 TX0 interrupt
  EUSCI_B0->TXBUF = 0;} // Load dummy value to transmit buffer
}
```

Listing 9.17 The Master Receiver I²C Example, Using Registers

```c
#include "msp.h"

#define S1 BIT1
#define SLAVE_ADDRESS 0x48

#define LED2_RED BIT0
#define LED2_GREEN BIT1
#define LED2_BLUE BIT2

 char RXData;

void main(void){

 WDT_A->CTL = WDT_A_CTL_PW | WDT_A_CTL_HOLD;
```

```c
        P2->DIR = LED2_RED|LED2_GREEN|LED2_BLUE;
        P2->OUT = 0x00;
        P1->DIR &= ~S1;
        P1->REN = S1;
        P1->OUT = S1;
        P1->IE = S1;
        P1->IES = S1;
        P1->IFG = 0x00;
        P1->SEL0 |= BIT6 | BIT7; // P1.6 and P1.7 as UCB0SDA and UCB0SCL

        EUSCI_B0->CTLW0 |= EUSCI_B_CTLW0_SWRST; // Hold EUSCI_B0 module in
            reset state
        EUSCI_B0->CTLW0 |= EUSCI_B_CTLW0_MODE_3|EUSCI_B_CTLW0_MST|
            EUSCI_B_CTLW0_SYNC;
        EUSCI_B0->CTLW0 |= EUSCI_B_CTLW0_UCSSEL_2; // Select SMCLK as EUSCI_B0
            clock
        EUSCI_B0->BRW = 0x001E; // Set BITCLK = BRCLK / (UCBRx+1) = 3 MHz / 30
            = 100 kHz
        EUSCI_B0->I2CSA = SLAVE_ADDRESS;
        EUSCI_B0->CTLW0 &= ~EUSCI_B_CTLW0_SWRST; // Clear SWRST to resume
            operation

        NVIC->ISER[1] = 0x00000008; // Port P1 interrupt is enabled in NVIC
        NVIC->ISER[0] = 0x00100000; // EUSCI_B0 interrupt is enabled in NVIC
        __enable_irq(); // All interrupts are enabled

        SCB->SCR |= SCB_SCR_SLEEPONEXIT_Msk; // Sleep on exit
        __sleep(); // enter LPM0
}

void EUSCIB0_IRQHandler(void){
 uint32_t status = EUSCI_B0->IFG; // Get EUSCI_B0 interrupt flag
 EUSCI_B0->IFG &=~ EUSCI_B_IFG_RXIFG0; // Clear EUSCI_B0 RX interrupt
     flag
 if(status & EUSCI_B_IFG_RXIFG0){ // Check if receive interrupt occurs
  RXData = EUSCI_B0->RXBUF;   // Load current RXData value to transmit
      buffer
  if(RXData == 'R') P2->OUT ^= LED2_RED; // Toggle P2.0 if 'R' is
      received
  else if(RXData == 'G') P2->OUT ^= LED2_GREEN; // Toggle P2.1 if 'G'
      is received
  else if(RXData == 'B') P2->OUT ^= LED2_BLUE; // Toggle P2.2 if 'B' is
      received
  EUSCI_B0->IE &= ~EUSCI_B_IE_RXIE0;} // Disable EUSCI_B0 RX interrupt
}

void PORT1_IRQHandler(void){
 uint32_t status = P1->IFG;
 P1->IFG &= ~S1;

 if(status & S1){
  EUSCI_B0->IE |= EUSCI_B_IE_RXIE0; // Enable EUSCI_B0 RX interrupt
  EUSCI_B0->CTLW0 |= EUSCI_B_CTLW0_TXSTT;}
}
```

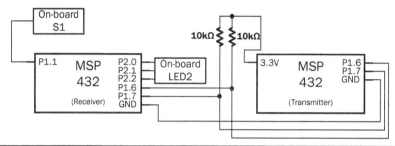

FIGURE 9.8 The connection diagram for the master receiver and slave transmitter I²C example.

function I2C_setMode and set the slave address by the function I2C_setSlaveAddress. In the port ISR, we enable the EUSCI_B0 I²C transmit interrupt using the function I2C_enableInterrupt. Before sending data to slave, we check whether the master has sent the stop signal and send a single byte to slave using I2C_masterIsStopSent and I2C_masterSendSingleByte. Finally, we clear the enabled interrupt flag and disable the EUSCI_A0 I²C transmit interrupt using the I2C_disableInterrupt function in the EUSCI_B0 ISR.

In Listing 9.19, again we set the peripheral pins. Afterwards, we initialize and enable the EUSCI_B0 module in I²C slave mode by the functions I2C_initSlave and I2C_enableModule. Then, we set the mode as receive by the function I2C_setMode. In the EUSCI_B0 ISR function, we clear the enabled interrupt flag, read received values using the function I2C_slaveGetData and toggle the red, green, and blue colors of LED2 according to the received data.

We next repeat the examples in Listings 9.16 and 9.17 using DriverLib functions. We provide the corresponding C codes in Listings 9.20 and 9.21, respectively.

In Listing 9.20, we first set the peripheral pins and port interrupt. Afterwards, we initialize and enable the EUSCI_B0 module in I²C slave mode and set the mode as transmitter. In the EUSCI_B0 ISR function, we clear the enabled interrupt flag, send the data to master, and disable the transmit interrupt.

In Listing 9.21, we set the peripheral pins. Afterwards, we initialize and enable the EUSCI_B0 module in the I²C master mode. Then, we set the mode as receive. In port ISR, we enable the EUSCI_B0 I²C receive interrupt. Afterwards, we start the master receiver using the I2C_masterReceiveStart function. We clear the enabled interrupt flag, read received values using the function I2C_masterReceiveSingle, and toggle the red, green, and blue colors of LED2 according to received data in the EUSCI_B0 ISR.

9.5 Application: Flame Detector

The aim of this application is to learn how to use the digital communication modes on the MSP432 microcontroller. As a real-life application, we will design a flame detector for smart homes. In this section, we provide the equipment list, layout of the circuit, and system design specifications.

9.5.1 Equipment List

We will use the digital BH1750 light sensor and MCP9808 temperature sensor in the application. BH1750 is a light-intensity sensor that gives digital output via its I²C interface. The BH1750 light sensor starts up at power down mode after power supply is

Digital Communication

Listing 9.18 The Master Transmitter I²C Example, Using DriverLib Functions

```c
#include <ti/devices/msp432p4xx/driverlib/driverlib.h>

#define SLAVE_ADDRESS 0x48

const eUSCI_I2C_MasterConfig i2cConfig =
{
 EUSCI_B_I2C_CLOCKSOURCE_SMCLK,        // SMCLK Clock Source
 3000000,                              // SMCLK = 3MHz
 EUSCI_B_I2C_SET_DATA_RATE_100KBPS,    // Desired I2C Clock of 400khz
 0,                                    // No byte counter threshold
 EUSCI_B_I2C_NO_AUTO_STOP              // No Autostop
};

 char TXData[10] = "RGBBGRBGGR";
 int i = 0;

void main(void){

 WDT_A_holdTimer();

 GPIO_setAsPeripheralModuleFunctionInputPin(GPIO_PORT_P1, GPIO_PIN6 |
     GPIO_PIN7, GPIO_PRIMARY_MODULE_FUNCTION);
 GPIO_setAsInputPinWithPullUpResistor(GPIO_PORT_P1, GPIO_PIN1);
 GPIO_clearInterruptFlag(GPIO_PORT_P1, GPIO_PIN1);
 GPIO_enableInterrupt(GPIO_PORT_P1, GPIO_PIN1);

 I2C_initMaster(EUSCI_B0_BASE, &i2cConfig);
 I2C_enableModule(EUSCI_B0_BASE);
 I2C_setMode(EUSCI_B0_BASE, EUSCI_B_I2C_TRANSMIT_MODE);
 I2C_setSlaveAddress(EUSCI_B0_BASE, SLAVE_ADDRESS);
 I2C_enableInterrupt(EUSCI_B0_BASE, EUSCI_B_I2C_NAK_INTERRUPT);

 Interrupt_enableInterrupt(INT_PORT1);
 Interrupt_enableInterrupt(INT_EUSCIB0);
 Interrupt_enableSleepOnIsrExit();
 Interrupt_enableMaster();

 PCM_setPowerState(PCM_LPM0_LDO_VCORE1);
}

void EUSCIB0_IRQHandler(void){
 uint32_t status = I2C_getEnabledInterruptStatus(EUSCI_B0_BASE);

 I2C_clearInterruptFlag(EUSCI_B0_BASE, status);

 if (status & EUSCI_B_I2C_TRANSMIT_INTERRUPT0)
   I2C_disableInterrupt(EUSCI_B0_BASE, EUSCI_B_I2C_TRANSMIT_INTERRUPT0);
}

void PORT1_IRQHandler(void){
 uint32_t status = GPIO_getEnabledInterruptStatus(GPIO_PORT_P1);
 GPIO_clearInterruptFlag(GPIO_PORT_P1, status);

 if(status & GPIO_PIN1){
  I2C_enableInterrupt(EUSCI_B0_BASE, EUSCI_B_I2C_TRANSMIT_INTERRUPT0);
  while (MAP_I2C_masterIsStopSent(EUSCI_B0_BASE) ==
      EUSCI_B_I2C_SENDING_STOP);
  I2C_masterSendSingleByte(EUSCI_B0_BASE,TXData[i++%10]);}
}
```

Listing 9.19 The Slave Receiver I²C Example, Using DriverLib Functions

```c
#include <ti/devices/msp432p4xx/driverlib/driverlib.h>

#define SLAVE_ADDRESS 0x48

 char RXData;
 int i = 0;

void main(void){

 WDT_A_holdTimer();

 GPIO_setAsPeripheralModuleFunctionInputPin(GPIO_PORT_P1, GPIO_PIN6 |
     GPIO_PIN7, GPIO_PRIMARY_MODULE_FUNCTION);
 GPIO_setAsOutputPin(GPIO_PORT_P2, GPIO_PIN0 | GPIO_PIN1 | GPIO_PIN2);
 GPIO_setOutputLowOnPin(GPIO_PORT_P2, GPIO_PIN0 | GPIO_PIN1 |
    GPIO_PIN2);

 I2C_initSlave(EUSCI_B0_BASE, SLAVE_ADDRESS,
     EUSCI_B_I2C_OWN_ADDRESS_OFFSET0, EUSCI_B_I2C_OWN_ADDRESS_ENABLE);
 I2C_enableModule(EUSCI_B0_BASE);
 I2C_setMode(EUSCI_B0_BASE, EUSCI_B_I2C_RECEIVE_MODE);
 I2C_enableInterrupt(EUSCI_B0_BASE, EUSCI_B_I2C_RECEIVE_INTERRUPT0);

 Interrupt_enableInterrupt(INT_EUSCIB0);
 Interrupt_enableSleepOnIsrExit();
 Interrupt_enableMaster();

 PCM_setPowerState(PCM_LPM0_LDO_VCORE1);
}

void EUSCIB0_IRQHandler(void){
 uint32_t status = I2C_getEnabledInterruptStatus(EUSCI_B0_BASE);
 I2C_clearInterruptFlag(EUSCI_B0_BASE, status);

 if (status & EUSCI_B_I2C_RECEIVE_INTERRUPT0){
  RXData = I2C_slaveGetData(EUSCI_B0_BASE);
  if(RXData == 'R') GPIO_toggleOutputOnPin(GPIO_PORT_P2, GPIO_PIN0);
  else if(RXData == 'G') GPIO_toggleOutputOnPin(GPIO_PORT_P2,
     GPIO_PIN1);
  else if(RXData == 'B') GPIO_toggleOutputOnPin(GPIO_PORT_P2,
     GPIO_PIN2);}
}
```

connected. The user should first send power-up command. Then, the start command should be sent to the device. Afterwards, the result can be read directly in the unsigned integer format. This value should be converted to luminance value [20].

MCP9808 is a temperature sensor that gives digital output via its I²C interface. Within the sensor, the temperature value can be read from the register (0x05) as 13-bit two's complement form. This value should be converted to degrees Celsius [21].

Listing 9.20 The Slave Transmitter I²C Example, Using DriverLib Functions

```c
#include <ti/devices/msp432p4xx/driverlib/driverlib.h>

#define SLAVE_ADDRESS 0x48

 char TXData[10] = "RGBBGRBGGR";
 int i = 0;

void main(void){

 WDT_A_holdTimer();

 GPIO_setAsPeripheralModuleFunctionInputPin(GPIO_PORT_P1, GPIO_PIN6 |
    GPIO_PIN7, GPIO_PRIMARY_MODULE_FUNCTION);

 I2C_initSlave(EUSCI_B0_BASE, SLAVE_ADDRESS,
    EUSCI_B_I2C_OWN_ADDRESS_OFFSET0, EUSCI_B_I2C_OWN_ADDRESS_ENABLE);
 I2C_enableModule(EUSCI_B0_BASE);
 I2C_enableInterrupt(EUSCI_B0_BASE, EUSCI_B_I2C_START_INTERRUPT );

 Interrupt_enableSleepOnIsrExit();
 Interrupt_enableInterrupt(INT_EUSCIB0);
 PCM_setPowerState(PCM_LPM0_LDO_VCORE1);
}
void EUSCIB0_IRQHandler(void){
 uint_fast16_t status = I2C_getEnabledInterruptStatus(EUSCI_B0_BASE);
 I2C_clearInterruptFlag(EUSCI_B0_BASE, status);

 if (status & EUSCI_B_I2C_START_INTERRUPT ){
  I2C_enableInterrupt(EUSCI_B0_BASE, EUSCI_B_I2C_TRANSMIT_INTERRUPT0);
  I2C_slavePutData(EUSCI_B0_BASE, TXData[i++%10]);}
 else if(status & EUSCI_B_I2C_TRANSMIT_INTERRUPT0){
  I2C_disableInterrupt(EUSCI_B0_BASE, EUSCI_B_I2C_TRANSMIT_INTERRUPT0);
  I2C_slavePutData(EUSCI_B0_BASE, 0);}
}
```

9.5.2 Layout

The layout of the flame detector is given in Fig. 9.9.

9.5.3 System Design Specifications

The design of the flame detector will be as follows. The fire will be detected with light and temperature sensors. Sensor outputs will be continuously checked. An alert will be given using LED1 connected to pin P1.0 on the MSP432 LaunchPad. To raise an alert, the sensor outputs should exceed 30% of the average of last 100 measurements over 100 seconds.

We could not add the C code of the project here due to page limitations. However, we provided the complete CCS project for this application in the companion website for the reader. We strongly suggest implementing it.

Listing 9.21 The Master Receiver I²C Example, Using DriverLib Functions

```c
#include <ti/devices/msp432p4xx/driverlib/driverlib.h>

#define SLAVE_ADDRESS       0x48

const eUSCI_I2C_MasterConfig i2cConfig =
{
 EUSCI_B_I2C_CLOCKSOURCE_SMCLK,        // SMCLK Clock Source
 3000000,                              // SMCLK = 3MHz
 EUSCI_B_I2C_SET_DATA_RATE_100KBPS,    // Desired I2C Clock of 400khz
 0,                                    // No byte counter threshold
 EUSCI_B_I2C_NO_AUTO_STOP              // No Autostop
};

 char RXData;
 int i=0;

void main(void){

 WDT_A_holdTimer();

 GPIO_setAsPeripheralModuleFunctionInputPin(GPIO_PORT_P1, GPIO_PIN6 |
    GPIO_PIN7, GPIO_PRIMARY_MODULE_FUNCTION);
 GPIO_setAsOutputPin(GPIO_PORT_P2, GPIO_PIN0 | GPIO_PIN1 | GPIO_PIN2);
 GPIO_setOutputLowOnPin(GPIO_PORT_P2, GPIO_PIN0 | GPIO_PIN1 | GPIO_PIN2);
 GPIO_setAsInputPinWithPullUpResistor(GPIO_PORT_P1, GPIO_PIN1);
 GPIO_clearInterruptFlag(GPIO_PORT_P1, GPIO_PIN1);
 GPIO_enableInterrupt(GPIO_PORT_P1, GPIO_PIN1);

 I2C_initMaster(EUSCI_B0_BASE, &i2cConfig);
 I2C_setSlaveAddress(EUSCI_B0_BASE, SLAVE_ADDRESS);
 I2C_setMode(EUSCI_B0_BASE, EUSCI_B_I2C_RECEIVE_MODE);
 I2C_enableModule(EUSCI_B0_BASE);

 Interrupt_enableSleepOnIsrExit();
 Interrupt_enableInterrupt(INT_PORT1);
 Interrupt_enableInterrupt(INT_EUSCIB0);
 Interrupt_enableMaster();

 PCM_setPowerState(PCM_LPM0_LDO_VCORE1);
}

void EUSCIB0_IRQHandler(void){
 uint_fast16_t status = I2C_getEnabledInterruptStatus(EUSCI_B0_BASE);
 I2C_clearInterruptFlag(EUSCI_B0_BASE, status);

 if (status & EUSCI_B_I2C_RECEIVE_INTERRUPT0){
  i++;
  RXData = I2C_masterReceiveSingle(EUSCI_B0_BASE);
  if(RXData == 'R') GPIO_toggleOutputOnPin(GPIO_PORT_P2, GPIO_PIN0);
   else if(RXData == 'G') GPIO_toggleOutputOnPin(GPIO_PORT_P2, GPIO_PIN1);
   else if(RXData == 'B') GPIO_toggleOutputOnPin(GPIO_PORT_P2, GPIO_PIN2);
  I2C_disableInterrupt(EUSCI_B0_BASE, EUSCI_B_I2C_RECEIVE_INTERRUPT0);}
}

void PORT1_IRQHandler(void){
 uint32_t status = GPIO_getEnabledInterruptStatus(GPIO_PORT_P1);
 GPIO_clearInterruptFlag(GPIO_PORT_P1, status);

 if(status & GPIO_PIN1){
   I2C_enableInterrupt(EUSCI_B0_BASE, EUSCI_B_I2C_RECEIVE_INTERRUPT0);
   I2C_masterReceiveStart(EUSCI_B0_BASE);}
}
```

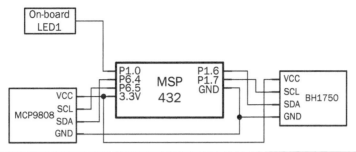

FIGURE 9.9 Layout of the flame detector.

9.6 Summary

MSP432 has digital communication capabilities. In this chapter, we explored these in detail. We started with the eUSCI_Ax and eUSCI_Bx modules available in the MSP432 microcontroller. The eUSCI_Ax supports UART and SPI communication modes. UART is the only asynchronous communication mode available in MSP432. SPI, on the other hand, is a synchronous and fast communication mode. eUSCI_Bx supports SPI and I^{2}C communication modes. Therefore, SPI is supported by the two USCI modules. I^{2}C is also synchronous. We explored each mode in detail in this chapter. We also provided sample C codes for the three communication modes. Finally, we provided an application on using the I^{2}C communication mode.

9.7 Exercises

In this section, we will not offer new problems. Instead, we will ask the reader to solve problems given in previous chapters using two MSP432 LaunchPad boards and establishing a digital communication link between them. Some sample problems are given below.

9.1 Solve Problem 5.5 using two MSP432 LaunchPad boards. The first MSP432 LaunchPad will be used for the button. The second will be used for the LEDs. Establish a digital communication link between these two boards using UART, SPI, and I^{2}C communication modes.

9.2 Solve Problem 5.6 using two MSP432 LaunchPad boards. The first MSP432 LaunchPad will be used for the button. The second will be used for the LEDs. Establish a digital communication link between these two boards using UART, SPI, and I^{2}C communication modes.

9.3 Solve Problem 6.2 using two MSP432 LaunchPad boards. The first MSP432 LaunchPad will be used for the button. The second will be used for the LEDs. Establish a digital communication link between these two boards using UART, SPI, and I^{2}C communication modes.

CHAPTER 10

Wireless Communication

Internet of things (IoT) applications have started changing all aspects of our lives for good. It is estimated that machines will use the Internet more than man in the near future. An IoT application needs two main tools such as microcontroller (with possible sensors on it) and wireless communication module to connect it to the Internet. Therefore, this chapter will focus on the wireless communication part of IoT. Here, we will focus on wireless local area network (WLAN), Bluetooth, ZigBee, RFID, GSM, and GPS. We will benefit from available modules to use these wireless communication systems. To do so, we will briefly introduce wireless communication first. Then, we will provide necessary background on each module followed by a dedicated application.

10.1 Background on Wireless Communication

As the name implies, wireless communication systems operate via transmission through space rather than a wired connection. This has the advantage of allowing users to send and receive data almost anywhere. Wireless communication systems differ according to the radio frequency (RF) and modulation techniques they are using [15]. Although these topics are not in the scope of this book, it is important to know some basic information about them. Therefore, we will briefly introduce them here.

Wireless communication systems use radio signals having various frequency values. The RF is the portion of the electromagnetic spectrum starting from about one kHz to three GHz. RF spectrum is allocated by government agencies in most countries. This means that using frequency bands without permission is not allowed. However, some frequency bands are unlicensed and can be used freely. These are referred as industrial, scientific, and medical (ISM) bands. As the name implies, they are reserved for the use of scientific, medical, and industrial applications. The ISM bands are defined by ITU radio regulations. By the way, ITU is the United Nations specialized agency for information and communication technologies. To note here, the ISM band allocation can vary in different countries due to local regulations. Only the 2.4 GHz band is standard all over the world.

We provide the operating frequency of popular wireless communication systems in Table 10.1. To note here, the common point of these systems is that they can be used in microcontroller applications. As can be seen in this table, wireless communication systems work in specific frequency bands. Besides, they can be classified according to their range such as short and long as can be seen in Table 10.1. WLAN, Bluetooth, ZigBee, and RFID are short-range systems. GSM and GPS are long-range systems.

Chapter Ten

Wireless System	Operation Frequency	Range
WLAN	2.4 GHz, 5 GHz	50–300 m
Bluetooth	2.4 GHz	2–100 m
ZigBee	868 MHz, 900 MHz, 2.4 GHz	50–1000 m
RFID	13.56 MHz	1–100 cm
GSM	800 MHz–2.4 GHz	35–70 km
GPS	1.2–1.5 GHz	20200 km

TABLE 10.1 Operation Frequency and Range of Wireless Communication Systems

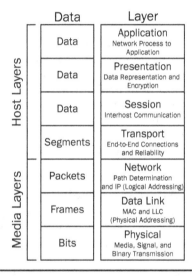

FIGURE 10.1 OSI model.

10.2 Wireless Local Area Network

Wireless local area network (WLAN) is the first wireless communication option to be considered in this chapter. Using it, we can connect two or more devices within a limited area. We first give some basic information on WLAN next. Then, we provide two sample applications using it.

10.2.1 Definition

WLAN is standardized by IEEE under number 802.11, which is often called as Wi-Fi [22]. All 802.11 Wi-Fi standards operate within the ISM frequency band. Hence, no license is required for them.

WLAN benefits from the open systems interconnection (OSI) model for communication. To be more specific, OSI defines how applications can communicate over a network. OSI consists of seven layers from low to high as physical, data link, network, transport, session, presentation, and application. We provide the schematic of these layers in Fig. 10.1. Here, lower levels are for hardware-based settings, whereas higher levels are for abstract communication protocols. More information on OSI can be found in [23].

Wireless Communication

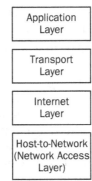

FIGURE 10.2 TCP/IP model.

Standard	Operation Frequency (GHz)	Maximum Bit Rate	Range (m)
802.11	2.4	2 Mbps	20–100
802.11a	5	54 Mbps	30–120
802.11b	2.4	11 Mbps	35–140
802.11g	2.4	54 Mbps	35–140
802.11n	2.4, 5	600 Mbps	70–250
802.11ac	5	7 Gbps	35–140

TABLE 10.2 Details of Physical Sublayers for IEEE 802.11 Standards

Transmission control protocol/Internet protocol (TCP/IP) is the network model used in the current Internet architecture. In a way, it is the implementation of the OSI model. TCP/IP consists of four layers from application to network access, as in Fig. 10.2. IEEE standard 802.11 defines a set of data link and physical layer protocols of OSI model or network interface of the TCP/IP model. More information on the TCP/IP model can be found in [24].

HTTP is the most commonly used application protocol of the OSI model. Basically, all websites and services use it to communicate with each other over Internet. HTTP is an application-level protocol since it sits on top of the TCP layer in the OSI model. HTTP works as follows. A client sends request to a Web server for Web elements such as Web pages, images, and text data. After the request is received, the server sends the Web elements requested by the client. Then, the connection between the client and server across Internet is ended. A new connection must be established for each request.

The IEEE standard 802.11 defines a series of encoding and transmission schemes for wireless communication at the physical sublayer. These are the 802.11, 802.11a, 802.11b, 802.11g, and 802.11n standards. The 802.11ac standard is the latest one providing raw data rates of up to 7 Gbps. Details of physical sublayers are given in Table 10.2.

10.2.2 WLAN Applications

We will benefit from Espressif's ESP8266 Wi-Fi chip for WLAN applications in this section. ESP8266 is an integrated Wi-Fi chip with an embedded microcontroller on it [25].

It can communicate with MSP432 through UART protocol. It has a specific AT command set that can be used in making a TCP/IP connection [26]. In this book, we use the ESP-01 module (having the ESP8266 chip on it) with baud rate 9600 bps.

10.2.2.1 Connecting to a Web Server

We will develop an intelligent umbrella holder in our first application. This device will connect to the `api.openweather.com` Web server to retrieve weather information in text form. Then, we parse the received text. If there is thunderstorm, rain, or snow our intelligent umbrella holder will warn us by onboard LEDs. To note here, the ESP-01 module should be connected to Internet before sending a request to the Web server. Hence, we need a Wi-Fi connected to Internet.

The connection between the MSP432 LaunchPad and ESP-01 module will be as follows. The MSP432 pins P3.2 (eUSCI A2 Rx) and P3.3 (eUSCI A2 Tx) will be connected to the ESP-01 module's TX and RX pins, respectively. The V_{CC} and GND pins of the ESP-01 module will be connected to the 3.3 V and GND pins of the MSP432 LaunchPad, respectively. The CH_PD pin of ESP-01 module will be also connected to the 3.3 V pin of the MSP432 LaunchPad.

We connect the MSP432 to Internet using the C code in Listing 10.1. In the first part of this C code, we define the variables to be used. Here, `ssid` and `pass` strings hold the name and password for the Wi-Fi access point; `server` and `url` strings hold the Web address to be connected. While compiling the code, please do not forget to write your SSID and password to the `ssid` and `pass` strings. The `RXData` string is used for receiving UART data from the ESP-01 module. The `TXData` string is used for transmitting UART data to the ESP-01 module. The `html` string is used for sending data to the Web server via the ESP-01 module.

In the main function in Listing 10.1, we first initialize the MSP432 microcontroller using the function `initMSP432` in Listing 10.2. Here, `uartConfig` is the UART configuration object to set the UART baud rate to 9600 bps at 12-MHz main clock. Then, we set up the eUSCI module in UART mode and eUSCI interrupts using UART functions in Sec. 9.2. We configure eUSCI A0 for host PC communication and eUSCI A2 for ESP-01 module communication. To note here, all data transfer between the Wi-Fi module and MSP432 is also sent to the host PC. Therefore, the reader can observe all AT commands submitted to form a Wi-Fi connection.

We initialize the ESP-01 module using the function `initESPclient` in Listing 10.3. In this function, we first configure the ESP-01 module in client mode using AT commands `RST` and `CWMODE`. Then, we configure and connect to the Wi-Fi using the AT command `CWJAP`. When the module is connected to the Wi-Fi, we get the IP address using the AT command `CIFSR`.

The function `initESPclient` consists of the `sendData` function in Listing 10.4. Within this function, we first clear the `RXData` string using the function `memset`. Then, we send data to the ESP-01 module using the function `UART_transmitData` in a while loop. Once all data is sent to the module, we wait for the ESP-01 module's response either with `acknowledge` or `error` messages during timeout. If we receive an error message, then we hold the program in a forever loop. Otherwise, the program continues.

The `sendData` function consists of the `delay_ms` function in Listing 10.5. This function gives approximately one-millisecond delay using the 12-MHz main clock.

We provide the eUSCI ISR in Listing 10.6. Here in eUSCIA0 ISR, we send data received from PC host to the ESP-01 module. Similarly, in eUSCIA2 ISR we send data

Listing 10.1 Usage of WLAN, Connecting to a Web Server

```c
#include <ti/devices/msp432p4xx/driverlib/driverlib.h>
#include <stdint.h>
#include <stdio.h>
#include <string.h>
#include <stdbool.h>

 char *ssid = "\"WiFiSSID\"";
 char *pass = "\"WiFiPass\"";
 char *server = "api.openweathermap.org";
 char *url = "/data/2.5/weather?q=istanbul&APPID=5
     c876b8580797df03e7137b4d79d14fb";

 char RXData[1000] = "";
 char TXData[1000] = "";
 int buf_ind = 0, i = 0;

 char html[500] = "";

void initMSP432(void);
void initESPclient(void);
void connectWebserver(void);
void sendData(char *data, char *ack, char *err, const int timeout);
void delay_ms(uint16_t ms);
char* itoa(int val);

void main(void) {

 WDT_A_holdTimer();

 GPIO_setAsOutputPin(GPIO_PORT_P2,GPIO_PIN0 | GPIO_PIN1| GPIO_PIN2);
 GPIO_setOutputLowOnPin(GPIO_PORT_P2, GPIO_PIN0 | GPIO_PIN1|
     GPIO_PIN2);

 initMSP432();
 initESPclient();

 while(1)
   connectWebserver();
}
void EUSCIA0_IRQHandler(void) {
     ...
}

void EUSCIA2_IRQHandler(void) {
     ...
}
```

received from the ESP-01 to host PC. We also save the received data from the ESP-01 module to the string `RXData`.

We connect to the Web server using the function `connectWebserver` in Listing 10.7. In this function, we establish a TCP/IP connection to the server using the AT command `CIPSTART`. Once we establish the connection, we send data using the AT command

Listing 10.2 initMSP432.c

```c
const eUSCI_UART_Config uartConfig =
{
 EUSCI_A_UART_CLOCKSOURCE_SMCLK,
 78,
 2,
 0,
 EUSCI_A_UART_NO_PARITY,
 EUSCI_A_UART_LSB_FIRST,
 EUSCI_A_UART_ONE_STOP_BIT,
 EUSCI_A_UART_MODE,
 EUSCI_A_UART_OVERSAMPLING_BAUDRATE_GENERATION
};

void initMSP432(void){

 GPIO_setAsPeripheralModuleFunctionInputPin(GPIO_PORT_P1,GPIO_PIN2 |
     GPIO_PIN3, GPIO_PRIMARY_MODULE_FUNCTION);
 GPIO_setAsPeripheralModuleFunctionInputPin(GPIO_PORT_P3,GPIO_PIN2 |
     GPIO_PIN3, GPIO_PRIMARY_MODULE_FUNCTION);

 PCM_setPowerState(PCM_AM_LDO_VCORE1);

 CS_setDCOCenteredFrequency(CS_DCO_FREQUENCY_12);

 UART_initModule(EUSCI_A0_BASE, &uartConfig);
 UART_initModule(EUSCI_A2_BASE, &uartConfig);
 UART_enableModule(EUSCI_A0_BASE);
 UART_enableModule(EUSCI_A2_BASE);
 UART_enableInterrupt(EUSCI_A0_BASE, EUSCI_A_UART_RECEIVE_INTERRUPT);
 UART_enableInterrupt(EUSCI_A2_BASE, EUSCI_A_UART_RECEIVE_INTERRUPT);

 Interrupt_enableInterrupt(INT_EUSCIA0);
 Interrupt_enableInterrupt(INT_EUSCIA2);
 Interrupt_enableMaster();
}
```

CIPSEND. Here, sent data is basically the URL address of the server that we receive weather information. Data also consists of the HTTP request method. In this application we will make a GET request. The URL is in the form of

```
/data/2.5/weather?q=your_city_name&APPID=your_ID
```

You can get a free ID by signing up to the website http://openweathermap.org. For more information on this website, please see [27, 28]. After sending a request to the server, we wait for a response from it. Then, we check whether the received data contains the AT command +IPD, which indicates that there is incoming data from the Web server. Once the server sends the response, we search for the weather condition ID. If there is a thunderstorm, rain, or snow, we toggle the red, green, and blue colors of LED2 connected to pins P2.0, P2.1, and P2.2 on the MSP432 LaunchPad accordingly. For more information on weather condition IDs, please see [29].

Listing 10.3 The `initESPclient` Function

```
void initESPclient(void){
 char ap[100] = "AT+CWJAP=";
 strcat(ap, ssid);strcat(ap, ",");
 strcat(ap, pass);strcat(ap, "\r\n");
 sendData("AT+RST\r\n","OK","ERROR",1000);
 delay_ms(2000);
 sendData("AT\r\n","OK","ERROR",1000);
 sendData("AT+CWMODE=1\r\n","OK","ERROR",1000);
 sendData(ap,"OK","ERROR",20000);
 sendData("AT+CIFSR\r\n","OK","ERROR",10000);
 delay_ms(3000);
}
```

Listing 10.4 The `sendData` Function

```
void sendData(char *data, char *ack, char *err, const int timeout){
 long int time = 0;
 memset(RXData, 0, sizeof(RXData));
 buf_ind = 0;

 while (*data != 0x00)
  UART_transmitData(EUSCI_A2_BASE, *data++);

 while( timeout > time){
  time++;
  delay_ms(1);
  if(strstr(RXData, ack)!= NULL) break;
  else if(strstr(RXData, err)!= NULL) while(1);}

 delay_ms(100);
}
```

Listing 10.5 The `delay_ms` Function

```
void delay_ms(uint16_t ms){
 uint16_t delay;
 volatile uint32_t i;
 for (delay = ms; delay >0 ; delay--)
  for (i=1200; i >0;i--);
}
```

10.2.2.2 Simple Web Server

In our second WLAN application, we create a simple HTTP Web server using MSP432 and the ESP-01 module. This server holds a simple Web page that shows the sensor values obtained in our leakage control application in Sec. 8.6. Hence, the user can observe the sensor values from a Web page.

The connection between the MSP432 LaunchPad and ESP-01 module will be as follows. The MSP432 pins P3.2 (eUSCI A2 Rx) and P3.3 (eUSCI A2 Tx) will be connected to the ESP-01 module's TX and RX pins, respectively. The V_{CC} and GND pins of the ESP-01

Listing 10.6 eUSCI ISR

```c
void EUSCIA0_IRQHandler(void) {
 uint8_t data;
 uint32_t status = UART_getEnabledInterruptStatus(EUSCI_A0_BASE);
 UART_clearInterruptFlag(EUSCI_A0_BASE, status);

 if(status & EUSCI_A_UART_RECEIVE_INTERRUPT_FLAG) {
  data = UART_receiveData(EUSCI_A0_BASE);
  UART_transmitData(EUSCI_A2_BASE, data);}
}

void EUSCIA2_IRQHandler(void) {
 uint8_t data;
 uint32_t status = UART_getEnabledInterruptStatus(EUSCI_A2_BASE);
 UART_clearInterruptFlag(EUSCI_A2_BASE, status);

 if(status & EUSCI_A_UART_RECEIVE_INTERRUPT_FLAG) {
  data = UART_receiveData(EUSCI_A2_BASE);
  UART_transmitData(EUSCI_A0_BASE, data);
  RXData[buf_ind++] = data;}
}
```

module will be connected to the 3.3 V and GND pins of the MSP432 LaunchPad, respectively. The CH_PD pin of ESP-01 module will be also connected to the 3.3 V pin of the MSP432 LaunchPad.

We provide the C code for the Web server application in Listing 10.8. In the first part of the code, we define the variables to be used in the code as in the previous application. Here, the sensor_value1 and sensor_value2 simulate the three sensor values obtained from the leakage control system application. Then, we initialize the MSP432 using the function initMSP432 in Listing 10.2. Afterwards, we initialize the ESP-01 module using the function initESPserver in Listing 10.9. In an infinite loop, we listen and handle the incoming client requests using the function handleClientWifi in Listing 10.10. Once the program runs, it will set up a Web server. You can observe connection and Wi-Fi status via host PC. Connect to the hotspot and open a Web page from a browser. Go to the IP address given in Wi-Fi status. Now, you can observe the status of sensor values.

We provide the initESPserver function in Listing 10.9. In this function, we configure the ESP-01 module in host mode using the AT commands RST and CWMODE. Then, we configure and create the Wi-Fi hotspot using the AT command CWSAP. When the Wi-Fi hotspot is created, we get the IP address using the AT command CIFSR. Finally, we configure the ESP-01 module as a Web server using the AT commands CIPMUX and CIPSERVER.

We provide the handleClientWifi function in Listing 10.10. Here, we wait to receive all data sent from the ESP-01 module and copy it to the string receiveddata. Then, we check whether the received data from the module consists of the AT command +IPD. This command indicates that there is incoming data from the Web client. Finally, we serve the html data (containing the sensor values) to be displayed on the client and close the connection using the AT commands CIPSEND and CIPCLOSE.

Listing 10.7 The connectWebserver Function

```
void connectWebserver(void){
 int received = 0;
 char receiveddata[1000] = "";

 strcpy(TXData,"AT+CIPSTART=\"TCP\",\"");
 strcat(TXData,server);
 strcat(TXData,"\",80\r\n");
 sendData(TXData,"OK","ERROR",5000);

 strcpy(html,"GET ");
 strcat(html,url);
 strcat(html," HTTP/1.1\r\nHost: ");
 strcat(html,server);
 strcat(html,"\r\nConnection: close\r\n\r\n");
 strcpy(TXData,"AT+CIPSEND=");
 strcat(TXData,itoa(strlen(html)));
 strcat(TXData,"\r\n");

 sendData(TXData,"OK","ERROR",5000);
 sendData(html,"OK","ERROR",5000);

 while(!received){
  if(buf_ind>0){
   delay_ms(1000);
   strncpy(receiveddata, RXData, buf_ind);
   memset(RXData, 0, sizeof(RXData));
   buf_ind = 0;}
  if(strstr(receiveddata, "+IPD") != NULL){
   if(strstr(receiveddata, "\"weather\":[{\"id\":5") != NULL){ // Rain
    GPIO_setOutputLowOnPin(GPIO_PORT_P2, GPIO_PIN0 | GPIO_PIN1|
        GPIO_PIN2);
    GPIO_setOutputHighOnPin(GPIO_PORT_P2,GPIO_PIN0);}
   else if(strstr(receiveddata, "\"weather\":[{\"id\":6") != NULL){ //
       Snow
    GPIO_setOutputLowOnPin(GPIO_PORT_P2, GPIO_PIN0 | GPIO_PIN1|
        GPIO_PIN2);
    GPIO_setOutputHighOnPin(GPIO_PORT_P2,GPIO_PIN1);}
   else if(strstr(receiveddata, "\"weather\":[{\"id\":2") != NULL){ //
       Thunderstorm
    GPIO_setOutputLowOnPin(GPIO_PORT_P2, GPIO_PIN0 | GPIO_PIN1|
        GPIO_PIN2);
    GPIO_setOutputHighOnPin(GPIO_PORT_P2,GPIO_PIN2);}
   else GPIO_setOutputHighOnPin(GPIO_PORT_P2, GPIO_PIN0 | GPIO_PIN1|
       GPIO_PIN2);
   received = 1;}}

 delay_ms(5000);
}
```

Listing 10.8 Usage of WLAN, Creating a Simple HTTP Server

```c
#include <ti/devices/msp432p4xx/driverlib/driverlib.h>
#include <stdint.h>
#include <stdio.h>
#include <string.h>
#include <stdbool.h>

 char *ssid = "\"msp432ap\"";
 char *pass = "\"12345678\"";
 char RXData[1000] = "";
 char TXData[1000] = "";
 int buf_ind = 0, i = 0;
 char *sensor_value1 = "300";
 char *sensor_value2 = "500";
 char *sensor_value3 = "800";
 char header[500];
 char html[500];

void initMSP432(void);
void initESPserver(void);
void sendData(char *data, char *ack, char *err, const int timeout);
void handleClientWifi(void);
void delay_ms(uint16_t ms);
char* itoa(int val);

void main(void) {

 WDT_A_holdTimer();

 initMSP432();
 initESPserver();

 while(1)
   handleClientWifi();
}

void EUSCIA0_IRQHandler(void) {
     ...
}

void EUSCIA2_IRQHandler(void) {
     ...
}
```

We provide the itoa function in Listing 10.11. In this function, we convert an integer value to a character array. To do so, we convert each digit in the integer to character in a for loop. Then, we save the result in a character array.

10.3 Bluetooth

Bluetooth is the second wireless communication option to be considered in this chapter. We first give some basic information on Bluetooth next. Then, we provide three sample applications using it.

Listing 10.9 The `initESPserver` Function

```
void initESPserver(void) {
 char ap[100] = "AT+CWSAP=";
 strcat(ap, ssid);
 strcat(ap, ",");
 strcat(ap, pass);
 strcat(ap, ",11,3\r\n");

 sendData("AT+RST\r\n","OK","ERROR",1000);
 delay_ms(2000);
 sendData("AT\r\n","OK","ERROR",1000);
 sendData("AT+CWMODE=2\r\n","OK","ERROR",1000);
 sendData(ap,"OK","ERROR",1000);
 sendData("AT+CIFSR\r\n","OK","ERROR",1000);
 sendData("AT+CIPMUX=1\r\n","OK","ERROR",1000);
 sendData("AT+CIPSERVER=1,80\r\n","OK","ERROR",1000);
 delay_ms(3000);
}
```

10.3.1 Definition

Bluetooth is a low-power, short-range wireless connectivity solution standardized by IEEE under number 802.15.1 [30]. IEEE no longer maintains Bluetooth. This is handled by the Bluetooth special interest group (SIG) [31]. Bluetooth operates at 2.4 to 2.485 GHz within the ISM frequency band. Hence, no license is required for its operation.

Bluetooth has two commonly used versions. These are Bluetooth basic rate/enhanced data rate (BR/EDR) and Bluetooth with low energy (LE). The former and latter were adopted as versions 2.0 / 2.1 and 4.0 / 4.1 / 4.2, respectively.

Bluetooth building blocks are defined by protocols in its core specification. These are the RF, link control (LC), link manager (LM), and logical link control and adaptation (L2CAP) protocols. These protocols are used in both Bluetooth BR/EDR and LE versions. The Bluetooth LE version also requires attribute (ATT), generic attribute profile (GATT), and generic access profile (GAP) protocols. The core specification blocks with additional LE protocols are as in Fig. 10.3. There are additional protocols defined in Bluetooth core specification. However, these are not essential. Also, there are non-core specifications defined by third-party groups.

Bluetooth divides data into packets and transmits each packet in one of its channels. The Bluetooth BR/EDR version has 79 designated channels each having 1-MHz bandwidth. The Bluetooth LE has 40 channels, each having 2-MHz bandwidth. Among these 40 channels, 37 are for data communication and 3 are for advertising.

A Bluetooth LE device may operate in different modes such as advertising, scanning, master device, and slave device. In the advertising mode, the Bluetooth LE device periodically transmits advertising information. If the device is in advertise-only mode, then it will not respond to any requests. However, if the device is in normal advertising mode, then it may respond to requests by other devices. When in scanning mode, the device listens advertising information transmitted by other devices and may request additional information if its active scan mode is enabled. If the device is in passive scan mode, then it will not request any additional information.

Once the connection is established between two Bluetooth LE devices, the connection initiator (scanner) becomes the master device. The advertiser becomes the slave

Listing 10.10 The `handleClientWifi` Function

```c
void handleClientWifi(void){
 char receiveddata[1000] = "";
 char id[1] = "";

 if(buf_ind>0){
  delay_ms(1000);
  strncpy(receiveddata, RXData, buf_ind);
  memset(RXData, 0, sizeof(RXData));
  buf_ind = 0;}

  if(strstr(receiveddata, "+IPD") != NULL){
   strncpy(id, strstr(receiveddata, "+IPD,")+5, 1);
   memset(receiveddata, 0, sizeof(receiveddata));

   strcpy(html,"HTTP/1.1 200 OK\r\nContent-Type: text/html\r\n\r\n");
   strcat(html,"<!DOCTYPE html><html><body><h1>MSP432 Webserver</h1>");
   strcat(html,"<p><h2>Home Status</h2></p>");
   strcat(html,"<br>Sensor Value 1: ");
   strcat(html, sensor_value1);
   strcat(html,"<br>Sensor Value 2: ");
   strcat(html, sensor_value2);
   strcat(html,"<br>Sensor Value 3: ");
   strcat(html, sensor_value3);
   strcat(html,"</body></html>");

   strcpy(TXData,"AT+CIPSEND=");
   strncat(TXData, id, 1);
   strcat(TXData,",");
   strcat(TXData,itoa(strlen(html)));
   strcat(TXData,"\r\n");
   sendData(TXData,"OK","ERROR",1000);
   sendData(html,"OK","ERROR",1000);

   strcpy(TXData,"AT+CIPCLOSE=");
   strncat(TXData, id, 1);
   strcat(TXData,"\r\n");
   sendData(TXData,"OK","ERROR",1000);}
}
```

device. Slave devices may have only one connection at a time, while master devices may have multiple connections with different slave devices simultaneously. This makes Bluetooth LE a perfect solution for IoT applications.

10.3.2 Bluetooth Applications

We provide three Bluetooth applications in this chapter. The first one is implemented on CSR's BC417 module (known as HC-05/06) for Bluetooth BR/EDR communication. The second and third applications are implemented on TI's CC2541 (know as HM-10) for the Bluetooth LE communication. HC-06 and HM-10 are integrated Bluetooth solutions with ready-to-use modules. They are connected to the microcontroller through its UART port and communicate with their AT command set [32, 33].

Listing 10.11 The `itoa` Function

```
char* itoa(int val){
 static char buf[32] = {0};
 int i = 0;

 if (val == 0){
  buf[i++] = '0';
  buf[i] = '\0';
  return buf;}

 i = 30;
 for(; val && i ; --i, val /= 10)
  buf[i] = "0123456789abcdef"[val % 10];

 return &buf[i+1];
}
```

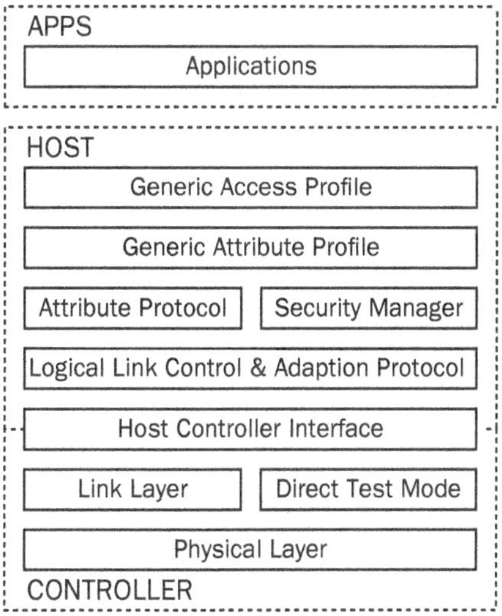

```
APPS
  Applications

HOST
  Generic Access Profile
  Generic Attribute Profile
  Attribute Protocol    Security Manager
  Logical Link Control & Adaption Protocol
  Host Controller Interface
  Link Layer           Direct Test Mode
  Physical Layer
CONTROLLER
```

FIGURE 10.3 Bluetooth building blocks.

10.3.2.1 Bluetooth BR/EDR Request Listener

In our first application, we will form an intelligent refrigerator module such that when eggs, milk, or fruits are below critical level in the refrigerator, it will request them directly from the market. We use the HC-06 module to form a Bluetooth request listener in our application. Through it, we will receive the status of the refrigerator. We will simulate the refrigerator with a Bluetooth enabled mobile device. We will send characters E, M, and F when eggs, milk, or fruits are below critical level, respectively. On the receiver side (MSP432 LaunchPad), there is a GSM module that sends messages to the market. For the sake of simplicity, we will simulate this operation by changing the color of LED2 on the MSP432 LaunchPad. Hence, we change the color when any of the characters E, M, and F come.

The connection between the MSP432 LaunchPad and HC-06 module will be as follows. The MSP432 pins P3.2 (eUSCI A2 Rx) and P3.3 (eUSCI A2 Tx) will be connected to

Listing 10.12 Bluetooth BR/EDR Request Listener

```c
#include <ti/devices/msp432p4xx/driverlib/driverlib.h>
#include <stdint.h>
#include <stdio.h>
#include <string.h>
#include <stdbool.h>

void initMSP432(void);
void handleClientBT(void);
void delay_ms(uint16_t ms);

 char RXData[1000] = "";
 char TXData[40] = "";
 int buf_ind = 0;

void main(void) {

 WDT_A_holdTimer();

 GPIO_setAsOutputPin(GPIO_PORT_P2,GPIO_PIN0 | GPIO_PIN1| GPIO_PIN2);
 GPIO_setOutputLowOnPin(GPIO_PORT_P2, GPIO_PIN0 | GPIO_PIN1|
     GPIO_PIN2);

 initMSP432();

 while(1)
    handleClientBT();
}

void EUSCIA0_IRQHandler(void) {
    ...
}

void EUSCIA2_IRQHandler(void) {
    ...
}
```

the HC-06 module's TX and RX pins, respectively. The V_{CC} and GND pins of the HC-06 module will be connected to the 3.3 V and GND pins of the MSP432 LaunchPad, respectively.

We provide the main program for our first Bluetooth application in Listing 10.12. In the first part of the code, we define the variables to be used. Here, the string RXData is used for receiving UART data from the HC-06 module. The string TXData is used for transmitting UART data to the module. In the main function, we first set up the red, green, and blue colors of LED2 connected to pins P2.0, P2.1, and P2.2 on the MSP432 LaunchPad as output and initialize the MSP432 using the function initMSP432 in Listing 10.2. In an infinite loop, we listen and handle the incoming client requests using the function handleClientBT in Listing 10.13. Once the MSP432 is powered up, a Bluetooth connection should be available. You can observe connection with Bluetooth-enabled mobile device (preferably Android based). Connect to the Bluetooth device and open a Bluetooth terminal. There are several Bluetooth terminal applications as freeware

Listing 10.13 The `handleClientBT` Function

```c
void handleClientBT(void){
 char receiveddata[1000] = "";

 if(buf_ind>0){
  delay_ms(1000);
  strncpy(receiveddata, RXData, buf_ind);
  memset(RXData, 0, sizeof(RXData));
  buf_ind = 0;}

 if(strstr(receiveddata, "E") != NULL || strstr(receiveddata, "e")
     != NULL){
  GPIO_setOutputHighOnPin(GPIO_PORT_P2, GPIO_PIN0);
  delay_ms(2000);
  GPIO_setOutputLowOnPin(GPIO_PORT_P2, GPIO_PIN0);}
 else if(strstr(receiveddata, "M") != NULL || strstr(receiveddata, "m")
     != NULL){
  GPIO_setOutputHighOnPin(GPIO_PORT_P2, GPIO_PIN1);
  delay_ms(2000);
  GPIO_setOutputLowOnPin(GPIO_PORT_P2, GPIO_PIN1);}
 else if(strstr(receiveddata, "F") != NULL || strstr(receiveddata, "f")
     != NULL){
  GPIO_setOutputHighOnPin(GPIO_PORT_P2, GPIO_PIN2);
  delay_ms(2000);
  GPIO_setOutputLowOnPin(GPIO_PORT_P2, GPIO_PIN2);}
}
```

for mobile phones. You can send characters (E, M, or F) through them as the simulation of our intelligent refrigerator to turn on red, green, and blue colors of LED2 on the MSP432 LaunchPad.

We provide the `handleClientBT` function in Listing 10.13. Here, we wait to receive all data sent from the HC-06 module and copy it to the string `receiveddata`. Then, we check whether the received data from the module contains characters E, M, or F. We toggle the red, green, and blue colors of LED2 for two seconds according to the received data.

10.3.2.2 Bluetooth LE Beacon

In our second application, we use the HM-10 module to create a Bluetooth advertiser (beacon). We can connect this advertiser to one of our critical household items. Hence, when it is lost, we can track it. The Bluetooth beacon data and signal level can be observed by a Bluetooth 4.0 enabled mobile device. Hence, the critical item attached with beacon can be located by looking at its data and signal level.

The connection between the MSP432 LaunchPad and HM-10 module will be as follows. The MSP432 pins P3.2 (eUSCI A2 Rx) and P3.3 (eUSCI A2 Tx) will be connected to the HM-10 module's TX and RX pins, respectively. The V_{CC} and GND pins of the HM-10 module will be connected to the 3.3 V and GND pins of the MSP432 LaunchPad, respectively.

We provide the C code for our second application in Listing 10.14. In the first part of this code, we define the variables to be used. Here, the string `RXData` is used for receiving UART data from the HM-10 module. In the main function, we first initialize the MSP432 using the function `initMSP432` in Listing 10.2. Then, we initialize the HM-10

Listing 10.14 Usage of the Bluetooth LE in Beacon Mode

```
#include <ti/devices/msp432p4xx/driverlib/driverlib.h>
#include <stdint.h>
#include <stdio.h>
#include <string.h>
#include <stdbool.h>

 char RXData[1000] = "";
 int buf_ind = 0;

void initMSP432(void);
void initHMbeacon(void);
void sendData(char *data, char *ack, char *err, const int timeout);
void delay_ms(uint16_t ms);

void main(void){

 WDT_A_holdTimer();

 initMSP432();
 initHMbeacon();

 Interrupt_enableSleepOnIsrExit();
 PCM_setPowerState(PCM_LPM0_LDO_VCORE1);
}

void EUSCIA0_IRQHandler(void){
        ...
}

void EUSCIA2_IRQHandler(void){
        ...
}
```

Listing 10.15 The initHMbeacon Function

```
void initHMbeacon(void){
 sendData("AT\r\n","OK","ERROR",1000);
 sendData("AT+RENEW\r\n","OK","ERROR",1000);
 sendData("AT+RESET\r\n","OK","ERROR",1000);
 sendData("AT\r\n","OK","ERROR",1000);
 sendData("AT+ADVI5\r\n","OK","ERROR",1000);
 sendData("AT+MARJ0x1234\r\n","OK","ERROR",1000);
 sendData("AT+MINO0xFA01\r\n","OK","ERROR",1000);
 sendData("AT+NAMEMSP432iBeacon\r\n","OK","ERROR",1000);
 sendData("AT+IBEA1\r\n","OK","ERROR",1000);
 sendData("AT+RESET\r\n","OK","ERROR",1000);
 delay_ms(3000);
}
```

module using the function initHMbeacon in Listing 10.15. Once the MSP432 is powered up, a Bluetooth connection will be available. You can observe the advertised data by a Bluetooth 4.0 enabled mobile device (preferably Android based). There are several freeware applications generally called beacon scanner for this purpose.

Listing 10.16 Usage of Bluetooth LE in Master Mode

```c
#include <ti/devices/msp432p4xx/driverlib/driverlib.h>
#include <stdint.h>
#include <stdio.h>
#include <string.h>
#include <stdbool.h>

 char RXData[1000] = "";
 int buf_ind = 0, i = 0;

void initMSP432(void);
void initHMmaster(void);
void sendData(char *data, char *ack, char *err, const int timeout);
void delay_ms(uint16_t ms);

void main(void){

 WDT_A_holdTimer();

 GPIO_setAsInputPinWithPullUpResistor(GPIO_PORT_P1, GPIO_PIN1 |
    GPIO_PIN4);
 GPIO_clearInterruptFlag(GPIO_PORT_P1, GPIO_PIN1 | GPIO_PIN4);
 GPIO_enableInterrupt(GPIO_PORT_P1, GPIO_PIN1 | GPIO_PIN4);

 initMSP432();
 initHMmaster();

 Interrupt_enableInterrupt(INT_PORT1);
 Interrupt_enableSleepOnIsrExit();
 PCM_setPowerState(PCM_LPM0_LDO_VCORE1);
}

void EUSCIA0_IRQHandler(void){
    ...
}

void EUSCIA2_IRQHandler(void){
    ...
}

void PORT1_IRQHandler(void){
    ...
}
```

We provide the initHMbeacon function in Listing 10.15. In this function, we configure the HM-10 module in beacon mode using AT commands. We first reset the device into factory defaults using the AT commands RENEW and RESET. Then, we configure the advertisement interval as 546.25 millisecond using the AT command ADVI. We configure major and minor fields using the AT commands MARJ and MINO. We set the advertising name using the AT command NAME. Finally, we configure the HM-10 module in beacon mode using the AT IBEA command and reset the module using the AT command RESET.

272 Chapter Ten

Listing 10.17 The `initHMmaster` Function

```
void initHMmaster(void){
 sendData("AT\r\n","OK","ERROR",1000);
 sendData("AT+RENEW\r\n","OK","ERROR",1000);
 sendData("AT+RESET\r\n","OK","ERROR",1000);
 sendData("AT+ROLE1\r\n","OK","ERROR",1000);
 sendData("AT+INQ\r\n","OK","ERROR",1000);
 delay_ms(1000);
 sendData("AT+INQ\r\n","OK","ERROR",1000);
 delay_ms(1000);
 sendData("AT+CONN1\r\n","OK","ERROR",1000);
 delay_ms(3000);
}
```

Listing 10.18 Port ISR for the MSP432 Device Connected to the Bluetooth LE in Master Mode

```
void PORT1_IRQHandler(void){
 uint32_t status = GPIO_getEnabledInterruptStatus(GPIO_PORT_P1);
 GPIO_clearInterruptFlag(GPIO_PORT_P1, status);

 if(status & GPIO_PIN1){
  UART_transmitData(EUSCI_A2_BASE, 'D');
  UART_transmitData(EUSCI_A0_BASE, 'D');}
 else if(status & GPIO_PIN4){
  UART_transmitData(EUSCI_A2_BASE, 'S');
  UART_transmitData(EUSCI_A0_BASE, 'S');}
}
```

10.3.2.3 Bluetooth LE Communication Application

In our third Bluetooth application, we use two HM-10 modules for Bluetooth LE communication between two MSP432 LaunchPads. Here, one MSP432 LaunchPad is located on a washing machine with digital sensors to detect detergent and softener levels. These sensors notify us when the detergent or softener level is critical. Here, we will simulate digital sensors with buttons. The first MSP432 sends characters D or S to the second MSP432. The second MSP432 LaunchPad has a GSM module attached to it. When it receives the character D or S from the first MSP432 LaunchPad (nearby the washing machine) it directly sends SMS to the market. We will simulate the GSM module with LEDs for the sake of simplicity. Hence, when the second MSP432 receives the character D or S, it turns on the corresponding LED. The MSP432 device on the washing machine will be set as master. The other MSP432 will be set as slave.

The connection between the MSP432 LaunchPad and HM-10 module will be as follows. The MSP432 pins P3.2 (eUSCI A2 Rx) and P3.3 (eUSCI A2 Tx) will be connected to the HM-10 module's TX and RX pins, respectively. The V_{CC} and GND pins of the HM-10 module will be connected to the 3.3 V and GND pins of the MSP432 LaunchPad, respectively.

We provide the C code to be used in the master device in Listing 10.16. In the first part of this code, we define the variables to be used. Here, the string `RXData` is used for receiving UART data from the HM-10 module. In the main function of the code, we first configure the buttons S1 and S2 connected to pins P1.1 and P1.4 on the MSP432 LaunchPad as input with pull-up resistors. We also enable the associated port interrupt. Then, we initialize the MSP432 using the function `initMSP432` in Listing 10.2. Afterwards,

Listing 10.19 Usage of Bluetooth LE in Slave Mode

```c
#include <ti/devices/msp432p4xx/driverlib/driverlib.h>
#include <stdint.h>
#include <stdio.h>
#include <string.h>
#include <stdbool.h>

 char RXData[1000] = "";
 int buf_ind = 0, i = 0;
void initMSP432(void);
void initHMslave(void);
void handleClientBLE(void);
void sendData(char *data, char *ack, char *err, const int timeout);
void delay_ms(uint16_t ms);

void main(void){

 WDT_A_holdTimer();

 GPIO_setAsOutputPin(GPIO_PORT_P2, GPIO_PIN0 | GPIO_PIN1| GPIO_PIN2);
 GPIO_setOutputLowOnPin(GPIO_PORT_P2, GPIO_PIN0 | GPIO_PIN1|
     GPIO_PIN2);

 initMSP432();
 initHMslave();

 while(1)
  handleClientBLE();
}

void EUSCIA0_IRQHandler(void){
    ...
}

void EUSCIA2_IRQHandler(void){
    ...
}
```

we initialize the HM-10 module in master mode using the function initHMmaster in Listing 10.17. Once the MSP432 is powered up, the HM-10 module will be connected to a Bluetooth 4.0 slave device it detects first.

We provide the initHMmaster function in Listing 10.17. In this function, we configure the HM-10 module in master mode using AT commands. Here, we first restore factory default settings and reset the module using AT commands RENEW and RESET. Then, we set the HM-10 module as master using the AT command ROLE. Afterwards, we scan for nearby slave devices using the AT command INQ and connect to the first slave device using the AT command CONN.

After connection, the master device will send the D or S characters to the slave device when the button S1 or S2 is pressed through a port ISR in Listing 10.18.

We provide the C code to be used in the slave device in Listing 10.19. In the main function, we first set up the red and green colors of LED2 connected to pins P2.0 and

Listing 10.20 The `initHMslave` Function

```c
void initHMslave(void){
 sendData("AT\r\n","OK","ERROR",1000);
 sendData("AT+RENEW\r\n","OK","ERROR",1000);
 sendData("AT+RESET\r\n","OK","ERROR",1000);
 delay_ms(3000);
}
```

Listing 10.21 The `handleClientBLE` Function

```c
void handleClientBLE(void){
 char receiveddata[1000] = "";

 if(buf_ind>0){
  delay_ms(1000);
  strncpy(receiveddata, RXData, buf_ind);
  memset(RXData, 0, sizeof(RXData));
  buf_ind = 0;}

 if(strstr(receiveddata, "D") != NULL || strstr(receiveddata, "d")
     != NULL){
  GPIO_setOutputHighOnPin(GPIO_PORT_P2, GPIO_PIN0);
  delay_ms(2000);
  GPIO_setOutputLowOnPin(GPIO_PORT_P2, GPIO_PIN0);}
 else if(strstr(receiveddata, "S") != NULL || strstr(receiveddata, "s")
     != NULL){
  GPIO_setOutputHighOnPin(GPIO_PORT_P2, GPIO_PIN1);
  delay_ms(2000);
  GPIO_setOutputLowOnPin(GPIO_PORT_P2, GPIO_PIN1);}
}
```

P2.1 on the MSP432 LaunchPad as output and initialize the MSP432 using the function `initMSP432` in Listing 10.2. Afterwards, we initialize the HM-10 module in slave mode using the function `initHMslave` in Listing 10.20. In an infinite loop, we listen and handle the incoming client requests using the function `handleClientBLE` in Listing 10.21.

We provide the `initHMslave` function in Listing 10.20. In this function, we configure the HM-10 module in slave mode using AT commands. In factory defaults, the HM-10 module is configured as a slave device. Hence, we restore factory default settings and reset the module using AT commands `RENEW` and `RESET`.

We provide the `handleClientBLE` function in Listing 10.21. Here, we wait to receive all data sent from the HM-10 module and copy it to the string `receiveddata`. Then, we check whether the received data consists of characters D or S. We toggle the red and green colors of LED2 for two seconds according to the received data. We have provided the `sendData` function in Listing 10.4.

10.4 ZigBee

ZigBee is the third wireless communication option to be considered in this chapter. We first give some basic information on ZigBee next. Then, we provide a sample application using it.

Wireless Communication 275

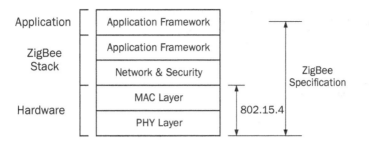

FIGURE 10.4 ZigBee stack.

10.4.1 Definition

Zigbee is a low-power, short-range wireless connectivity solution based on the IEEE 802.15.4 standard. ZigBee alliance is responsible for developing the ZigBee protocol [34]. ZigBee operates at unlicensed bands including 2.4 GHz, 900 MHz, and 868 MHz.

The IEEE 802.15.4 standard defines the physical and media access control (MAC) layers for ZigBee. ZigBee protocols are standardized by the ZigBee specification. They define upper layers of the protocol stack including application as shown in Fig. 10.4. In other words, ZigBee uses the IEEE 802.15.4 standard for wireless communication and adds logical network, security, and application software upon it. Its standards are called input device, remote control, retail services, smart energy, light link, home automation, building automation, telecommunication services, health care, and 2030.5 (IP).

The IEEE 802.15.4 standard offers two types of physical layers. One of them operates at 2.4 GHz and can be used to attain higher throughput. The other layer operates at 816/915 MHz and can be used to attain larger coverage area. This reduces the number of nodes to be used in a given area.

There are two device types in IEEE 802.15.4: full function devices (FFD) and reduced function devices (RFD). FFD are equipped with full set of MAC layer functions and can be end-point devices or coordinator devices for ZigBee. RFD are equipped with reduced MAC layer functions and act as only end-point devices for ZigBee. Only one coordinator device can be active in a ZigBee network.

The most important characteristic of ZigBee is that it can support different network topologies such as star, mesh, or tree. In star topology, devices can talk to each other directly in master-slave settings. Here, the master can only be FFD and slave can be FFD or RFD. In the tree topology, FFD can talk to other FFD within their range and send message to other FFD outside their range. The mesh topology is similar to tree topology. However, multiple connection paths from one device to another can be created using it.

10.4.2 ZigBee Application

We will use Xbee series 1 modules in our ZigBee application. These modules are ready to be used in mesh configuration. Each module can be connected to MSP432 through its UART port and communicates with AT commands [35].

In our ZigBee application, we will use two Xbee modules for ZigBee mesh communication. Through this setup, we will send data from one MSP432 to another. Here, one MSP432 is located nearby a window and has a sensor to detect whether the window is open or closed. The other MSP432 has a GSM module. When the second MSP432 receives the character W, it sends an SMS message to the police. Here, we will simulate

Listing 10.22 Usage of ZigBee in Mesh Mode, Transmitter Side

```c
#include <ti/devices/msp432p4xx/driverlib/driverlib.h>
#include <stdint.h>
#include <stdio.h>
#include <string.h>
#include <stdbool.h>

 char RXData[1000] = "";
 int buf_ind = 0, i = 0;

void initMSP432(void);
void initXBee(void);
void sendData(char *data, char *ack, char *err, const int timeout);
void delay_ms(uint16_t ms);

void main(void) {

 WDT_A_holdTimer();

 GPIO_setAsInputPinWithPullUpResistor(GPIO_PORT_P1, GPIO_PIN1);
 GPIO_clearInterruptFlag(GPIO_PORT_P1, GPIO_PIN1);
 GPIO_enableInterrupt(GPIO_PORT_P1, GPIO_PIN1);

 initMSP432();
 initXBee();

 Interrupt_enableInterrupt(INT_PORT1);
 Interrupt_enableSleepOnIsrExit();
 PCM_setPowerState(PCM_LPM0_LDO_VCORE1);
}

void EUSCIA0_IRQHandler(void) {
    ...
}

void EUSCIA2_IRQHandler(void) {
    ...
}

void PORT1_IRQHandler(void) {
    ...
}
```

the window sensor by a button. We will also simulate the SMS with LED for the sake of simplicity.

The connection between the MSP432 LaunchPad and Xbee series 1 module will be as follows. The MSP432 pins P3.2 (eUSCI A2 Rx) and P3.3 (eUSCI A2 Tx) will be connected to the Xbee series 1 module's TX and RX pins, respectively. The V_{CC} and GND pins of the Xbee series 1 module will be connected to the 3.3 V and GND pins of the MSP432 LaunchPad, respectively.

We provide the C code for the transmitter part of our ZigBee application in Listing 10.22. In the first part of this code, we define the variables to be used. Here, the string RXData is used for receiving UART data from the Xbee module. In the main

Listing 10.23 Port ISR for the MSP432 Device

```c
void PORT1_IRQHandler(void){
 uint32_t status = GPIO_getEnabledInterruptStatus(GPIO_PORT_P1);

 GPIO_clearInterruptFlag(GPIO_PORT_P1, status);

 if(status & GPIO_PIN1){
  UART_transmitData(EUSCI_A2_BASE, 'W');
  UART_transmitData(EUSCI_A0_BASE, 'W');}
}
```

Listing 10.24 The initXBee Function

```c
void initXBee(void){
 sendData("+++","OK","ERROR",1000);
 delay_ms(3000);
 sendData("AT\r\n","OK","ERROR",1000);
 sendData("ATID3331\r\n","OK","ERROR",1000);
 sendData("ATRE\r\n","OK","ERROR",1000);
 delay_ms(3000);
}
```

function of the C code, we first configure the button S1 connected to pin P1.1 on the MSP432 LaunchPad as input with the pull-up resistor and enable the port interrupt. Then, we initialize the microcontroller using the function initMSP432 in Listing 10.2. Afterwards, we initialize the Xbee module using the function initXBee in Listing 10.23. Once the MSP432 is powered up, this Xbee module will send data to all nearby Xbee modules configured with the same personal network area identification (PAN ID). We provide the port ISR in Listing 10.23. Here, we send the character W to both PC host and Xbee module when the button S1 is pressed through a port ISR.

We provide the initXBee function in Listing 10.24. In this function, we configure the Xbee module using AT commands. Here, we first enable the AT mode using the command +++. Then, we set the PAN ID and reset the Xbee module using AT commands ID and RE.

We provide the C code for the receiver part of the application in Listing 10.25. In the main function of the code, we first set up LED1 connected to pin P1.0 on the MSP432 LaunchPad as output and initialize the MSP432 microcontroller using the function initMSP432 in Listing 10.2. Afterwards, we initialize the Xbee module using the function initHMslave in Listing 10.23. In an infinite loop, we listen and handle the incoming client requests using the function handleClientXbee in Listing 10.26.

We provide the handleClientXbee function in Listing 10.23. Here, we wait to receive all data sent from the Xbee module and copy it to the string receiveddata. Then, we check whether the received data from the module contains the character W. We toggle LED1 for two seconds according to the received data. We provide the sendData function in Listing 10.4.

Listing 10.25 Usage of ZigBee in Mesh Mode, Receiver Side

```
#include <ti/devices/msp432p4xx/driverlib/driverlib.h>
#include <stdint.h>
#include <stdio.h>
#include <string.h>
#include <stdbool.h>

 char RXData[1000] = "";
 int buf_ind = 0, i = 0;

void initMSP432(void);
void initXBee(void);
void handleClientXbee(void);
void sendData(char *data, char *ack, char *err, const int timeout);
void delay_ms(uint16_t ms);

void main(void) {

 WDT_A_holdTimer();

 GPIO_setAsOutputPin(GPIO_PORT_P1, GPIO_PIN0);
 GPIO_setOutputLowOnPin(GPIO_PORT_P1, GPIO_PIN0);

 initMSP432();
 initXBee();

 while(1)
   handleClientXbee();
}
void EUSCIA0_IRQHandler(void) {
        ...
}
void EUSCIA2_IRQHandler(void) {
        ...
}
```

10.5 RFID

RFID is the fourth wireless communication option to be considered in this chapter. We first give some basic information on RFID next. Then, we provide a sample application using it.

10.5.1 Definition

The RFID system is composed of an RFID reader and tag. The RFID reader sends signal to the tag via antenna. The tag responds to this signal with its unique information. Hence, the communication between them is established.

RFID tags are either active or passive. Active RFID tags contain their own power source. It gives them the ability to send data with the range of up to 100 meters. Passive RFID tags do not have their own power source. Instead, they are powered by the electromagnetic energy transmitted from the RFID reader. The transmitted energy must be

Listing 10.26 The `handleClientXbee` Function

```c
void handleClientXbee(void){
 char receiveddata[1000] = "";

 if(buf_ind>0){
  delay_ms(1000);
  strncpy(receiveddata, RXData, buf_ind);
  memset(RXData, 0, sizeof(RXData));
  buf_ind = 0;}

 if(strstr(receiveddata, "W") != NULL || strstr(receiveddata, "w")
     != NULL){
  GPIO_setOutputHighOnPin(GPIO_PORT_P1, GPIO_PIN0);
  delay_ms(2000);
  GPIO_setOutputLowOnPin(GPIO_PORT_P1, GPIO_PIN0);}
}
```

high enough to power the tags. Hence, passive RFID tags can send data with a range up to 25 meters. Passive RFID tags primarily operate at three frequency ranges. These are low frequency (LF) 125–134 kHz, high frequency (HF) 13.56 MHz, and ultra high frequency (UHF) 856 to 960 MHz.

The passive RFID tag and reader communicate in a way known as the RFID coupling mechanism. The form of coupling depends on the intended application. Hence, the type of RFID coupling used will affect the frequency of the system. The main RFID coupling techniques are backscatter, inductive, and capacitive.

RFID backscatter coupling uses the RF power transmitted by the reader to energise the tag. The tag reflects back some of the power transmitted by the reader. This way, it also sends back information to the reader. RFID backscatter coupling uses UHF.

RFID inductive coupling uses transfer of energy from the reader to tag via the mutual inductance between the two circuits in them. To use the RFID inductive coupling, both the tag and reader should have induction or antenna coils. When the tag is placed close enough to the reader, the field from the reader coil will couple to the coil from the tag. A voltage will be induced in the tag that will be rectified and used to power the tag circuitry. RFID inductive coupling normally uses LF or HF.

RFID capacitive coupling uses capacitive effects to provide the coupling between the tag and reader. The system is often used in smart cards. When a smart card is inserted into a reader, the card will be in very close proximity to the reader. Hence, capacitive coupling can be done. Rather than having coils or antennas, capacitive coupling uses electrodes (the plates of the capacitor to provide the required coupling).

10.5.2 RFID Application

We will use the RMD6300 RFID card reader module in our application. This module is designed for reading code from the EM4100 compatible read-only tags and read/write card. EM4100 is the RFID protocol used for 125-kHz RFID communication. It contains 64-bit data where nine bits are used for the header, 14 bits are used for parity, 40 bits are used for data, and one bit is used for stop. In our RFID application, we will use the RFID card reader module and tag to open the door. Here LED1 connected to pin P1.1 on the MSP432 LaunchPad will simulate the door lock.

Listing 10.27 Usage of RFID card reader to toggle LED1

```c
#include <ti/devices/msp432p4xx/driverlib/driverlib.h>
#include <stdint.h>
#include <stdio.h>
#include <string.h>
#include <stdbool.h>

 char RXData[1000] = "";
 char ID[13] = "0D0040A514FC";
 int buf_ind = 0, i = 0;

void initMSP432(void);
void handleClientRfid(void);
void delay_ms(uint16_t ms);

void main(void) {

 WDT_A_holdTimer();

 GPIO_setAsOutputPin(GPIO_PORT_P1, GPIO_PIN0);
 GPIO_setOutputLowOnPin(GPIO_PORT_P1, GPIO_PIN0);

 initMSP432();

 while(1)
   handleClientRfid();
}

void EUSCIA0_IRQHandler(void) {
     ...
}

void EUSCIA2_IRQHandler(void) {
     ...
}
```

The connection between the MSP432 LaunchPad and RMD6300 module will be as follows. The MSP432 pins P3.2 (eUSCI A2 Rx) and P3.3 (eUSCI A2 Tx) will be connected to the RMD6300 module's TX and RX pins, respectively. The V_{CC} and GND pins of the RMD6300 module will be connected to the 5-V and GND pins of the MSP432 LaunchPad, respectively.

We provide the C code for our RFID application in Listing 10.27. In the first part of this code, we define the variables to be used. Here, the string RXData is used for receiving UART data from the RFID module. The string ID is used for the RFID card ID. Do not forget to write your RFID card ID to this string. In the main function, we first set up LED1 connected to pin P1.0 on the MSP432 LaunchPad as output and initialize the MSP432 microcontroller using the function initMSP432 in Listing 10.2. In an infinite loop, we listen and handle the incoming requests using the function handleClientRfid in Listing 10.28. Once the MSP432 is powered up, you can toggle LED1 by bringing the RFID tag closer to the RFID card reader.

We provide the handleClientRfid function in Listing 10.28. Here, we wait to receive all data sent from the RFID module and copy it to the string receiveddata.

Listing 10.28 The `handleClientRfid` Function

```
void handleClientRfid(void){
 char receiveddata[1000] = "";

 if(buf_ind>0){
 delay_ms(1000);
 strncpy(receiveddata, RXData, buf_ind);
 memset(RXData, 0, sizeof(RXData));
 buf_ind = 0;}

 if(strstr(receiveddata, ID) != NULL){
 memset(receiveddata, 0, sizeof(receiveddata));
 memset(RXData, 0, sizeof(RXData));
 buf_ind = 0;
 GPIO_toggleOutputOnPin(GPIO_PORT_P1, GPIO_PIN0);
 delay_ms(2000);
 GPIO_toggleOutputOnPin(GPIO_PORT_P1, GPIO_PIN0);}
}
```

Then, we check whether the received data from the module consists of the RFID card ID. If this is the case, we turn on LED1 for two seconds.

10.6 GSM

GSM is the fifth wireless communication option to be considered in this chapter. GSM stands for global system for mobile communication. We extensively use this communication mode in our mobile phones. We first give some basic information on GSM next. Then, we provide two sample applications using it.

10.6.1 Definition

GSM services operate at 850-, 900-, 1800-, and 1900-MHz frequency bands depending on the country and service provider. Unfortunately, GSM operations are fairly complex. Therefore, we ask the reader to check available resources to fully grasp it. We will use a module for our GSM applications. This module will handle almost all GSM requirements. Hence, we will not deal with low-level GSM operations.

10.6.2 GSM Applications

We will use the GSM click module in our applications [36]. This module uses Telit's GL865-QUAD GSM IC for communication [37]. This IC supports GSM/GPRS communication with quad-band coverage with 850-, 900-, 1800-, or 1900-MHz frequencies. Hence, it can be used worldwide. GSM click has an SMA antenna connector and SIM card socket. It is ready to be used with UART interface and communicates with MSP432 via Telit's AT command set [38]. Please do not forget to disable the PIN protection of your SIM card via SIM settings of your mobile phone before using it with the GSM click module.

10.6.2.1 SMS Receive Application

In our first application, we will use the GSM click module to receive an SMS message to turn on the heater. Here, LED1 connected to pin P1.0 on the MSP432 LaunchPad will simulate the heater.

Listing 10.29 Usage of the GSM Click Module to Receive SMS Message

```c
#include <ti/devices/msp432p4xx/driverlib/driverlib.h>
#include <stdint.h>
#include <stdio.h>
#include <string.h>
#include <stdbool.h>

 char RXData[1000] = "";
 int buf_ind = 0, i = 0;

void initMSP432(void);
void initGSM(void);
void handleClientGSM(void);
void sendData(char *data, char *ack, char *err, const int timeout);
void delay_ms(uint16_t ms);

void main(void) {

 WDT_A_holdTimer();

 GPIO_setAsOutputPin(GPIO_PORT_P1, GPIO_PIN0);
 GPIO_setOutputLowOnPin(GPIO_PORT_P1, GPIO_PIN0);

 initMSP432();
 initGSM();

 while(1)
   handleClientGSM();
}

void EUSCIA0_IRQHandler(void) {
     ...
}

void EUSCIA2_IRQHandler(void) {
     ...
}
```

The connection between the MSP432 LaunchPad and GSM click module will be as follows. The MSP432 pins P3.2 (eUSCI A2 Rx) and P3.3 (eUSCI A2 Tx) will be connected to the GSM click module's TX and RX pins, respectively. The V_{CC} and GND pins of the GSM click module will be connected to the 3.3-V and GND pins of the MSP432 LaunchPad, respectively.

We provide the C code for our first GSM application in Listing 10.29. In the first part of the code, we define the variables to be used. Here, the string RXData is used for receiving UART data from the GSM click module. In the main function, we first set up LED1 as output. We initialize the MSP432 microcontroller using the function initMSP432 in Listing 10.2. Then, we initialize the GSM click module using the function initGSM in Listing 10.30. In an infinite loop, we listen and handle incoming requests using the function handleClientGSM in Listing 10.31. Once MSP432 is powered up, the user can toggle LED1 by sending strings heater on or heater off via SMS message.

Listing 10.30 The `initGSM` Function

```
void initGSM(void) {
 sendData("AT\r\n","OK","ERROR",1000);
 sendData("ATZ\r","OK","ERROR",1000);
 sendData("AT#SLED=2\r\n","OK","ERROR",1000);
 sendData("AT+CNMI=1,2,0,0,0\r\n","OK","ERROR",1000);
 sendData("AT+CMGF=1\r\n","OK","ERROR",1000);
 delay_ms(3000);
}
```

Listing 10.31 The `handleClientGSM` Function

```
void handleClientGSM(void) {
 char receiveddata[1000] = "";

 if(buf_ind>0){
  delay_ms(1000);
  strncpy(receiveddata, RXData, buf_ind);
  memset(RXData, 0, sizeof(RXData));
  buf_ind = 0;}

 if(strstr(receiveddata, "heater on") != NULL)
  GPIO_setOutputHighOnPin(GPIO_PORT_P1, GPIO_PIN0);
 else if(strstr(receiveddata, "heater off") != NULL)
  GPIO_setOutputLowOnPin(GPIO_PORT_P1, GPIO_PIN0);
}
```

We provide the `initGSM` function in Listing 10.30. In this function, we configure the GSM click module using its AT commands. Here, we first reset the module and toggle the status LED (green LED) on the GSM click module using AT commands `ATZ` and `SLED`. Then, we enable SMS receive indication using the AT command `CNMI`. Finally, we set the message format as text using the AT command `CMGF`.

We provide the `handleClientGSM` function in Listing 10.31. In this function, we wait to receive all data sent from the GSM click module and copy it to the string `receiveddata`. Then, we check whether the received data from the GSM click module contains the string `heater on` or `heater off`. Afterwards, we toggle LED1 according to the received data.

10.6.2.2 SMS Send Application

In our second application, we will send SMS message to the market if the detergent or softener level gets low in our intelligent washing machine. Hence, this is the continuation of our previous application on Bluetooth LE. Here, we will simulate the low-level signals by pressing buttons S1 and S2 connected to pins P1.1 and P1.4 on the MSP432 LaunchPad. Therefore, the predefined SMS message will be selected by pressing the buttons. To note here, this application can also be modified for the intelligent refrigerator in Sec. 10.3, the window control application in Sec. 10.4, or the garage gate opener in Sec. 10.7.

The connection between the MSP432 LaunchPad and GSM click module will be as follows. The MSP432 pins P3.2 (eUSCI A2 Rx) and P3.3 (eUSCI A2 Tx) will be connected to the GSM click module's TX and RX pins, respectively. The V_{CC} and GND pins of the GSM click module will be connected to the 3.3-V and GND pins of the MSP432 LaunchPad, respectively.

Listing 10.32 Usage of the GSM Click Module to Transmit SMS Message

```c
#include <ti/devices/msp432p4xx/driverlib/driverlib.h>
#include <stdint.h>
#include <stdio.h>
#include <string.h>
#include <stdbool.h>

char SMS_Softener[] = "Softener";
char SMS_Detergent[] = "Detergent";
char PN[20] = "..."; //please add your phone number here

 char RXData[1000] = "";
 int buf_ind = 0, i = 0;

void initMSP432(void);
void initGSMsend(void);
void sendData(char *data, char *ack, char *err, const int timeout);
void delay_ms(uint16_t ms);

void main(void) {

 WDT_A_holdTimer();

 GPIO_setAsInputPinWithPullUpResistor(GPIO_PORT_P1, GPIO_PIN1|
     GPIO_PIN4);
 GPIO_clearInterruptFlag(GPIO_PORT_P1, GPIO_PIN1|GPIO_PIN4);
 GPIO_enableInterrupt(GPIO_PORT_P1, GPIO_PIN1|GPIO_PIN4);

 initMSP432();
 initGSMsend();

 Interrupt_enableInterrupt(INT_PORT1);
while(1);
}

void EUSCIA0_IRQHandler(void) {
    ...
}

void EUSCIA2_IRQHandler(void) {
    ...
}

void PORT1_IRQHandler(void) {
    ...
}
```

We provide the C code for our second GSM application in Listing 10.32. In the first part of this code, we define the variables to be used. Here, the string RXData is used for receiving UART data from the GSM click module. The strings SMS_Detergent and SMS_Softener are used for transmitting UART data to the GSM click module. The string PN is used for phone number to send the SMS. Do not forget to enter a receiver phone number to this string. In the main function, we first initialize the MSP432 microcontroller using the function initMSP432 in Listing 10.2. Then, we initialize the GSM

Listing 10.33 The initGSMsend Function

```
void initGSMsend(void){
 sendData("AT\r","OK","ERROR",1000);
 sendData("ATZ\r","OK","ERROR",1000);
 sendData("AT#SLED=2\r","OK","ERROR",1000);
 sendData("AT+CMGF=1\r","OK","ERROR",1000);
 delay_ms(3000);
}
```

Listing 10.34 Port ISR for the MSP432 Device Connected to the GSM Click Module

```
void PORT1_IRQHandler(void){
 uint32_t status = GPIO_getEnabledInterruptStatus(GPIO_PORT_P1);

 GPIO_clearInterruptFlag(GPIO_PORT_P1, status);
 char SMS[100] = "AT+CMGS=\"";

 if(status & GPIO_PIN1){;
  UART_transmitData(EUSCI_A0_BASE, 'D');
  strcat(SMS, PN);
  strcat(SMS, "\"\r");
  strcat(SMS, SMS_Detergent);
  strcat(SMS, "\r\x1A\r\n");}
 else if(status & GPIO_PIN4){
  UART_transmitData(EUSCI_A0_BASE, 'S');
  strcat(SMS, PN);
  strcat(SMS, "\"\r");
  strcat(SMS, SMS_Softener);
  strcat(SMS, "\r\x1A\r\n");}
  sendData(SMS,"OK","ERROR", 5000);
}
```

click module using the function initGSMsend in Listing 10.33. Once the MSP432 is powered up, it will send the content of SMS_Detergent or SMS_Softener to the receiver number when the button S1 or S2 connected to pins P1.1 and P1.4 on the MSP432 LaunchPad is pressed.

We provide the initGSMsend function in Listing 10.33. In this function, we configure the GSM click module using its AT commands. Here, we first reset the module and toggle the status LED (green LED) on the GSM click module using AT commands ATZ and SLED. Then, we set the message format as text using the AT command CMGF.

The GSM click module will send the SMS_Detergent or SMS_Softener messages when the button S1 or S2 is pressed through a port ISR. We provide the port ISR in Listing 10.34. Here we send the SMS message using the AT command CMGS.

10.7 GPS

GPS is the sixth and last wireless communication option to be considered in this chapter. GPS stands for the global positioning system using satellites. The GPS system does not require the user to transmit any data. Instead, each satellite transmits a unique signal that allows the GPS module to decode and compute the precise location of the satellite. Next, we give some basic information on GPS. Then, we provide a sample application using it.

286 Chapter Ten

10.7.1 Definition

GPS is a free satellite-based navigation system developed by U.S. Department of Defence. Initially, GPS contained 24 satellites orbiting around the Earth. Now, the satellite number has increased to 32. Each GPS satellite has an extremely accurate clock. The GPS module receives a timestamp from each visible satellite, along with data on where in the sky it is located. From this information, the GPS receiver knows its distance to each satellite. Using trilateration, the GPS receiver can calculate its position on Earth [39]. If the receiver can lock onto at least three satellites, it can accurately calculate its 2D position (latitude and longitude) using trilateration. Locking onto four satellites leads to 3D position (latitude, longitude, and altitude).

GPS data is displayed in different message formats usually over a serial interface. Almost all GPS receivers output NMEA data developed by the National Marine Electronics Association [40]. This standard is formatted in lines of data called sentences. Each sentence contains various data bits (characters) organized in comma-separated format. A sample NMEA sentence in GPGLL format is as follows.

```
$GPGLL,4003.9039,N,10512.5793,W,235317,A
```

The above GPGLL sentence can be decoded as follows. The second part `4003.9039,N` is the latitude as 40 degrees, 03.9040 minutes, north. The third part `10512.5793,W` is the longitude as 105 degrees, 12.5792 minutes, west. The fourth part `235317` is the Greenwich mean time 23:53:17. The character `A` stands for the validation to indicate that this is a valid data.

The GPS receiver continuously sends NMEA sentences via its UART interface. The user should capture this data starting with its NMEA sentence identifier (i.e. GPGLL). Then, the sentence can be parsed based on the commas in it. To note here, each NMEA sentence has constant number of commas. Hence, data can be retrieved by looking between two commas. NMEA sentences also contain the data validation identifier that indicates whether the GPS receiver module has locked on three or more satellites.

10.7.2 GPS Application

In our GPS application, we will use the Neo-6M module to get the user position [41]. Based on it, the developed module will send an SMS message to open the garage gate. The Neo-6M module is ready to be used with UART interface. Moreover, it supports NMEA sentences. Within the application, when GPS module finds the user location within range, LED1 connected to pin P1.0 on the MSP432 LaunchPad will turn on. Here, LED1 simulates the SMS message send action.

The connection between the MSP432 LaunchPad and Neo-6M module will be as follows. The MSP432 pins P3.2 (eUSCI A2 Rx) and P3.3 (eUSCI A2 Tx) will be connected to the Neo-6M module's TX and RX pins, respectively. The V_{CC} and GND pins of the Neo-6M module will be connected to the 3.3-V and GND pins of the MSP432 LaunchPad, respectively.

We provide the C code for our GPS application in Listing 10.35. In the first part of this code, we define the variables to be used. Here, the string `RXData` is used for receiving UART data from the Neo-6M GPS module. Variables `Lat` and `Long` are used for storing the user location. Do not forget to enter your approximate location into these strings. You can find your location by using your mobile phone's GPS and a map application (such as GoogleMaps). In the main function, we first set up LED1 connected to pin P1.0 on the MSP432 LaunchPad as output and initialize the MSP432 microcontroller using

Listing 10.35 Usage of the GPS Module

```c
#include <ti/devices/msp432p4xx/driverlib/driverlib.h>
#include <stdint.h>
#include <stdio.h>
#include <string.h>
#include <stdbool.h>

 char RXData[1000] = "";

 char Lat_degmin[]="4100.2"; // 41A°00'2x.xx"N
 char Lon_degmin[]="02911.5"; // 29A°11'3x.xx"E

 int buf_ind = 0, i = 0;

void initMSP432(void);
void handleClientGPS(void);
void delay_ms(uint16_t ms);

void main(void) {

 WDT_A_holdTimer();

 GPIO_setAsOutputPin(GPIO_PORT_P1, GPIO_PIN0 );
 GPIO_setOutputLowOnPin(GPIO_PORT_P1, GPIO_PIN0 );

 initMSP432();

 while(1)
   handleClientGPS();
}

void EUSCIA0_IRQHandler(void) {
     ...
}

void EUSCIA2_IRQHandler(void) {
     ...
}
```

the function `initMSP432` in Listing 10.2. In an infinite loop, we listen and handle the incoming requests using the `handleClientGPS` function in Listing 10.31.

We provide the `handleClientGPS` function in Listing 10.36. In this function, we check whether the received data from the GPS module contains the string GPGLL and data validation character indicated by A. Then, we check whether the location of our device is close to our home by comparing latitude and longitude values. If this is the case, then we turn on LED1.

10.8 Summary

IoT is becoming a fact of our lives. Therefore, we focused on six wireless communication methods related to IoT applications in this chapter. While explaining these methods, we did not take a theoretical approach. Instead, we benefit from available modules in the market. Our main aim was showing how each communication method can be used in

Listing 10.36 The `handleClientGPS` Function

```
void handleClientGPS(void){
 char receiveddata[1000] = "";

 if(buf_ind>0){
  delay_ms(100);
  strncpy(receiveddata, RXData, buf_ind);
  memset(RXData, 0, sizeof(RXData));
  buf_ind = 0;}

 if(strstr(receiveddata, "$GPGLL") != NULL && strstr(receiveddata,
    ",A,") != NULL){
  if(strstr(receiveddata, Lat_degmin) != NULL && strstr(receiveddata,
     Lon_degmin) != NULL){
   GPIO_setOutputHighOnPin(GPIO_PORT_P1, GPIO_PIN0 );
   delay_ms(2000);
   GPIO_setOutputLowOnPin(GPIO_PORT_P1, GPIO_PIN0 );}}
}
```

connection with the MSP432 microcontroller. To do so, we introduced sample applications related to our main theme that is smart homes. We believe this approach will be more useful to the reader when he or she plans to use a wireless communication method in an actual application.

10.9 Exercises

In this section, we will not offer new problems. Instead, we will ask the reader to solve problems given in previous chapters using two MSP432 LaunchPad boards and establishing a wireless communication link between them. Some sample problems are given below.

10.1 Solve Problem 5.5 using two MSP432 LaunchPad boards. The first MSP432 LaunchPad will be used for the button. The second will be used for the LEDs. Establish a wireless link between these two boards using the following:
  a. Wi-Fi communication
  b. Bluetooth LE communication
  c. ZigBee communication

10.2 Solve Problem 5.6 using two MSP432 LaunchPad boards. The first MSP432 LaunchPad will be used for the button. The second will be used for the LEDs. Establish a wireless link between these two boards using the following:
  a. Wi-Fi communication
  b. Bluetooth LE communication
  c. ZigBee communication

10.3 Solve Problem 6.2 using two MSP432 LaunchPad boards. The first MSP432 LaunchPad will be used for the button. The second will be used for the LEDs. Establish a wireless link between these two boards using the following:
  a. Wi-Fi communication
  b. Bluetooth LE communication
  c. ZigBee communication

CHAPTER 11

Flash Memory and RAM Operations

This chapter is on flash memory and RAM operations on the MSP432 microcontroller. We can reach flash memory content and modify it via coding. This will be the first item to be considered in the chapter. As for RAM operations, we will focus on the usage of memory protection unit and advanced encryption standard accelerator (AES256) module available in MSP432. The memory protection unit will allow us to control and secure the sensitive data in RAM. The AES256 will allow us to encrypt data in RAM. We will provide sample applications on using the memory protection unit and AES256 module. Finally, we will provide advanced applications on the concepts introduced in this chapter.

11.1 Flash Memory and Controller

Flash memory is the place the code is saved in the microcontroller. As a reminder, flash is nonvolatile. Hence, it can keep the saved data even if energy is not provided. Therefore, it can be taken as another form of ROM. However, the flash can easily be programmed by feeding a suitable voltage to it.

MSP432 flash memory is divided into two sections: main and information. The executable code and constant values are kept in the main section. The calibration data, serial number, and similar factory settings are kept in the information section. To note here, there is no physical difference between these two sections.

The main section of the flash memory for the MSP432 microcontroller is 256 kB. It spans the memory addresses between 0x00000000 and 0x0003FFFF. This space is divided into two identical banks as Bank0 and Bank1 each having 128 kB memory space. Bank0 spans the memory addresses between 0x00000000 and 0x0001FFFF. Bank1 spans the memory addresses between 0x00020000 and 0x0003FFFF.

The information section of the flash memory for the MSP432 microcontroller is 16 kB. It spans the memory addresses between 0x00200000 and 0x00203FFF. This space is also divided into two identical banks as Bank0 and Bank1 each having 8 kB memory space. Bank0 spans the memory addresses between 0x00200000 and 0x00201FFF. Bank1 spans the memory addresses between 0x00202000 and 0x00201FFF.

The MSP432 has an internal flash controller (FLCTL) module for erasing, writing, and reading operations. The CPU can write to a single byte, word (4 bytes) or full-word (16 bytes) location of the flash memory. In the erasing operation, minimum erasable memory region is 4 kB. Here, each 4-kB memory region is called sector. Sector and mass erase (to erase entire flash memory) options can be used when erasing the flash memory.

There are also advanced write and erase operations available for MSP432. For further information on these, please see [4].

The CPU can read data from flash with data access widths of 8, 16, or 32 bits. There is also an additional read buffering option that allows reading data as 128-bit blocks. This option is initially disabled. In the reading operation, CPU frequency may be higher than the maximum reading frequency of the flash controller. This can cause corrupted data reading from flash. To prevent this, the FLCTL stalls the read access for a defined number of wait states. The number of wait states can be 0, 1, 2, or 3 for different read modes and CPU frequencies. For further information on this issue, please see [4].

11.1.1 Flash Controller Usage via DriverLib Functions

The FLCTL can be used via DriverLib functions. We provide the necessary functions for this purpose below. The complete function list for the FLCTL can be found in [11].

```
void FlashCtl_getMemoryInfo (uint32_t addr, uint32_t *sectorNum,
    uint32_t *bankNum)
/*
Calcultes the flash bank and sector number for given memory address.
    The memory address should be in main memory.
addr: memory address to calculate flash bank and sector number.
sectorNum: pointer to save sector number.
bankNum: pointer to save bank number.
*/

bool FlashCtl_unprotectSector (uint_fast8_t memorySpace, uint32_t
    sectorMask)
/*
Disables the write protection on given sector, on given memory. It
    returns true if protection is disabled successfully.
memorySpace: memory space to disable write protection. It can be one of
    main memory Bank0, main memory Bank1, information memory Bank0 or
    information memory Bank1. Please see DriverLib user guide for macro
    definitions.
sectorMask: flash sector to disable write protection. It can be one of
    flash sector 0 to 31. Please see DriverLib user guide for macro
    definitions.
*/

bool FlashCtl_protectSector (uint_fast8_t memorySpace, uint32_t
    sectorMask)
/*
Enables the write protection on a given sector or memory. It returns
    true if protection is enabled successfully.
memorySpace: memory space to enable write protection. It can be one of
    main memory Bank0 or Bank1, information memory Bank0 or Bank1.
    Please see DriverLib user guide for macro definitions.
sectorMask: flash sector to enable write protection. It can be one of
    flash sector 0 to 31. Please see DriverLib user guide for macro
    definitions.
*/

bool FlashCtl_performMassErase (void)
```

```
/*
Erases the whole flash memory. It returns true if memory is erased
    successfully.
*/

bool FlashCtl_eraseSector (uint32_t addr)
/*
Erases the memory sector. It returns true if sector is erased
    successfully.
addr: start address of the memory sector to be erased. Minimum memory
    size to be erased is 4 kB. If the given address is not in 4 kB
    memory boundary, then the entire sector will be erased.
*/

bool FlashCtl_programMemory (void *src, void *dest, uint32_t length)
/*
Writes data to flash memory. It is a blocking function. It disables the
    master interrupt during operation and then re-enables it
    automatically to prevent write errors.
src: pointer to data to be saved into flash.
dest: pointer to memory location in flash where data is to be saved.
length: length of data to be saved in flash.
*/

void FlashCtl_setWaitState (uint32_t bank, uint32_t waitState)
/*
Changes the number of wait states used by the flash controller for read
    operations. When changing the clock frequency of CPU, this
    function must be used to allow for readable flash memory.
bank: Flash bank to set wait state for. It can be one of FLASH_BANK0 or
    FLASH_BANK1.
waitState: The number of wait states to set. It can be 0, 1, 2 or 3.
    Please see DriverLib user guide and datasheet for more detail.
*/

void FlashCtl_enableInterrupt (uint32_t flags)
/*
Enables the flash controller interrupts.
flags: interrupt to be enabled. Please see DriverLib user guide for
    macro definitions.
*/

void FlashCtl_disableInterrupt (uint32_t flags)
/*
Disables the flash controller interrupts.
flags: interrupt to be disabled. Please see DriverLib user guide for
    macro definitions.
*/

void FlashCtl_clearInterruptFlag (uint32_t flags)
/*
Clears the flash controller interrupts flags.
flags: interrupt to be cleared. Please see DriverLib user guide for
    macro definitions.
*/
```

```
uint32_t FlashCtl_getEnabledInterruptStatus (void)
/*
Returns the interrupt status flag masked with enabled interrupt mask.
*/
```

11.1.2 Coding Practices for the Flash Controller

We provide an example on the usage of the FLCTL in this section. This example demonstrates how to read data from flash memory using pointer operations. Besides, it shows how to erase and write data to flash using DriverLib functions. We provide the corresponding C code in Listing 11.1. Here, we first set the core voltage level 1 to operate at high speed as indicated in Sec. 7.3.1. Then, we set the flash wait states of Bank0 and Bank1 to two using the function `FlashCtl_setWaitState`. Afterwards, we set the DCO frequency to 48 MHz to read, erase, and write to flash memory faster. We first read from flash memory using pointer operations. Here, we read the consecutive 32 bytes starting from the flash address `0x0003E000`. To note here, this address is picked arbitrarily. Then, we disable write protection on sector 30 (the address `0x0003E000`) using the function `FlashCtl_unprotectSector`. Afterwards, we clear the sector using the function `FlashCtl_eraseSector`. We write new data to flash memory starting from the memory address `0x0003E000` using the function `FlashCtl_programMemory` and enable write protection using the function `FlashCtl_protectSector`. Finally, we read the consecutive 32 bytes starting from memory address `0x0003E000` to check the written values.

11.2 Memory Protection Unit

Memory protection unit (MPU) is a special module used for managing access properties for selected memory regions. Through it the user can set different read/write permissions for selected memory parts having user-defined size. If the user tries to access the protected memory without permission, MPU detects a fault. Hence, a fault interrupt is generated. This allows the programmer to apply specific operations when access to protected memory without permission is initiated.

The MPU has control over eight different memory regions. The memory size for these can be selected as 2^x bytes, where x can get values between 5 and 31. If the memory size for a region is equal to or larger than 256 bytes, then it can be divided into eight subregions with equal size. Any of these subregions can be arbitrarily disabled or enabled.

In Cortex-M4 processors, there are two access levels: privileged and unprivileged. The code can use all CPU instructions and has access to all peripheral devices and memory regions in the privileged access. On the other hand, the code cannot use some CPU instructions and cannot access some peripheral devices (like NVIC, system timer) in the unprivileged access. Also, some memory regions might be restricted to the unprivileged access. The program can have either privileged or unprivileged access in normal code execution. However, when an interrupt comes, the normal code execution is stopped and an exception handler takes over the operation. In this situation, the code must have privileged access.

There are six different data access permission options defined for the privileged and unprivileged accesses as given in Table 11.1. Besides, there is also an instruction access permission in which the selected region can be enabled or disabled for code execution.

Listing 11.1 Usage of the Flash Controller Module

```c
#include <ti/devices/msp432p4xx/driverlib/driverlib.h>

#define DataAddress 0x3E000

 char writeData[32] = "MSP432 UART DMA Example";
 char readData[32];
 int i;

void main(void){

 WDT_A_holdTimer();

 PCM_setPowerState(PCM_AM_LDO_VCORE1);
 FlashCtl_setWaitState(FLASH_BANK0, 2);
 FlashCtl_setWaitState(FLASH_BANK1, 2);
 CS_setDCOCenteredFrequency(CS_DCO_FREQUENCY_48);

 for(i=0;i<32;i++)
  readData[i]= *(uint8_t*)(DataAddress + i);

 FlashCtl_unprotectSector(FLASH_MAIN_MEMORY_SPACE_BANK1,
     FLASH_SECTOR30);
 FlashCtl_eraseSector(DataAddress);
 FlashCtl_programMemory(writeData, (void*)DataAddress, 32);
 FlashCtl_protectSector(FLASH_MAIN_MEMORY_SPACE_BANK1,FLASH_SECTOR31);

 for(i=0;i<32;i++)
  readData[i]= *(uint8_t*)(DataAddress + i);

 PCM_setPowerState(PCM_LPM0_LDO_VCORE1);
}
```

Privileged Mode	Unprivileged Mode
No access	No access
Read/Write	No access
Read/Write	Read only
Read/Write	Read/Write
Read only	No access
Read only	Read only

TABLE 11.1 Data Access Permission Options

If the program tries to access a memory location that is modified by the MPU without given permissions, a memory management fault is generated. This fault sets a specific bit in a system control block (SCB). Afterwards, a MPU fault interrupt is generated. The SCB is a special peripheral unit used in Cortex-M4F microprocessors. For further information on it, please see [4].

11.2.1 Memory Protection Unit Usage via DriverLib Functions

The MPU can be controlled via DriverLib functions. We provide the necessary functions to be used for this purpose below. The complete function list for the MPU can be found in [11].

```
void MPU_enableModule (uint32_t mpuConfig)
/*
Enables the MPU and configures its behavior in privileged mode and hard
    fault or NMI.
mpuConfig: configuration parameter determining the MPU behavior in
    privileged mode and hard fault or NMI. Please see DriverLib user
    guide for macro definitions.
*/

void MPU_disableModule (void)
/*
Disables the MPU.
*/

void MPU_enableRegion (uint32_t region)
/*
Enables MPU for given region. The memory region should be configured
    before calling this function.
region: memory region to be enabled. It can be one of 0 to 7.
*/

void MPU_disableRegion (uint32_t region)
/*
Disables MPU for a given region.
region: memory region to be disabled. It can be one of 0 to 7.
*/

void MPU_setRegion (uint32_t region, uint32_t addr, uint32_t flags)
/*
Configures the selected memory region.
region: memory region to be configured.
addr: base address of the memory region to be configured.
flags: configuration parameters including region size, execute
    permission, read/write permissions, disabled sub-regions, initial
    status. Please see DriverLib user guide and datasheet for more
    detail.
*/

void MPU_enableInterrupt (void)
/*
Enables the MPU interrupt.
*/

void MPU_disableInterrupt (void)
/*
Disables the MPU interrupt.
*/
```

Listing 11.2 Usage of the Memory Protection Unit

```c
#include <ti/devices/msp432p4xx/driverlib/driverlib.h>

uint32_t Data[32];
uint32_t ReadData[32];
const uint32_t flagSet = MPU_RGN_SIZE_32B | MPU_RGN_PERM_NOEXEC |
    MPU_RGN_PERM_PRV_RO_USR_RO | MPU_RGN_ENABLE;
uint32_t i=0;

void main(void){

 WDT_A_holdTimer();

 GPIO_setAsOutputPin(GPIO_PORT_P1, GPIO_PIN0);
 GPIO_setOutputLowOnPin(GPIO_PORT_P1, GPIO_PIN0);
 GPIO_setAsInputPinWithPullUpResistor(GPIO_PORT_P1, GPIO_PIN1);
 GPIO_clearInterruptFlag(GPIO_PORT_P1, GPIO_PIN1);
 GPIO_enableInterrupt(GPIO_PORT_P1, GPIO_PIN1);

 MPU_setRegion(0, (uint32_t)&Data[8] , flagSet);

 Interrupt_enableInterrupt(INT_PORT1);
 Interrupt_enableInterrupt(FAULT_MPU);
 Interrupt_enableMaster();

 for(i=0;i<32;i++)
  Data[i] = 255;

 MPU_enableModule(MPU_CONFIG_PRIV_DEFAULT);

 for(i=0;i<32;i++)
  ReadData[i] = Data[i];

 i=0;

 while(1){
  PCM_setPowerState(PCM_LPM0_LDO_VCORE1);
  Data[i] = ++i;}
}
void PORT1_IRQHandler(void){
    GPIO_clearInterruptFlag(GPIO_PORT_P1, GPIO_PIN1);
}
void MemManage_Handler(void){
    GPIO_setOutputHighOnPin(GPIO_PORT_P1, GPIO_PIN0);
    while(1);
}
```

11.2.2 Coding Practices for the Memory Protection Unit

We provide an example on the usage of MPU in this section. Through it we demonstrate data read and write operations with and without MPU protection. We provide the corresponding C code in Listing 11.2. Here, we first set LED1 connected to pin P1.0 on the MSP432 LaunchPad as output. We also set the button S1 connected to pin P1.1 on the MSP432 LaunchPad as input with the pull-up resistor and enable the peripheral interrupt. We configure the MPU for memory region using the function `MPU_setRegion`. Here, we set flags to protect 32 bytes of data, with read-only access for both privileged and unprivileged modes, without code execution in the region. We also set the flag to start memory protection initially. We set the base address as the eighth element of data array. Then, we write 255 to this address without memory protection. Afterwards, we enable the MPU using the function `MPU_enableModule` and read data from the array. In a while loop, we put the CPU to low-power mode. When the button S1 is pressed, CPU exits from the low-power mode and writes data to the array. When the protected memory location (the eighth element of the data array) is reached, an error is generated and MPU interrupt is generated. In the MPU ISR, we turn on LED1 and trap the code in an infinite loop.

11.3 Advanced Encryption Standard Accelerator Module

The advanced encryption standard accelerator (AES256) module can be used to encrypt and decrypt data according to advanced encryption standard. The AES256 module can encrypt or decrypt 128-bit data with 128-, 192-, or 256-bit keys. The data and key must be loaded to related registers either in byte or on half-word basis.

In the encryption process, 128-bit data is encrypted with the selected key. In the decryption process, there are two options. First, the encrypted data can be decrypted with the same key. Second, the selected key is encrypted first. Then, this encrypted key is used to decrypt the encrypted data. The AES256 module has four cipher modes each having a different approach to encrypt or decrypt data with the selected key. These are electronic codebook (ECB), cipher block chaining (CBC), output feedback (OFB), and cipher feedback (CFB). For further information on these, please see [4].

11.3.1 AES256 Module Usage via DriverLib Functions

The AES256 module can be controlled via DriverLib functions. We provide the necessary functions to be used for this purpose below. Here, `moduleInstance` is the AES256 module to be configured. The complete function list for the AES256 module can be found in [11].

```
bool AES256_setCipherKey (uint32_t moduleInstance, const uint8_t *
   cipherKey, uint_fast16_t keyLength)
/*
Sets encryption (cipher) key into the AES256 module.
cipherKey: pointer to the cipher key to be loaded.
keyLength: length of the cipher key which can be 128, 192 or 256 bits.
   Please see DriverLib user guide for macro definitions.
*/

void AES256_encryptData (uint32_t moduleInstance, const uint8_t *data,
   uint8_t *encryptedData)
/*
```

```
    Encrypts the 128-bit data and stores it in an array. The cipher key
        must be loaded before using this function.
    data: pointer to data to be encrypted.
    encryptedData: pointer to array in which the encrypted data is saved.
    */

    void AES256_decryptData (uint32_t moduleInstance, const uint8_t *data,
        uint8_t *decryptedData)
    /*
    Decrypts the 128-bit data and stores it in an array. The cipher key
        must be loaded before using this function.
    data: pointer to data to be decrypted.
    decryptedData: pointer to array in which the decrypted data is saved.
    */

    bool AES256_setDecipherKey (uint32_t moduleInstance, const uint8_t *
        cipherKey, uint_fast16_t keyLength)
    /*
    Sets decryption (decipher) key into the AES256 module.
    cipherKey: pointer to the decipher key to be loaded.
    keyLength: length of the decipher key which can be 128, 192, or 256
        bits. Please see DriverLib user guide for macro definitions.
    */

    void AES256_clearInterruptFlag (uint32_t moduleInstance)
    /*
    Clears the AES256 interrupt flag.
    */

    uint32_t AES256_getInterruptFlagStatus (uint32_t moduleInstance)
    /*
    Returns the status of the AES256 interrupt flag.
    */

    void AES256_enableInterrupt (uint32_t moduleInstance)
    /*
    Enables the AES256 interrupt.
    */

    void AES256_disableInterrupt (uint32_t moduleInstance)
    /*
    Disables the AES256 interrupt.
    */
```

11.3.2 Coding Practices for the AES256 Module

We provide an example on the usage of the AES256 module in this section. Through it we demonstrate data encryption and decryption using DriverLib functions. We provide the corresponding C code in Listing 11.3. Here, we first set LED1 connected to pin P1.0 on the MSP432 LaunchPad as output. Then, we set the key for encryption using the function AES256_setCipherKey. We encrypt data in blocks of 128-bits using the function AES256_encryptData. Afterwards, we set the key for encryption using

Listing 11.3 Usage of the AES256 Module

```c
#include <ti/devices/msp432p4xx/driverlib/driverlib.h>

#define SIZE 48

uint8_t Data[SIZE] = "MSP432 AES256 Encrypted Data Example..";
uint8_t EncryptedData[SIZE];
uint8_t DecryptedData[SIZE];
int i = 0;
uint8_t CipherKey[16] = { 0, 1, 2, 3, 4, 5, 6, 7, 8, 9, 10, 11, 12,
    13, 14, 15 };

void main(void){

 WDT_A_holdTimer();

 GPIO_setAsOutputPin(GPIO_PORT_P1, GPIO_PIN0);
 GPIO_setOutputLowOnPin(GPIO_PORT_P1, GPIO_PIN0);

 AES256_setCipherKey(AES256_BASE, CipherKey, AES256_KEYLENGTH_128BIT);

 for(i=0;i<SIZE/16;i++)
  AES256_encryptData(AES256_BASE, Data+i*16, EncryptedData+i*16);

 AES256_setDecipherKey(AES256_BASE, CipherKey,
     AES256_KEYLENGTH_128BIT);

 for(i=0;i<SIZE/16;i++)
  AES256_decryptData(AES256_BASE, EncryptedData+i*16, DecryptedData+i
      *16);

 GPIO_setOutputHighOnPin(GPIO_PORT_P1, GPIO_PIN0);

 PCM_setPowerState(PCM_LPM0_LDO_VCORE0);
}
```

the `AES256_setDecipherKey` function. We decrypt the encrypted data in blocks of 128-bits using the function `AES256_decryptData`. Finally, we turn on LED1 to indicate the process has completed and set the CPU in the low-power mode.

11.4 Application: Advanced Home Entrance System

The aim in this application is learning how to use the flash controller and MPU on the MSP432 microcontroller. As a real-life application, we will improve our home entrance system introduced in Chap. 6. The equipment list to be used and the layout in the application are the same. Therefore, we don't provide them here.

The system design specifications for this application are the same as in the home entrance system introduced in Chap. 6. However, the password will be stored in flash. When the microcontroller restarts, there will be no need to store the password again. Also the password in memory will be protected. Hence, the user cannot change it without entering the correct password first.

We could not add the C code of the project here due to page limitations. However, we provided the complete CCS project for this application in the companion website for the reader. We strongly suggest implementing it.

11.5 Application: Code Updater

The aim in this application is learning how to use the flash controller on the MSP432 microcontroller. As a real-life application, we will have a code updater. We may want to update some part of the code after it is debugged and embedded on the flash memory. For example, the threshold values in the leakage control system, introduced in Chap. 8, may depend on the environment. Thus, the system may need to be calibrated after production. This can be done using our code updater system without reprogramming the MSP432. Here, the user will send the threshold values via UART communication. Then, these values will be updated in flash memory. The threshold values sent from UART can be identified by sending *a* and *b* before sending threshold values.

We could not add the C code of the project here due to page limitations. However, we provided the complete CCS project for this application in the companion website for the reader. We strongly suggest implementing it.

11.6 Summary

Flash memory can be reached by writing a C code on MSP432. This allows modifying some parts of the code or static data in an application. Therefore, we first focused on flash memory operations in this chapter. We also introduced the MPU and AES256 modules to add extra protection on data. The memory protection unit allows us to control and secure the sensitive data. The AES256 module allows encrypting and decrypting data. We provided sample applications on the usage of these modules via DriverLib functions. We also provided two advanced applications on the concepts introduced in this chapter. These may help the reader to add extra security to his or her developed systems.

11.7 Exercises

11.1 Write a C code to read 128 bytes of data from flash memory with the starting address 0x0008F000. Keep the read data in an array. Form a function to encrypt the read data array with 256-bit encryption.

11.2 Write a C code to decrypt 128 bytes of data (in an array) with a 256-bit decryption. Add a module to your code to write 128 bytes of data to flash memory with the starting address 0x00060000.

11.3 Use two MSP432 LaunchPads to execute the code blocks in Problems 11.1 and 11.2 as follows:
   a. The first MSP432 LaunchPad will be used for sending the 128 bytes of encrypted data when the button S1 is pressed.
   b. The second MSP432 LaunchPad will be used for receiving the encrypted data. When reception is complete, the second MSP432 will decrypt the data and save it to flash memory.
   c. Establish a digital communication link between these two boards using UART, SPI, and I^2C communication modes.

CHAPTER 12
Direct Memory Access

Large amount of data may need to be copied from one location to another in the microprocessor for some applications. Although the copy operation can be performed by using the CPU, this is not desirable since the CPU cannot handle other tasks at the same time. Therefore, modern microcontrollers have a direct memory access (DMA) module for such operations. This module handles data transfer operations without interrupting the CPU. MSP432 has an eight-channel DMA module for this purpose. In this chapter, we will explore DMA operations using this module starting with its properties. We will use DriverLib functions for DMA applications since their register-level implementations are extremely complex. Afterwards, we will give example projects on the usage of the DMA module. Finally, we will provide a real-life application on DMA module usage.

12.1 The DMA Module

DMA is a special module having control over microprocessor's memory bus. With this feature, data blocks can be transferred between memory and peripherals (or memory to memory) without interfering the CPU. Meanwhile, the CPU can handle other tasks or be kept in a low-power mode. Hence, the DMA module provides significant benefits to reduce the CPU load and power consumption if used properly.

The MSP432 DMA module is graphically shown in Fig. 12.1. As can be seen in this figure, the heart of the DMA module is the ARM PL230 microDMA controller (DMAC). This controller gets transfer properties from a special register and performs DMA operations (cycle) automatically. These will be explained in detail in the following sections.

12.1.1 Transfer Types

The DMA module supports three different transfer types as peripheral-to-memory, memory-to-peripheral, and memory-to-memory. There are eight different DMA channels to be used in these transfers. Eight different DMA sources can be selected for each channel. The complete list of DMA sources for each channel is as in Table 12.1. In this table, Source 0 is reserved for all channels. It can be used for memory-to-memory transfer type.

12.1.2 Fundamental Properties

The microDMA controller needs data and address increment size for the source and destination. Furthermore, arbitration rate, transfer size, and cycle operating mode are

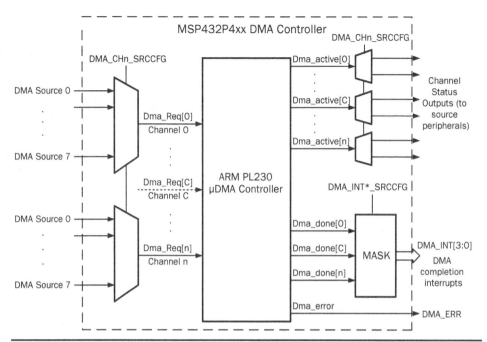

Figure 12.1 Block diagram of the DMA module.

DMA Channel	Source 1	Source 2	Source 3	Source 4	Source 5	Source 6	Source 7
Channel 0	eUSCI_A0 TX	eUSCI_B0 TX0	eUSCI_B3 TX1	eUSCI_B2 TX2	eUSCI_B1 TX3	TA0CCR0	AES256_Trigger0
Channel 1	eUSCI_A0 RX	eUSCI_B0 RX0	eUSCI_B3 RX1	eUSCI_B2 RX2	eUSCI_B1 RX3	TA0CCR2	AES256_Trigger1
Channel 2	eUSCI_A1 TX	eUSCI_B1 TX0	eUSCI_B0 TX1	eUSCI_B3 TX2	eUSCI_B2 TX3	TA1CCR0	AES256_Trigger2
Channel 3	eUSCI_A1 RX	eUSCI_B1 RX0	eUSCI_B0 RX1	eUSCI_B3 RX2	eUSCI_B2 RX3	TA1CCR2	Reserved
Channel 4	eUSCI_A2 TX	eUSCI_B2 TX0	eUSCI_B1 TX1	eUSCI_B0 TX2	eUSCI_B3 TX3	TA2CCR0	Reserved
Channel 5	eUSCI_A2 RX	eUSCI_B2 RX0	eUSCI_B1 RX1	eUSCI_B0 RX2	eUSCI_B3 RX3	TA2CCR2	Reserved
Channel 6	eUSCI_A3 TX	eUSCI_B3 TX0	eUSCI_B2 TX1	eUSCI_B1 TX2	eUSCI_B0 TX3	TA3CCR0	DMAE0 (Pin P7.0)
Channel 7	eUSCI_A3 RX	eUSCI_B3 RX0	eUSCI_B2 RX1	eUSCI_B1 RX2	eUSCI_B0 RX3	TA3CCR2	ADC14

Table 12.1 DMA Sources for All Channels

needed to perform one DMA transfer cycle. The DMA cycle starts with a request from the source and ends after the predefined number of transfers are complete. The microDMA controller parameters can be updated after each cycle.

The microDMA controller supports three different data size as 8-bit (one byte), 16-bit (half-word), and 32-bit (one word). Source and destination data size must be equal. If this is not the case, then the microDMA controller uses the source data size instead of the destination data size. The address increment size is used to adjust the period of the address locations to be processed. This allows reaching nonsuccessive data locations. In other words, data may be skipped during transfer.

Different address increment sizes can be selected for each data size. For example, 8-, 16-, or 32-bit address increment size can be used for eight-bit data size. In a similar manner, 16- or 32-bit address increment size can be used for the 16-bit data size. Only 32-bit address increment size can be used for the 32-bit data size. There is also an option for no increment for each data size value. When this option is selected, the DMA cycle is performed without skipping any data.

The arbitration rate is labeled by R. The microDMA controller arbitrates after 2^R DMA transfers for high-priority DMA requests. Here, Channel 0 has the highest priority. Channel 7 has the lowest priority. R can get values between 0 and 10 to obtain arbitration rates between 1 and 1024 transfers. The transfer size is labeled by N and can get values between 1 and 1024.

12.1.3 Cycle Operating Modes

The microDMA controller supports eight different cycle operating modes. These decide on how the controller performs a DMA cycle. We will only explain invalid cycle, basic cycle, auto-request cycle, and ping-pong cycle modes in this book. There are also memory scatter-gather using the primary data structure, memory scatter-gather using the alternate data structure, peripheral scatter-gather using the primary data structure, and peripheral scatter-gather using the alternate data structure modes. For further information on these, please see [4].

The invalid cycle mode is the base mode for other cycle modes. After the microDMA controller completes a DMA cycle, cycle type is set to invalid automatically to prevent repeating the same DMA cycle. In the basic cycle mode, the microDMA controller must receive a request from the same channel after arbitration to continue the DMA cycle. Let's take two generic channels as A and B. One DMA cycle is composed of the following operations in the basic cycle mode:

1. The microDMA controller starts the DMA cycle with a request from the channel A.
2. If $N \leq 2^R$, the microDMA controller performs N DMA transfers, then continues from step 6.
3. The microDMA controller arbitrates after 2^R DMA transfers and N value is updated to show the remaining DMA transfer number.
4. If the channel B has a DMA request and has higher priority, then the microDMA controller services it.
5. If the channel A has a DMA request, the microDMA controller continues from step 2.
6. The DMA cycle ends and a DMA interrupt is generated (if enabled beforehand).

In the auto-request cycle mode, the microDMA controller only receives a single request at the beginning of the DMA cycle. The controller serves the request without any other request from the same channel. Again, let's take two generic channels as A and B. One DMA cycle for the auto-request cycle mode is as follows:

1. The microDMA controller starts the DMA cycle with a request from the channel A.
2. If $N \leq 2^R$, the microDMA controller performs N DMA transfers, then continues from step 6.
3. The microDMA controller arbitrates after 2^R DMA transfers and N value is updated to show the remaining DMA transfer number.
4. If the channel B has a DMA request and has higher priority, the microDMA controller services it.

5. If the channel A has a higher priority, the microDMA controller continues from step 2.
6. DMA cycle ends and a DMA interrupt is generated (if enabled beforehand).

In the ping-pong cycle mode, the microDMA controller uses two data structures: primary and alternate. First, the controller performs one DMA cycle using the primary data structure. Afterwards, it performs the next DMA cycle using the alternate data structure. Then, the microDMA controller switches back to the primary data structure for the next DMA cycle. This switching process continues until the controller reads an invalid data structure or related DMA channel is disabled.

12.1.4 DMA Interrupts

The microDMA controller can generate up to four different interrupt requests. These are DMA_INT0, DMA_INT1, DMA_INT2, and DMA_INT3. All interrupt requests from eight channels are controlled in the DMA_INT0 using their interrupt flag. DMA_INT1, DMA_INT2, and DMA_INT3 are special. Each can be used for an interrupt from a single channel to obtain a faster interrupt response.

12.2 DMA Module Usage via DriverLib Functions

The DMA module can be controlled via DriverLib functions. We provide the necessary functions to be used for this purpose below. The complete function list for the DMA module can be found in [11].

```
void CS_setExternalClockSourceFrequency (uint32_t lfxt_XT_CLK_
    frequency, uint32_t hfxt_XT_CLK_frequency)
/*
Sets the external crystal frequency values for LFXT and HFXT
    oscillators.
This function must be called if an external crystal LFXT or HFXT is
    used. It does not change any register.
lfxt_XT_CLK_frequency:  the LFXT frequency in Hz.
hfxt_XT_CLK_frequency: the HFXT frequency in Hz.
*/

void DMA_enableModule (void)
/*
Enables the DMA module.
*/

void DMA_disableModule (void)
/*
Disables the DMA module.
*/

void DMA_enableChannel (uint32_t channelNum)
/*
Enables the DMA channel to be used. The channel is automatically
    disabled after the DMA transfer is complete.
```

```
channelNum: channel number to be enabled. It can be one of channel 0
    to 7. Please see the DriverLib user guide for macro definitions.
*/

void DMA_disableChannel (uint32_t channelNum)
/*
Disables the DMA channel.
channelNum: channel number to be disabled. It can be one of channel 0
    to 7. Please see the DriverLib user guide for macro definitions.
*/

void DMA_setControlBase (void *controlTable)
/*
Sets the base address of control table used by the enabled DMA
    channels. The base address must be configured before any channel
        functions are used. The control table holds configuration
        parameters of DMA channel. Size of the table depends on the
        channel number and transfer mode. Control table must be aligned
            in 1024-byte memory boundary. This can be done by the code
            line #pragma DATA_ALIGN(controlTable, 1024)
controlTable: pointer of the control table.
*/

void DMA_enableChannelAttribute (uint32_t channelNum, uint32_t attr)
/*
Enables attributes of the selected channel.
channelNum: channel number to be configured. It can be one of channel 0
    to 7. Please see the DriverLib user guide for macro definitions.
attr: attributes to be enabled. It can be the logical OR form of burst
    mode selection, alternate control structure selection, high
    priority, and hardware request masks. Please see the DriverLib user
        guide for macro definitions.
*/

void DMA_disableChannelAttribute (uint32_t channelNum, uint32_t attr)
/*
Disables attributes of selected channel.
channelNum: channel number to be configured. It can be one of channel 0
    to 7. Please see the DriverLib user guide for macro definitions.
attr: attributes to be disabled. It can be the logical OR form of burst
    mode selection, alternate control structure selection, high
    priority, and hardware request masks. Please see the DriverLib user
        guide for macro definitions.
*/

uint32_t DMA_getChannelAttribute (uint32_t channelNum, uint32_t attr)
/*
Returns attributes of the selected channel.
channelNum: channel number to be read. It can be one of channel 0 to 7.
    Please see the DriverLib user guide for macro definitions.
*/

void DMA_setChannelControl (uint32_t channelStructIndex, uint32_t
    control)
/*
```

*Configures the channel control structure.
channelStructIndex: channel control structure to be configured. It can
    be the logical OR form of channel number with primary or
    alternative structures, UDMA_PRI_SELECT or UDMA_ALT_SELECT.
control: control parameter to be set. It can be the logical OR form of
    data sizes, source and destination address increment types, and
    data transfer length. Please see the DriverLib user guide for macro
     definitions.
/

void DMA_setChannelTransfer (uint32_t channelStructIndex, uint32_t
    mode, **void** *srcAddr, **void** *dstAddr, uint32_t transferSize)
/*
Configures the channel transfer parameters.
channelStructIndex: channel control structure to be configured. It can
    be the logical OR form of channel number with primary or
    alternative structures, UDMA_PRI_SELECT or UDMA_ALT_SELECT.
mode: transfer mode to be used and can be one of stop, basic, auto,
    ping-pong, memory scatter-gather, and peripheral scatter-gather
    values. Please see the DriverLib user guide for macro definitions.
*srcAddr: pointer to the source data.
*dstAddr: pointer to the destination data.
transferSize: size of data to be transferred.
/

void DMA_assignChannel (uint32_t mapping)
/*
Assigns a peripheral to the DMA channel.
mapping: peripheral-channel assignment macro. Please see the DriverLib
    user guide for macro definitions.
/

void DMA_requestSoftwareTransfer (uint32_t channel)
/*
Initializes DMA transfer by software at selected DMA channel.
channel: DMA channel number to be initialized. It can be one of channel
     0 to 7. Please see the DriverLib user guide for macro definitions.
/

void DMA_assignInterrupt (uint32_t interruptNumber, uint32_t channel)
/*
Assigns selected DMA channel to the selected DMA interrupt handler.
interruptNumber: interrupt handler to be assigned. It can be one of
    DMA_INT1, DMA_INT2, and DMA_INT3.
channel: DMA channel number to be assigned. It can be one of channel 0
    to 7. Please see the DriverLib user guide for macro definitions.
/

void DMA_enableInterrupt (uint32_t interruptNumber)
/*
Enables the selected DMA interrupt.
interruptNumber: interrupt to be enabled. It can be one of DMA_INT0,
    DMA_INT1, DMA_INT2, DMA_INT3, and DMA_INTERR.
/

```
void DMA_disableInterrupt (uint32_t interruptNumber)
/*
Disables the selected DMA interrupt.
interruptNumber: interrupt to be disabled. It can be one of DMA_INT0,
    DMA_INT1, DMA_INT2, DMA_INT3, and DMA_INTERR.
*/

uint32_t DMA_getInterruptStatus (void)
/*
Returns the status of DMA interrupt mask.
*/
```

12.3 Coding Practices for the DMA Module

We provide sample C codes on the usage of the DMA module in this section. These examples contain memory-to-memory, peripheral-to-memory, and memory-to-peripheral data transfers. They also show basic, auto, and ping-pong modes with different data sizes. We use LED1 connected to pin P1.0, and red and green colors of LED2 connected to pins P2.0 and P2.1 on the MSP432 LaunchPad as output in the following C codes.

12.3.1 Memory-to-Memory Transfer

The first DMA usage example is on transferring data from one memory location to another. To do so, we configure the DMA module to use the primary structure, auto transfer mode, and eight-bit data size. The aim here is transferring 32 elements between two memory locations. We provide the corresponding C code in Listing 12.1. Here, we first set LED1 connected to pin P1.0 on the MSP432LaunchPad as output and clear the destination array. Before using any DMA functions, we define the controlTable as global array and align it on 1024-byte memory boundary. Then, we enable the DMA module and set the base address of the controlTable using DMA_enableModule and DMA_setControlBase functions. Afterwards, we configure the channel control structure and transfer parameters using DMA_setChannelControl and DMA_setChannelTransfer functions. Then, we enable the DMA Channel 0 and assign the DMA interrupt 1 to it using DMA_enableChannel and DMA_assignInterrupt functions. Finally, we initialize the transfer on DMA channel 0 using the function DMA_requestSoftwareTransfer and let the CPU stay in low-power mode. In the DMA ISR, we disable the DMA channel using the function DMA_disableChannel and turn on LED1 to indicate that DMA transfer has ended.

12.3.2 Transfer Between Memory and Peripheral Devices

The second example is on transferring data both from memory to peripheral and peripheral to memory using DMA. To do so, we use the UART module as the peripheral device such that it transfers data between the host PC and MSP432 LaunchPad bidirectionally. In a normal UART operation, data is transmitted byte by byte. If needed, an interrupt is generated after each transferred byte. In this example, we transmit 32 bytes at once using the DMA module then generate an interrupt. Meanwhile, the CPU stays in low-power mode. While adjusting the DMA module, the transfer size (set as 32 bytes) and the address are set. The microDMA controller handles the rest of the operations. The same operations are done while receiving data from the host PC.

Listing 12.1 Usage of the DMA Module in Memory-to-Memory Mode

```c
#include <ti/devices/msp432p4xx/driverlib/driverlib.h>

#pragma DATA_ALIGN(controlTable, 32)

uint8_t controlTable[32];
uint8_t source[32] =
    {0,1,2,3,4,5,6,7,8,9,8,7,6,5,4,3,2,1,0,1,2,3,4,5,6,7,8,9,8,7,6,5};
uint8_t destination[32];
int i;

void main(void) {

WDT_A_holdTimer();

GPIO_setAsOutputPin(GPIO_PORT_P1, GPIO_PIN0);
GPIO_setOutputLowOnPin(GPIO_PORT_P1, GPIO_PIN0);

for(i=0; i<32; i++)
 destination[i] = 0;

DMA_enableModule();
DMA_setControlBase(controlTable);
DMA_setChannelControl(UDMA_PRI_SELECT, UDMA_SIZE_8 | UDMA_SRC_INC_8 |
    UDMA_DST_INC_8 | UDMA_ARB_32);
DMA_setChannelTransfer(UDMA_PRI_SELECT, UDMA_MODE_AUTO, source,
    destination, 32);
DMA_enableChannel(0);
DMA_assignInterrupt(DMA_INT1, 0);

Interrupt_enableInterrupt(INT_DMA_INT1);
Interrupt_enableSleepOnIsrExit();
Interrupt_enableMaster();

DMA_requestSoftwareTransfer(0);

PCM_setPowerState(PCM_LPM0_LDO_VCORE0);
}
void DMA_INT1_IRQHandler(void) {
 DMA_disableChannel(0);
 GPIO_setOutputHighOnPin(GPIO_PORT_P1, GPIO_PIN0);
}
```

We configure the DMA module to use the primary structure, basic transfer mode, and eight-bit data size. We provide the corresponding C code in Listing 12.2. Here, we first set the red and green colors of LED2 connected to pins P2.0 and P2.1 on the MSP432 LaunchPad as output. Then, we configure and initialize the peripheral pins and eUSCI A0 module in UART mode at 115200 bps. Before using any DMA functions, we define `controlTable` as the global array and align it on 1024-byte memory boundary. Then, we enable the DMA module and set the base address of the `controlTable`. Afterwards, we use the function `DMA_assignChannel` to assign eUSCI A0 TX and RX to DMA Channels 0 and 1, respectively. Then, we disable all attributes to DMA

Listing 12.2 Usage of the DMA Module in Peripheral-to-Memory and Memory-to-Peripheral Modes

```c
#include <ti/devices/msp432p4xx/driverlib/driverlib.h>

const eUSCI_UART_Config uartConfig =
{
 EUSCI_A_UART_CLOCKSOURCE_SMCLK,      // SMCLK Clock Source
 6,                                    // BRDIV = 13
 8,                                    // UCxBRF = 0
 32,                                   // UCxBRS = 37
 EUSCI_A_UART_NO_PARITY,               // No Parity
 EUSCI_A_UART_LSB_FIRST,               // MSB First
 EUSCI_A_UART_ONE_STOP_BIT,            // One stop bit
 EUSCI_A_UART_MODE,                    // UART mode
 EUSCI_A_UART_OVERSAMPLING_BAUDRATE_GENERATION  // Oversampling
};

#pragma DATA_ALIGN(controlTable, 1024)

 uint8_t controlTable[64];
 char TXData[32] = "MSP432 UART DMA Example 12345678";
 char RXData[32];

void main(void){

 WDT_A_holdTimer();

 GPIO_setAsOutputPin(GPIO_PORT_P2, GPIO_PIN0 | GPIO_PIN1);
 GPIO_setOutputLowOnPin(GPIO_PORT_P2, GPIO_PIN0 | GPIO_PIN1);
 GPIO_setAsPeripheralModuleFunctionInputPin(GPIO_PORT_P1, GPIO_PIN2 |
     GPIO_PIN3, GPIO_PRIMARY_MODULE_FUNCTION);

 CS_setDCOCenteredFrequency(CS_DCO_FREQUENCY_12);

 UART_initModule(EUSCI_A0_BASE, &uartConfig);
 UART_enableModule(EUSCI_A0_BASE);

 DMA_enableModule();
 DMA_setControlBase(controlTable);
 DMA_assignChannel(DMA_CH0_EUSCIA0TX);
 DMA_assignChannel(DMA_CH1_EUSCIA0RX);
 DMA_disableChannelAttribute(DMA_CH0_EUSCIA0TX, UDMA_ATTR_ALTSELECT |
     UDMA_ATTR_USEBURST | UDMA_ATTR_HIGH_PRIORITY | UDMA_ATTR_REQMASK);
 DMA_disableChannelAttribute(DMA_CH1_EUSCIA0RX, UDMA_ATTR_ALTSELECT |
     UDMA_ATTR_USEBURST | UDMA_ATTR_HIGH_PRIORITY | UDMA_ATTR_REQMASK);
 DMA_setChannelControl(UDMA_PRI_SELECT | DMA_CH0_EUSCIA0TX, UDMA_SIZE_8
     | UDMA_SRC_INC_8 | UDMA_DST_INC_NONE | UDMA_ARB_1);
 DMA_setChannelControl(UDMA_PRI_SELECT | DMA_CH1_EUSCIA0RX, UDMA_SIZE_8
     | UDMA_SRC_INC_NONE | UDMA_DST_INC_8 | UDMA_ARB_1);
 DMA_setChannelTransfer(UDMA_PRI_SELECT | DMA_CH0_EUSCIA0TX,
     UDMA_MODE_BASIC, TXData, (void*)
     UART_getTransmitBufferAddressForDMA(EUSCI_A0_BASE), 32);
 DMA_setChannelTransfer(UDMA_PRI_SELECT | DMA_CH1_EUSCIA0RX,
     UDMA_MODE_BASIC, (void*) UART_getReceiveBufferAddressForDMA(
     EUSCI_A0_BASE), RXData, 32);
 DMA_assignInterrupt(DMA_INT1, DMA_CHANNEL_0);
```

```c
    DMA_assignInterrupt(DMA_INT2, DMA_CHANNEL_1);

    Interrupt_enableInterrupt(INT_DMA_INT1);
    Interrupt_enableInterrupt(INT_DMA_INT2);
    Interrupt_enableSleepOnIsrExit();

    DMA_enableChannel(DMA_CHANNEL_0);
    DMA_enableChannel(DMA_CHANNEL_1);

    PCM_setPowerState(PCM_LPM0_LDO_VCORE0);
}

void DMA_INT1_IRQHandler(void) {
    GPIO_setOutputHighOnPin(GPIO_PORT_P2, GPIO_PIN0);
    DMA_disableChannel(DMA_CHANNEL_0);
    DMA_disableInterrupt(INT_DMA_INT1);
}

void DMA_INT2_IRQHandler(void) {
    DMA_clearInterruptFlag(INT_DMA_INT2);
    GPIO_setOutputHighOnPin(GPIO_PORT_P2, GPIO_PIN1);
}
```

Channels 0 and 1 using the function `DMA_disableChannelAttribute`. We configure the channel control structure and transfer parameters and enable the DMA Channels 0 and 1. We assign them to DMA interrupts 1 and 2, and let the CPU stay in low-power mode. In the DMA interrupt 1 ISR, we disable the DMA Channel 0 using the function `DMA_disableChannel` and turn on the red color of LED2 to indicate that data transfer to host PC has ended. In the DMA interrupt 2 ISR, we disable the DMA Channel 1 and turn on the green color of LED2 to indicate that data transfer from the host PC has completed.

12.3.3 Peripheral to Memory Transfer

The third DMA example is on transferring data from the ADC14 module (as a peripheral device) to memory location. We set the DMA module to receive 64 samples and write them to the given address. The ADC14 module is triggered by the timer. Data reception is done by the DMA module within the ADC ISR. If we want to process data in real-time in a digital system, this should be done within one sampling period. Otherwise, we may miss the incoming data. To avoid this, we can use the ping-pong mode. Hence, data at the pong buffer can be processed, when the ping buffer is filled. This increases the data processing time to sampling period × buffer size. In our application, the DMA module is set to process 64 samples (32 for ping and 32 for pong buffers). The sampling rate is set as 16 Hz by the timer module. Therefore, the module gets data for two seconds and enter an DMA ISR afterwards. Within the ISR, the ping and pong states are changed. Then, the process is repeated again.

As in previous examples, we configure the DMA module to use the primary and alternate structures, ping-pong transfer mode, and eight-bit data size. We provide the corresponding C code in Listing 12.3. Here, we first set LED1 connected to pin P1.0 on the MSP432 LaunchPad as output, and configure and enable the Timer_A0 module to trigger the ADC at 16 Hz. We also set the ADC module to use its internal temperature

Listing 12.3 Usage of the DMA Module in Peripheral-to-Memory Ping-Pong Mode

```c
#include <ti/devices/msp432p4xx/driverlib/driverlib.h>

const Timer_A_UpModeConfig upModeConfig =
{
 TIMER_A_CLOCKSOURCE_ACLK,
 TIMER_A_CLOCKSOURCE_DIVIDER_1,
 2048,
 TIMER_A_TAIE_INTERRUPT_DISABLE,
 TIMER_A_CCIE_CCR0_INTERRUPT_DISABLE,
 TIMER_A_DO_CLEAR
};

const Timer_A_CompareModeConfig compareConfig =
{
 TIMER_A_CAPTURECOMPARE_REGISTER_1,
 TIMER_A_CAPTURECOMPARE_INTERRUPT_DISABLE,
 TIMER_A_OUTPUTMODE_SET_RESET,
 2048
};

#pragma DATA_ALIGN(controlTable, 1024)

 uint8_t controlTable[32];
 uint16_t destination_ping[32];
 uint16_t destination_pong[32];
 int i;
 float result[64];
 volatile int pingpong = 0;

void main(void){

 WDT_A_holdTimer();

 GPIO_setAsOutputPin(GPIO_PORT_P1, GPIO_PIN0);
 GPIO_setOutputLowOnPin(GPIO_PORT_P1, GPIO_PIN0);

 CS_setReferenceOscillatorFrequency(CS_REFO_32KHZ);
 CS_initClockSignal(CS_ACLK, CS_REFOCLK_SELECT, CS_CLOCK_DIVIDER_1);

 Timer_A_configureUpMode(TIMER_A0_BASE, &upModeConfig);
 Timer_A_initCompare(TIMER_A0_BASE, &compareConfig);

 ADC14_enableModule();
 ADC14_initModule(ADC_CLOCKSOURCE_MCLK, ADC_PREDIVIDER_1,
     ADC_DIVIDER_1, ADC_TEMPSENSEMAP);
 ADC14_configureSingleSampleMode(ADC_MEM0, true);
 ADC14_configureConversionMemory(ADC_MEM0,
     ADC_VREFPOS_AVCC_VREFNEG_VSS, ADC_INPUT_A22, false);
 ADC14_setSampleHoldTrigger(ADC_TRIGGER_SOURCE1, false);
 ADC14_enableSampleTimer(ADC_MANUAL_ITERATION);
 ADC14_enableConversion();

 REF_A_setReferenceVoltage(REF_A_VREF2_5V);
 REF_A_enableReferenceVoltage();
```

```c
    REF_A_enableTempSensor();

    DMA_enableModule();
    DMA_setControlBase(controlTable);
    DMA_disableChannelAttribute(DMA_CH7_ADC14, UDMA_ATTR_ALTSELECT |
        UDMA_ATTR_USEBURST | UDMA_ATTR_HIGH_PRIORITY | UDMA_ATTR_REQMASK);
    DMA_setChannelControl(UDMA_PRI_SELECT | DMA_CH7_ADC14, UDMA_SIZE_16 |
        UDMA_SRC_INC_NONE | UDMA_DST_INC_16 | UDMA_ARB_1);
    DMA_setChannelTransfer(UDMA_PRI_SELECT | DMA_CH7_ADC14,
        UDMA_MODE_PINGPONG, (void*) &ADC14->MEM[0], destination_ping, 32);
    DMA_setChannelControl(UDMA_ALT_SELECT | DMA_CH7_ADC14, UDMA_SIZE_16 |
        UDMA_SRC_INC_NONE | UDMA_DST_INC_16 | UDMA_ARB_1);
    DMA_setChannelTransfer(UDMA_ALT_SELECT | DMA_CH7_ADC14,
        UDMA_MODE_PINGPONG, (void*) &ADC14->MEM[0], destination_pong, 32);
    DMA_assignChannel(DMA_CH7_ADC14);
    DMA_enableChannel(DMA_CHANNEL_7);
    DMA_assignInterrupt(INT_DMA_INT1, DMA_CHANNEL_7);
    DMA_enableInterrupt(INT_DMA_INT1);

    Interrupt_enableInterrupt(INT_DMA_INT1);
    Interrupt_enableMaster();

    Timer_A_startCounter(TIMER_A0_BASE, TIMER_A_UP_MODE);

    while(1){
     PCM_setPowerState(PCM_LPM0_LDO_VCORE0);
     if (pingpong == 1){
      for(i=0;i<32;i++)
       result[i] = (float) ((int32_t)(destination_ping[i] - 4000) * (85 -
           30)) / (2000) + 30.0f;}
     else{
      for(i=0;i<32;i++)
       result[i+32] = (float) ((int32_t)(destination_pong[i] - 4000) *
           (85 - 30)) / (2000) + 30.0f;}
     GPIO_toggleOutputOnPin(GPIO_PORT_P1, GPIO_PIN0);}
    }

    void DMA_INT1_IRQHandler(void){
     if (DMA_getChannelAttribute(DMA_CHANNEL_7) & UDMA_ATTR_ALTSELECT){
      DMA_setChannelControl(UDMA_PRI_SELECT | DMA_CH7_ADC14, UDMA_SIZE_16 |
          UDMA_SRC_INC_NONE | UDMA_DST_INC_16 | UDMA_ARB_1);
      DMA_setChannelTransfer(UDMA_PRI_SELECT | DMA_CH7_ADC14,
          UDMA_MODE_PINGPONG, (void*) &ADC14->MEM[0], destination_ping,
          32);
      pingpong=1;}
     else{
      DMA_setChannelControl(UDMA_ALT_SELECT | DMA_CH7_ADC14, UDMA_SIZE_16 |
          UDMA_SRC_INC_NONE | UDMA_DST_INC_16 | UDMA_ARB_1);
      DMA_setChannelTransfer(UDMA_ALT_SELECT | DMA_CH7_ADC14,
          UDMA_MODE_PINGPONG, (void*) &ADC14->MEM[0], destination_pong,
          32);
      pingpong=0;}
    }
```

sensor. Before using any DMA functions, we define `controlTable` as global array and align it on 1024-byte memory boundary. Then, we enable the DMA module and set the base address of `controlTable`. Then, we disable all attributes of the DMA Channel 7, configure the channel control structure and transfer parameters, and assign the ADC to DMA Channel 7. We also enable and assign the DMA Channel 7 to DMA interrupt 1. We calculate the temperature value from sampled ADC values in a while loop according to the ping-pong flag. In the DMA interrupt 1 ISR, we check the channel attributes to get the used structure. If the alternate structure is currently being used and transfer from primary structure has ended, we reconfigure the primary structure and set the ping-pong flag. If the primary structure is currently being used and transfer from alternate structure has ended, we reconfigure the alternate structure and reset the ping-pong flag.

12.4 Application: Baby Monitor System

The aim of this application is to learn how to use the DMA module on the MSP432 microcontroller. As a real-life application, we will design a baby monitor system. In this section, we provide the equipment list, layout of the circuit, and system design specifications.

12.4.1 Equipment List

We will use the Groove sound sensor module to detect the surrounding sound signal. We will also use LED1 connected to pin P1.0 on the MSP432 LaunchPad in the baby monitor system.

12.4.2 Layout

The layout of the baby monitor system is given in Fig. 12.2.

12.4.3 System Design Specifications

The design of our baby monitor system will be as follows. The sound level from the Groove sensor will be sampled by the ADC using the DMA module. The sampling rate will be 10 kHz. When the average of last 1024 samples exceeds a threshold level, then LED1 connected to P1.0 pin on the MSP432 LaunchPad will turn on. The CPU will be at a suitable low-power mode during this operation. To realize the baby monitor, we should add a second MSP432 connected to the first one via Bluetooth connection. Hence, the sound level measuring and LED1 alarm parts can be in two different locations.

We could not add the C code of the project here due to page limitations. However, we provided the complete CCS project for this application in the companion website for the reader. We strongly suggest implementing it.

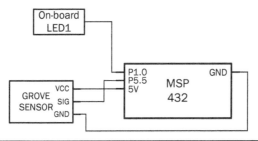

FIGURE 12.2 Layout of the baby monitor system.

12.5 Summary

The DMA module can be used when large data blocks need to be transferred between two locations within the microcontroller. The aim here is performing this task without interfering the CPU. MSP432 has an eight-channel DMA module for this purpose. In this chapter, we explored DMA operations using this module. We benefit from DriverLib functions while using the DMA module. We also provided a real-life application on the usage of the DMA module. We believe this chapter clarifies the DMA module. Hence, the reader can benefit from it whenever needed.

12.6 Exercises

In this section, we will not offer new problems. Instead, we will ask the reader to reconsider the exercises given in previous chapters by adding DMA to them. Some sample problems are given below:

12.1 Rewrite the code given in Listing 8.13 by using the DMA module.

12.2 Repeat the examples in Sec. 9.3.3 by using DMA modules in Listings 9.12 and 9.13.

12.3 Repeat the examples in Sec. 9.4.3 by using DMA modules in Listings 9.18 and 9.19.

12.4 Repeat the examples in Sec. 9.4.3 by using DMA modules in Listings 9.20 and 9.21.

CHAPTER 13
Real-time Operating System

As the operations to be performed on the microcontroller become more complex, it becomes difficult to use standard coding methods. There, the real-time operating system (RTOS) comes into play. RTOS provides a standardized environment to manage complex operations on the microcontroller with efficient use of its resources. We will introduce RTOS concepts in this chapter starting with fundamental definitions. Then, we will analyze TI-RTOS that will be used throughout the chapter. We will have an example-based approach to explain the RTOS concepts on TI-RTOS. We will also consider Energia from an RTOS perspective. Finally, we will provide an application to show the usefulness of RTOS.

13.1 What Is RTOS?

Up to now, we only considered fairly simple operations on the MSP432 microcontroller. A sequential C or assembly code was sufficient to satisfy the requirements in these operations. This is called bare-metal programming to indicate that the developed code directly interacts with the microcontroller resources. Unfortunately, this approach becomes infeasible when the operations to be performed become complex. For these, the efficient usage of microcontroller resources becomes the main concern. Hence, there should be a system to manage and organize these complex operations. RTOS is introduced for this purpose.

13.1.1 RTOS Fundamentals

RTOS needs several tools to organize complex operations to be performed on the microcontroller. The first and most important of these is the kernel. Then comes the drivers and instrumentation tools. To note here, different RTOS types may contain extra tools besides the mentioned ones. However, we will only focus on the kernel, drivers, and instrumentation tools in the following sections since they are the most used ones. Before going further, let's briefly explain RTOS fundamentals.

The work to be done should be divided into function-like blocks called threads in RTOS. The difference of the thread from a function is that it may have its own stack and control block. Besides, the thread is basically a function. There are four thread groups as hardware interrupt (Hwi), software interrupt (Swi), task, and idle loop. Each group has its specific usage area. The RTOS kernel is responsible for scheduling the execution of threads. Meanwhile, communication between these threads may become a necessity. The kernel also has special features such as semaphore, event, mailbox, and queue for this purpose. Time-based operations may also be needed in an RTOS application. The kernel also has special features to handle them.

There are two different RTOS types based on the scheduling mechanism they use. The first scheduling mechanism is called time sharing in which each thread is executed on a regular clock signal. The second scheduling mechanism is called event-driven or preemptive in which each thread has a priority assigned to it. The highest priority thread is always served first. This scheduling mechanism is more suitable for embedded systems where operations within them are executed under real-time constraints.

13.1.2 Advantages and Disadvantages of RTOS

RTOS offers several advantages when used in complex projects. Lethaby [42] summarizes these in five parts as preemptive multitasking design paradigm, pre-tested and pre-integrated communications stacks and drivers, application portability, system-level debug and analysis tools, and more efficient use of CPU resources. These advantages allow a large project to be developed by a diverse group of designers such that each focuses on specific threads. Besides, the developed project can be tested efficiently in system level. Hence, possible errors and malfunctions can be localized easily.

Although RTOS has several noteworthy advantages, it also has some disadvantages. The first of these is the increase in code size since RTOS needs extra predefined functions to operate. Besides, RTOS itself needs valuable CPU resources for the same reason. These are analyzed in detail in the TI-RTOS workshop series website [43]. We strongly suggest the reader to visit this website to get a feeling on the resource usage of RTOS. The second disadvantage of RTOS is that it is based on a different programming paradigm. It may be difficult for the programmer to get used to it. Therefore, RTOS may not be the best solution for all applications. In other words, single purpose and simple applications with low code size generally do not need RTOS.

13.1.3 RTOS Types

There are several RTOS types available in the market to be used in projects. Some of these are free of charge, while others require royalty fee. TI offers TI-RTOS and FreeRTOS to be used in CCS. Therefore, we will only consider TI-RTOS due to its ease of use in CCS. However, the concepts to be introduced in the following sections can also be applied on other RTOS types as well.

13.2 TI-RTOS

TI-RTOS does not come as a preinstalled application in CCS. It is available under the SimpleLink SDK. The user should add TI-RTOS to CCS from the Resource Explorer window as shown in Fig. 13.1.

TI-RTOS has three main components as kernel, middleware and drivers, and instrumentation. These are shown in Fig. 13.2. In this book, we will only explain kernel, instrumentation, and drivers in detail. For further information on the other parts of TI-RTOS, please see [44].

13.2.1 Kernel

TI interchangeably calls the kernel as SYS/BIOS or BIOS in its documents and CCS. To be consistent throughout the book, we will always call it as kernel. We will explain all kernel components in Sec. 13.3.

Real-time Operating System 317

FIGURE 13.1 CCS resource explorer window to install TI-RTOS.

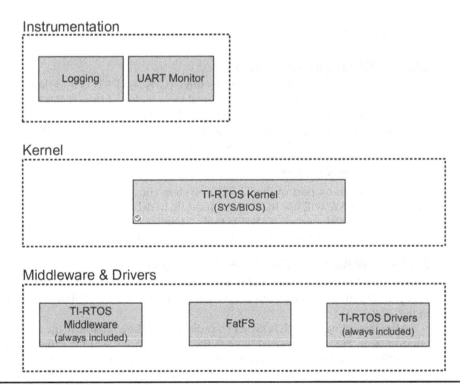

FIGURE 13.2 TI-RTOS components.

318 Chapter Thirteen

Figure 13.3 Creating a new TI-RTOS project.

13.2.2 Middleware and Drivers

Middleware and drivers section contains three parts as TI-RTOS middleware, TI-RTOS drivers, and FatFS. The first two parts are always included in a TI-RTOS project. Middleware includes TCP/IP and USB stacks. TI-RTOS drivers contain special functions for using microcontroller peripherals with TI-RTOS. We will explain them in detail in Sec. 13.4. FatFS can be included to operate on SD card file system.

13.2.3 Instrumentation

The instrumentation part of the TI-RTOS system consists of two parts as logging and UART monitor. We will use the logging module to debug and analyze the performance of the project developed via TI-RTOS tools. Therefore, we will explain in Sec. 13.5 how to use it. For further information on the UART monitor, please see [44].

13.2.4 Creating a TI-RTOS-Based CCS Project

We will benefit from a predefined project titled `hello` in developing TI-RTOS projects. This project can be reached from the resource explorer window following the folder list in Fig. 13.3.

After importing the TI-RTOS `hello` project, it will be available in the Project Explorer window as in Fig. 13.4. The `hello.c` is the main source file of the project. The content of this file is as in Listing 13.1. We will modify it throughout the chapter.

Let's briefly analyze the contents of the source file `hello.c`. Installed first two header files are for the XDCtools (preinstalled in CCS). These are necessary to use and configure TI-RTOS components. The third header file is for adding TI-RTOS support files

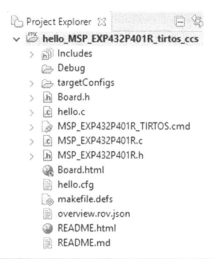

FIGURE 13.4 Project Explorer window of the TI-RTOS project.

Listing 13.1 The Contents of the `hello.c` File

```
#include <xdc/std.h>
#include <xdc/runtime/System.h>
#include <ti/sysbios/BIOS.h>

void main(void){

 System_printf("hello world\n");

 BIOS_exit(0);
}
```

to the project. Within main, there is a function `System_printf()`. When this function is called, its content is stored in a buffer. Then, data in this buffer is transferred to the CCS console window when the function `System_flush()` or `BIOS_exit()` is called. The last line in the code contains the function `BIOS_exit(0)` which ends the BIOS usage.

There is one more important file in the project list called `hello.cfg`. This file holds information on the configuration settings for the TI-RTOS project. We can open this file by right-clicking on it and selecting *Open With → XGCONF*. A welcome screen will be seen as in Fig. 13.5. We will use this file to add or modify TI-RTOS kernel components in the following section.

As we build and run the project, we should see the sentence "hello world" on the console window. This indicates that everything is in order. Hence, we are ready to use TI-RTOS.

13.3 TI-RTOS Kernel

As mentioned previously, kernel is the most important component of RTOS. TI groups kernel elements into five parts as threads, synchronization, memory management, diagnostics, and startup. Each part is also divided into subgroups. We provide the schematic view of the TI-RTOS kernel in Fig. 13.6. We will consider threads, synchronization, and

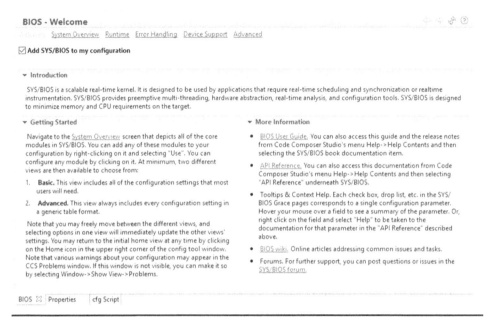

FIGURE 13.5 Welcome screen of the `hello.cfg` configuration file.

memory management groups in this section. We will consider the diagnostics group in Sec. 13.5. We will not consider the startup group in this book. For further information on it, please see [44].

13.3.1 Threads

Thread is the smallest code block that can be executed independently in an RTOS. TI-RTOS kernel provides four different thread types. These are hardware interrupt (Hwi), software interrupt (Swi), task, and idle loop. These threads are prioritized as provided. Hence, an Hwi will always be executed first within the preemptive scheduler. This is followed by the Swi, task, and idle loop. We will consider each thread type separately in the following sections.

13.3.1.1 Idle Loop

The idle loop is the background thread of the TI-RTOS kernel with lowest priority. Any Hwi, Swi, or task can preempt the idle loop during the process. The idle loop may contain one or more threads each associated with a function. These idle threads have the same priority level. Hence, they are run in the same order in which they were created in a continuous loop. An idle thread must be executed completely to execute the next idle thread.

We can create a thread in the idle loop using the configuration file mentioned in Sec. 13.2.4. To do so, open the `hello.cfg` file by right-clicking on it and select *Open With → XGCONF*. In the "BIOS - System Overview" window, click on the "Idle" box. A new window should appear as in Fig. 13.7.

As can be seen in Fig. 13.7, there are predefined threads called "User idle function 0", "User idle function 1", and so on. Do not forget that these threads will be executed in the order from 0 to the last one in the idle loop. We can associate a user-defined function to

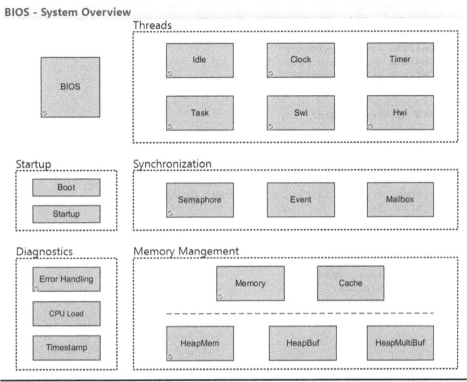

FIGURE 13.6 Schematic view of the TI-RTOS kernel components.

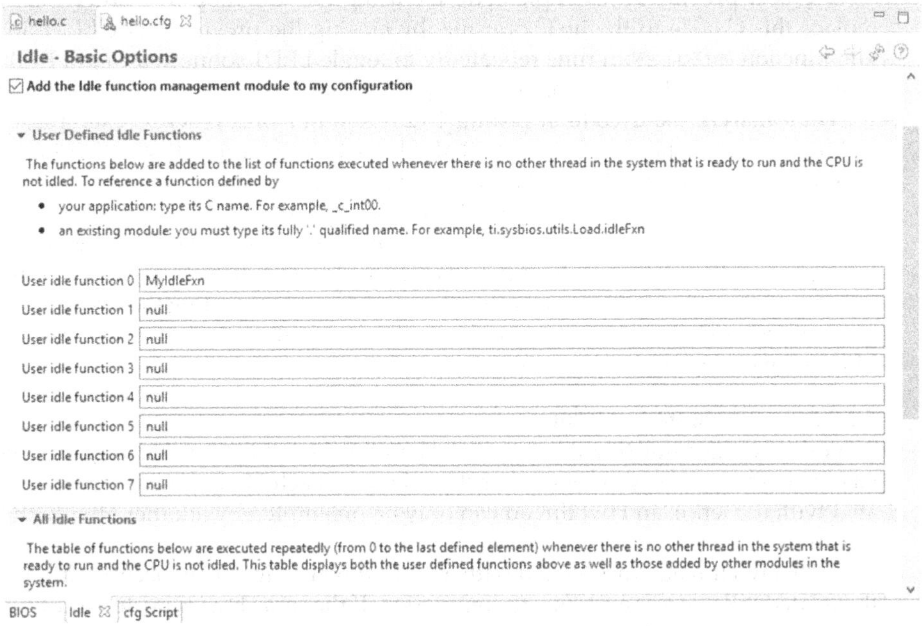

FIGURE 13.7 Creating a thread in the idle loop.

Listing 13.2 The C Code for the Idle Loop Example

```c
#include <xdc/std.h>
#include <xdc/runtime/System.h>
#include <ti/sysbios/BIOS.h>
#include <ti/drivers/GPIO.h>
#include "Board.h"

void MyIdleFxn(void);

void main(void){

 Board_initGPIO();

 System_printf("Starting Idle Thread Example\n");
 System_flush();

 BIOS_start();
}
void MyIdleFxn(void){
 GPIO_toggle(Board_LED0);
  __delay_cycles(24000000);
}
```

each thread by writing its name to the corresponding box. For our example, we declared that the function `MyIdleFxn` is associated with the zeroth thread in the idle loop.

We next provide an example code for using the idle loop in Listing 13.2. As we replace this C code to the `hello.c` file, by erasing the previous code block in it, the idle function `MyIdleFxn` runs repeatedly to toggle LED1 connected to pin P1.0 on the MSP432 LaunchPad approximately once per second.

Let's analyze the C code in Listing 13.2. As in our first TI-RTOS example, we first include necessary header files to add the kernel to our project. We also include the `board.h` file to use the MSP432 LaunchPad hardware. Within the main function, we initialize the GPIO pins by the function `Board_initGPIO()`. Then, we use the `System_printf` and `System_flush` functions to post the string "Starting Idle Thread Example" on the CCS console. Afterwards, we start the kernel usage by the function `BIOS_start()`. Since there is no other thread defined preempting the idle loop, the program executes the `MyIdleFxn` function associated with the zeroth idle thread. Hence, LED1 toggles approximately once per second.

13.3.1.2 Hardware Interrupt

A hardware interrupt (Hwi) supports a thread initiated by external signals. The Hwi has the highest priority among other thread types. Each Hwi also has a priority associated with it. Hence, an Hwi thread can only be preempted by another Hwi with higher priority.

Each Hwi thread is associated with an ISR. When an interrupt occurs, CPU jumps to the related ISR address and runs to completion. If the same Hwi is posted multiple times during this process, the ISR runs only once. To prevent this, the ISR execution must be kept as short as possible.

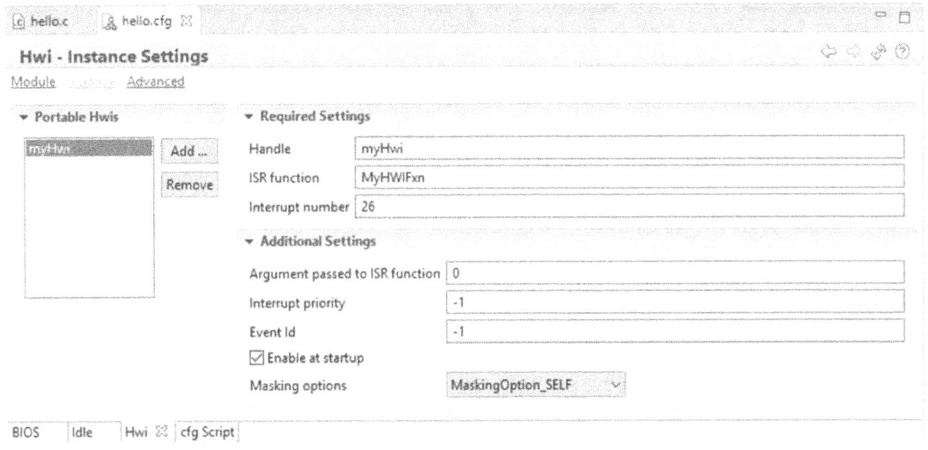

FIGURE 13.8 Creating an Hwi thread.

Each Hwi thread uses the system stack when it is triggered. We already know that a higher priority Hwi can preempt lower priority Hwi. Here, nesting comes into play. This process can cause increasing stack size. Hence, it must be used with precaution.

We can create an Hwi thread statically using the configuration file mentioned in Sec. 13.2.4. To do so, open the `hello.cfg` file by right-clicking on it and select *Open With → XGCONF*. In the "BIOS - System Overview" window, click on the "Hwi" box. In the opened window, tick the box titled "Add the portable Hwi management module to my configuration". Then, select the "Instance" option on the top of the window. A new window should appear as in Fig. 13.8.

To create a new Hwi in the window in Fig. 13.8, we should click on the "Add" button and set the properties of the Hwi thread in the "Required Settings" section. Here, we should provide a name to our thread handle, set the ISR function name to be used with this handle, and set the interrupt number associated with the interrupt to be serviced. This number should be obtained from the hardware documentation reference [4]. There are also additional settings in the same window. These are *Augment passed to ISR function*, *Interrupt priority*, *Event Id*, *Enable at startup*, and *Masking options*. Through these, we can adjust the Hwi thread properties further. These parameters can also be left with their initial values. For further information on them, please see [44].

Let's modify our TI-RTOS project by adding a hardware timer interrupt as Hwi. Within the project, we configure the hardware timer for generating interrupts with one second interval. We call the related function as `Timer_A_delay` and provide it in Listing 13.3.

Using this function, we toggle LED1 connected to pin P1.0 on the MSP432 Launch-Pad in our TI-RTOS project. Therefore, we create an Hwi called `myHwi`, and associate it with an ISR function `MyHWIFxn`. We set the interrupt number as 26 since we are using the hardware timer. To note here, we make all these adjustments on the configuration file as given in Fig. 13.8. We also use the idle loop to print the number of toggles in the CCS console.

We provide the C source code of our project in Listing 13.4. The main difference in this C code is the addition of the `MyHWIFxn` function for the Hwi thread. Besides, there is a `Timer_A_delay` function to set timer interrupts every second.

Listing 13.3 The `Timer_A_delay` Function

```
void Timer_A_delay(void){
 Timer_A_UpModeConfig initUpParam =
 {
  TIMER_A_CLOCKSOURCE_ACLK,
  TIMER_A_CLOCKSOURCE_DIVIDER_1,
  0x8000,
  TIMER_A_TAIE_INTERRUPT_DISABLE,
  TIMER_A_CCIE_CCR0_INTERRUPT_ENABLE,
  TIMER_A_DO_CLEAR,
 };

 CS_initClockSignal(CS_ACLK, CS_REFOCLK_SELECT, CS_CLOCK_DIVIDER_1);
 CS_initClockSignal(CS_SMCLK, CS_MODOSC_SELECT, CS_CLOCK_DIVIDER_2);
 CS_initClockSignal(CS_MCLK, CS_MODOSC_SELECT, CS_CLOCK_DIVIDER_2);
 Timer_A_configureUpMode(TIMER_A1_BASE, &initUpParam);
 Timer_A_clearCaptureCompareInterrupt(TIMER_A1_BASE,
     TIMER_A_CAPTURECOMPARE_REGISTER_0);
 Timer_A_startCounter(TIMER_A1_BASE, TIMER_A_UP_MODE);
}
```

We can also create an Hwi dynamically in C code. We provide such an example in Listing 13.5. This code performs the same operation as in our static Hwi example.

In Listing 13.5, `myHwi` is the handle name of the created Hwi. `MyHWIFxn` is the ISR function triggered by the Hwi. `hwiParams` is the structure name that contains Hwi instance parameters. Here, 26 is the hardware timer interrupt number obtained from the datasheet [4]. `arg` is the instance parameter that will be passed to `MyHWIFxn` when it is triggered. `enableInt` is the instance parameter to enable or disable the interrupt when the Hwi is created. If this parameter is set as `FALSE`, then the interrupt is disabled. Otherwise, the interrupt is enabled. Finally, `eb` is the error block that can be used to handle errors that may occur during Hwi creation.

13.3.1.3 Software Interrupt

The software interrupt (Swi) has the second highest priority after Hwi. An Swi thread can only be preempted by Hwi or another Swi with higher priority. If two or more Swi threads have the same priority, then they will be executed in the order of creation. An Swi is triggered by the TI-RTOS software function `Swi_post(Swi name)`. Each Swi thread is associated with a function. As the Swi is triggered, CPU jumps to this function and runs it to completion. Like an Hwi, if the same Swi is posted multiple times during execution, it only runs once.

The main aim of the Swi usage is post-processing for the Hwi. In other words, when an interrupt comes to trigger the Hwi, it calls the associated Swi to perform data processing. Hence, the Hwi can be ready for the next interrupt signal. Since an Swi can be preempted by hardware, it will not block the incoming Hwi.

Each Swi thread uses the system stack when it is triggered. We already know that Hwis and higher priority Swis can preempt lower priority Swis. This nesting process can cause increasing stack size. Hence, it must be used with precaution.

There are five software functions to post an Swi. These are `Swi_post`, `Swi_inc`, `Swi_dec`, `Swi_or`, and `Swi_andn`. These functions can be used in connection with the trigger value. The trigger (instance parameter) is introduced to use the five Swi functions

Listing 13.4 The C Code for the Static Hwi Usage Example

```c
#include <ti/devices/msp432p4xx/driverlib/driverlib.h>
#include <xdc/std.h>
#include <xdc/runtime/System.h>
#include <ti/sysbios/BIOS.h>
#include <ti/sysbios/hal/Hwi.h>
#include <ti/drivers/GPIO.h>
#include "Board.h"

void MyHWIFxn(UArg arg);
void MyIdleFxn(void);

void Timer_A_delay(void);

 int print = 0;
 int cnt = 0;

void main(void) {

 Board_initGeneral();
 Board_initGPIO();

 Timer_A_delay();

 System_printf("Starting HWI Example\n");
 System_flush();

 BIOS_start();
}

void MyHWIFxn(UArg arg){
 Timer_A_clearCaptureCompareInterrupt(TIMER_A1_BASE,
     TIMER_A_CAPTURECOMPARE_REGISTER_0);
 GPIO_toggle(Board_LED0);
 cnt ++;
 print = 1;
}

void MyIdleFxn(void) {
 if(print == 1){
   System_printf("LED Toggle cnt = %d \n", cnt);
   System_flush();
   print = 0;}
}

void Timer_A_delay(void) {
 ...
}
```

for special purposes. The value of the trigger is set initially. When the Swi is triggered and scheduled to be executed it resets to its initial value. The value before the reset operation can be obtained using the function `Swi_getTrigger` only in the Swi function. Let's focus on the five software functions to post an Swi and how the trigger value can be used with them.

Listing 13.5 The C Code for the Dynamic Hwi Usage Example

```c
#include <ti/devices/msp432p4xx/driverlib/driverlib.h>
#include <xdc/std.h>
#include <xdc/runtime/System.h>
#include <xdc/runtime/Error.h>
#include <ti/sysbios/BIOS.h>
#include <ti/sysbios/hal/Hwi.h>
#include <ti/drivers/GPIO.h>
#include "Board.h"

 Hwi_Params hwiParams;
 Hwi_Handle myHwi;
 Error_Block eb;

void MyHWIFxn(UArg arg);
void MyIdleFxn(void);
void Timer_A_delay(void);

 int print = 0;
 int cnt = 0;

void main(void){

 Board_initGeneral();
 Board_initGPIO();

 Error_init(&eb);
 Hwi_Params_init(&hwiParams);
 hwiParams.arg = 0;
 hwiParams.enableInt = TRUE;
 myHwi = Hwi_create(26, MyHWIFxn, &hwiParams, &eb);
 if (myHwi == NULL) System_abort("Hwi create failed");

 Timer_A_delay();

 System_printf("Starting HWI Example\n");
 System_flush();

 BIOS_start();
}

void MyHWIFxn(UArg arg){
 Timer_A_clearCaptureCompareInterrupt(TIMER_A1_BASE,
     TIMER_A_CAPTURECOMPARE_REGISTER_0);
 GPIO_toggle(Board_LED0);
 cnt ++;
 print = 1;
}

void MyIdleFxn(void){
 if(print == 1){
  System_printf("LED Toggle cnt = %d \n", cnt);
  System_flush();
  print = 0;}
}

void Timer_A_delay(void){
 ...
}
```

The function `Swi_post(mySwi)` posts the Swi `mySwi`. It does not change the trigger value while posting the Swi.

The function `Swi_inc(mySwi)` increases the trigger value by one and posts the Swi `mySwi`. The purpose of this function is as follows. We have mentioned previously that when an Swi is posted again before the previous one has ended, only the first Swi will be executed. With the help of the function `Swi_getTrigger(mySwi)`, the total number of Swi that is not executed can be obtained. Therefore, when the Swi is posted again, the operation can be repeated by this number.

The function `Swi_dec(mySwi)` decreases the trigger value by one and posts the Swi `mySwi` when the trigger becomes 0. We can give an example on the usage of this function as follows. Assume that the Hwi will get data from the ADC14 module and then post the Swi for processing the obtained data. However, the Swi should not be executed until 10 successive ADC data are obtained. To do so, we should set the initial trigger value as 10 and use this function. Hence, when the Hwi is triggered 10 times, it will post the Swi. Afterwards, the trigger value will turn back to 10 and the same procedure will be repeated again with every 10 trigger.

The function `Swi_or(mySwi, mask)` performs the logical OR operation between the trigger value and the mask. It then posts the Swi `mySwi`. This way, requests coming from different sources can be handled in the same Swi by using a switch statement. We can give an example on the usage of this function as follows. Assume that the two interrupts coming from different buttons use the same Swi. However, each interrupt requires a different operation to be executed. To handle this, we first set the initial trigger value as 0. Afterwards, we use the functions `Swi_or(mySwi, 1)` and `Swi_or(mySwi, 2)` within the ISR for the first and second buttons, respectively. When an interrupt comes from the first button, the trigger value is ORed by 1 and becomes 1. When an interrupt comes from the second button, the trigger value is ORed by 2 and becomes 2. For both cases, the Swi `mySwi` is posted. Within the Swi function, we read the trigger value by the function `Sw_getTrigger(mySwi)`. Then, we perform different operations using the trigger value in a switch-case or if-else structure.

The function `Swi_andn(mySwi, mask)` performs the logical AND operation between the trigger value and the mask. It then posts the Swi `mySwi` when the trigger value becomes 0. The purpose of this function is as follows. Assume that two different sources should finish their work to post the Swi. We set the initial trigger value as 3. When the first source finishes its work, it uses `Swi_andn(mySwi, 2)`. The AND operation is applied to trigger value (3) and mask (2). As a result, the trigger value becomes 2. Since the trigger value is not 0, the Swi is not posted. When the second source finishes its work, it uses `Swi_andn(mySwi, 1)`. The AND operation is applied to trigger value (2) and mask (1). The trigger value becomes 0 and Swi is posted. Afterwards, the trigger turns back to its initial value that is 3. The same procedure is repeated again.

We can create an Swi thread statically using the configuration file mentioned in Sec. 13.2.4. To do so, open the `hello.cfg` file by right-clicking on it and select *Open With → XGCONF*. In the "BIOS - System Overview" window, click on the "Swi" box. In the opened window, tick the box titled "Add the software interrupt threads module to my configuration". Here, the user can select the total number of priorities that can be used in Swi. Then, select the "Instance" option on the top of the window. A new window should appear as in Fig. 13.9.

To create a new Swi in the window in Fig. 13.9, we should click on the "Add" button and set the properties of the Swi thread in the "Required Settings" section. Here, we

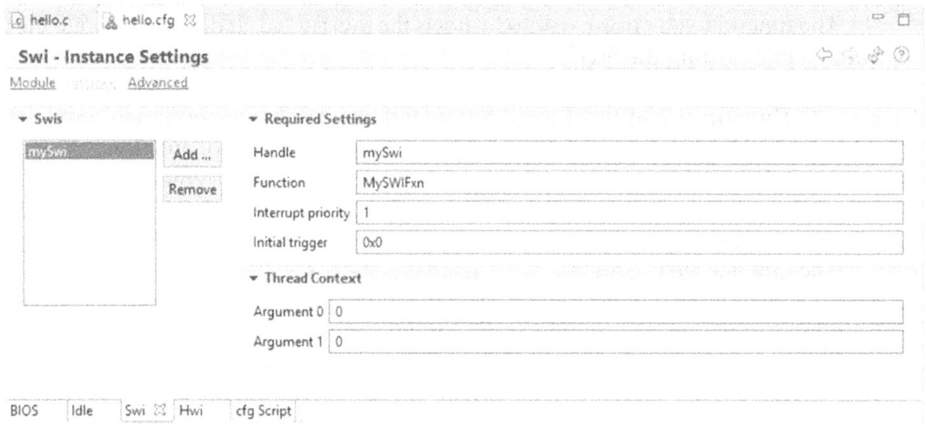

FIGURE 13.9 Creating an Swi thread.

should provide a name to our thread handle, set the function name to be used with this handle, set the priority for the created Swi, and initial trigger value. There are also additional settings in the same window under "Thread Content". Using these, arguments can be passed to the thread.

Let's modify our TI-RTOS project by adding a hardware timer interrupt now as Hwi and Swi. Within the project, we configure the hardware timer for generating interrupts with one second interval using the function `Timer_A_delay` (given in Listing 13.3). Different from the hardware interrupt example, we toggle LED1, connected to pin P1.0 on the MSP432 LaunchPad, in the Swi function in this example. Therefore, we create an Swi called `mySwi` and associate with it a function `MySWIFxn`. To note here, we make all these adjustments in the configuration file as given in Fig. 13.9.

We provide the C source code of our project in Listing 13.6. The main difference in this code is the addition of the `MySWIFxn` function for the Swi thread. Here, we also declare the Swi handle as `extern` to indicate that it is generated in the external configuration file.

We can also create an Swi dynamically in C code. We provide such an example in Listing 13.7. This code performs the same operation as in our static Swi example.

In Listing 13.7, `mySwi` is the handle name of the created Swi. `MySWIFxn` is the function triggered by Swi. `swiParams` is the structure name that contains Swi instance parameters. `arg0` and `arg1` are the instance parameters that will be passed to `MySWIFxn` when it is triggered. `priority` is the instance parameter to determine the Swi priority. This parameter can get values from 0 (lowest) to 15 (highest). `trigger` is the instance parameter to determine the special trigger value. Finally, `eb` is the error block to handle errors that may occur during Swi creation.

To delete a dynamically created Swi, `Swi_delete()` function is used. Do not forget that it can only be called from the task level. Preemption by a Swi with higher priority can be temporarily prevented using `Swi_disable()` function. This function disables all Swi preemptions. `Swi_restore()` function is used to enable all preemption process again. These functions can only be called in idle loop, task, or Swi function. Swis are enabled or disabled as a group. They cannot be enabled or disabled individually.

Listing 13.6 The C Code for the Static Swi Usage Example

```c
#include <ti/devices/msp432p4xx/driverlib/driverlib.h>
#include <xdc/std.h>
#include <xdc/runtime/System.h>
#include <ti/sysbios/BIOS.h>
#include <ti/sysbios/hal/Hwi.h>
#include <ti/sysbios/knl/Swi.h>
#include <ti/drivers/GPIO.h>
#include "Board.h"

 extern Swi_Handle mySwi;

void MySWIFxn(UArg arg0, UArg arg1);
void MyHWIFxn(UArg arg);
void MyIdleFxn(void);
void Timer_A_delay(void);

 int print = 0;
 int cnt = 0;

void main(void){

 Board_initGeneral();
 Board_initGPIO();

 Timer_A_delay();

 System_printf("Starting SWI Example\n");
 System_flush();

 BIOS_start();
}
void MyHWIFxn(UArg arg){
 Timer_A_clearCaptureCompareInterrupt(TIMER_A1_BASE,
     TIMER_A_CAPTURECOMPARE_REGISTER_0);
 Swi_post(mySwi);
}
void MySWIFxn(UArg arg0, UArg arg1){
 GPIO_toggle(Board_LED0);
 cnt ++;
 print = 1;
}
void MyIdleFxn(void){
 if(print == 1){
  System_printf("LED Toggle cnt = %d \n", cnt);
  System_flush();
  print = 0;}
}

void Timer_A_delay(void){
 ...
}
```

Listing 13.7 The C Code for the Dynamic Swi Usage Example

```c
#include <ti/devices/msp432p4xx/driverlib/driverlib.h>
#include <xdc/std.h>
#include <xdc/runtime/System.h>
#include <xdc/runtime/Error.h>
#include <ti/sysbios/BIOS.h>
#include <ti/sysbios/hal/Hwi.h>
#include <ti/sysbios/knl/Swi.h>
#include <ti/drivers/GPIO.h>
#include "Board.h"

 Error_Block eb;
 Hwi_Params hwiParams;
 Hwi_Handle myHwi;
 Swi_Params swiParams;
 Swi_Handle mySwi;

void MySWIFxn(UArg arg0, UArg arg1);
void MyHWIFxn(UArg arg);
void MyIdleFxn(void);
void Timer_A_delay(void);

 int print = 0;
 int cnt = 0;

void main(void){

  Board_initGeneral();
  Board_initGPIO();

  Error_init(&eb);
  Hwi_Params_init(&hwiParams);
  hwiParams.arg = 0;
  hwiParams.enableInt = TRUE;
  myHwi = Hwi_create(26, MyHWIFxn, &hwiParams, &eb);
  if (myHwi == NULL) System_abort("Hwi create failed");

  Error_init(&eb);
  Swi_Params_init(&swiParams);
  swiParams.arg0 = 0;
  swiParams.arg1 = 0;
  swiParams.trigger = 0;
  swiParams.priority = 1;
  mySwi = Swi_create(MySWIFxn, &swiParams, &eb);
  if (mySwi == NULL) System_abort("Swi create failed");

  Timer_A_delay();

  System_printf("Starting SWI Example\n");
  System_flush();

  BIOS_start();
}

void MyHWIFxn(UArg arg){
```

```
    Timer_A_clearCaptureCompareInterrupt(TIMER_A1_BASE,
        TIMER_A_CAPTURECOMPARE_REGISTER_0);
    Swi_post(mySwi);
}

void MySWIFxn(UArg arg0, UArg arg1){
 GPIO_toggle(Board_LED0);
 cnt ++;
 print = 1;
}

void MyIdleFxn(void){
 if(print == 1){
   System_printf("LED Toggle cnt = %d \n", cnt);
   System_flush();
   print = 0;}
}

void Timer_A_delay(void){
 ...
}
```

13.3.1.4 Task

The task is a special thread having higher priority than the idle loop and lower priority than Hwi and Swi. There are 16 different priority levels from 0 (lowest) to 15 (highest) available for each task. The task with higher priority will always preempt the one with lower priority. There is also another priority level defined as −1 for putting the task to inactive state.

Tasks are special because they have an additional state called *wait* or *block*. We already know that an Hwi or Swi should run to completion. However, tasks can wait during execution until a particular event occurs within the system. This is possible since each task has its own separate stack. The size of the stack can be adjusted. However, it must be used carefully to prevent stack overflow. For further information on this issue, please see [44].

There are five different execution states defined for a task. These are listed below:

- Running: It is the state the task is executed. There can be only one task in this state at any time.
- Ready: It is the state the task is scheduled for execution according to its priority level. The running task goes to this state if it is preempted or it yields to other tasks with the same priority using the function `Task_yield`.
- Blocked: It is the state the task waits until necessary resources are available. The running task goes to this state if `Semaphore_pend` or `Task_sleep` functions are used.
- Terminated: It is the state reached when the task is deleted using the function `Task_delete` or when the task finishes its job completely.
- Inactive: It is the state the task has a priority level −1. This priority can be set when the task is created or by calling the `Task_setPri` function. This function can also be used to change the priority level on the run.

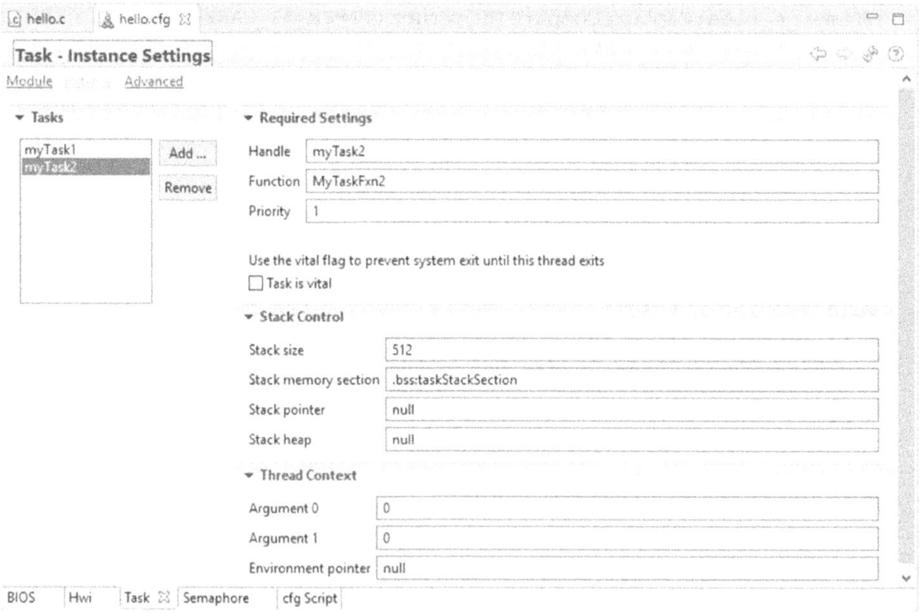

332 Chapter Thirteen

FIGURE 13.10 Creating a task thread.

We can create a task thread statically using the configuration file in Sec. 13.2.4. To do so, open the `hello.cfg` file by right-clicking on it and select *Open With → XGCONF*. In the "BIOS - System Overview" window, click on the "Task" box. In the opened window, tick the box titled "Add the Task threads module to my configuration". Here, the user can select the total number of priorities and the default stack size that can be used in the task. Then, select the "Instance" option on the top of the window. A new window should appear as in Fig. 13.10.

To create a new task in the window in Fig. 13.10, we should click on the "Add" button and set the properties of the task thread in the "Required Settings" section. Here, we should provide a name to our thread handle, set the function name to be used with this handle, and set the priority for the created task.

Let's modify our recent TI-RTOS project by adding handling post Hwi operations using two tasks instead of a Swi and idle loop. Within the project, we configure the hardware timer for generating interrupts with one-second interval using the function `Timer_A_delay` (given in Listing 13.3). Different from the software interrupt example, we toggle LED1, connected to pin P1.0 on the MSP432 LaunchPad, in one task function in this example. Besides, the console writing operation is done in another task function instead of the idle loop. Therefore, we create two tasks called `myTask1` and `myTask2` with priorities 2 and 1, respectively. We associate each task handle with functions `MyTaskFxn1` and `MyTaskFxn2`, respectively. To note here, we make all these adjustments on the configuration file as given in Fig. 13.10. We also set the stack size for each task as 512 bytes.

We had to use the semaphores with the mentioned tasks. We will explain in detail in Sec. 13.3.3 how semaphores work. However, to use them in this example, please click on the Semaphores box in the "BIOS - System Overview" window. In the opened window, tick the box titled "Add the Semaphore management module to my configuration".

Then, add two semaphores as `mySem1` and `mySem2` in the instances window as in Fig. 13.13.

We provide the C source code of our project in Listing 13.8. Here, two tasks replace the Swi and idle loop threads in the previous example. We declare the task and semaphore handles as `extern` to indicate that they are generated in the external configuration file.

In Listing 13.8, `myTask1` and `myTask2` are initially set to inactive state. We configure the hardware timer for generating interrupts with one-second interval. Hwi sets `myTask1` to ready state to trigger the `MyTaskFxn1` function. This task function is used to toggle LED1 and set `myTask1` to blocked state again. It also sets `myTask2` to ready state. Hence, this task prints LED toggle count values in the console window. To note here, this code can be fully understood after introducing semaphores in Sec. 13.3.3.

We can also create tasks dynamically in C code. We provide such an example in Listing 13.9. This code performs the same operation as in our static task example.

In Listing 13.9, `myTask1` and `myTask2` are the handle names of the created tasks. `MyTaskFxn1` and `MyTaskFxn2` are the functions associated with the mentioned tasks, respectively. `taskParams` is the structure name that contains task instance parameters. `arg0` and `arg1` are the instance parameters that will be passed to task functions when they are triggered. `priority` is the instance parameter to determine the task priority. `stack` is the instance pointer to use a predefined array for task stack. `stackSize` is the instance parameter to determine the size of the task stack. Finally, `eb` is the error block to handle errors that may occur during task creation.

13.3.2 Timing

TI-RTOS kernel provides time-related modules to be used in operations. These are the clock, timer, seconds, and timestamp. We will not deal with the seconds module in this book. For further information on this module, please see [44]. We will explain the remaining time-related modules in separate sections next.

13.3.2.1 The Clock Module

The clock module can be used to create a periodic system tick. In fact, this module is actually a special Swi with the highest priority level 15. However, this priority can be configured. The clock module uses one of the four Timer_A modules (introduced in Chap. 7) to generate its system tick. Initially, system tick period is set to one millisecond.

We can include the clock module to our TI-RTOS project statically using the configuration file mentioned in Sec. 13.2.4. To do so, open the `hello.cfg` file by right-clicking on it and select *Open With → XGCONF*. In the "BIOS - System Overview" window, click on the "Clock" box. In the opened window, tick the box titled "Add the Clock support module to my configuration". Then, select the "Instance" option on the top of the window. A new window should appear as in Fig. 13.11.

To create a new clock module (Swi) thread in the window in Fig. 13.11, we should click on the "Add" button and set its properties in the "Required Settings" section. Here, we should provide a name to our thread handle, set the function name to be used with this handle, set the initial timeout value (which cannot be 0), and set the period of the system tick. We can also tick the box titled "Start at boot time when instance is created". Hence, the clock module starts working when it is created. If this box is ticked, we should also set the initial timeout value.

We can reconfigure our LED toggle project to work with the clock module. To do so, we should only have a clock module and idle loop. Therefore, we should create a clock

Listing 13.8 The C Code for the Static Task Usage Example

```c
#include <ti/devices/msp432p4xx/driverlib/driverlib.h>
#include <xdc/std.h>
#include <xdc/runtime/System.h>
#include <ti/sysbios/BIOS.h>
#include <ti/sysbios/hal/Hwi.h>
#include <ti/sysbios/knl/Task.h>
#include <ti/sysbios/knl/Semaphore.h>
#include <ti/drivers/GPIO.h>
#include "Board.h"

 extern Task_Handle myTask1, myTask2;
 extern Semaphore_Handle mySem1, mySem2;

void MyTaskFxn1(UArg arg0, UArg arg1);
void MyTaskFxn2(UArg arg0, UArg arg1);
void MyHWIFxn(UArg arg);
void Timer_A_delay(void);

 int print = 0;
 int cnt = 0;

void main(void){

 Board_initGeneral();
 Board_initGPIO();

 Timer_A_delay();

 System_printf("Starting Task Example\n");
 System_flush();

 BIOS_start();
}

void MyHWIFxn(UArg arg){
 Timer_A_clearCaptureCompareInterrupt( TIMER_A1_BASE,
     TIMER_A_CAPTURECOMPARE_REGISTER_0 );
 Semaphore_post(mySem1);
}

void MyTaskFxn1(UArg arg0, UArg arg1){
 while(1){
  Semaphore_pend(mySem1, BIOS_WAIT_FOREVER);
  GPIO_toggle(Board_LED0);
  cnt ++;
  Semaphore_post(mySem2);}
}

void MyTaskFxn2(UArg arg0, UArg arg1){
 while(1){
  Semaphore_pend(mySem2, BIOS_WAIT_FOREVER);
  System_printf("LED Toggle cnt = %d \n", cnt);
  System_flush();}
}

void Timer_A_delay(void){
 ...
}
```

Listing 13.9 The C Code for the Dynamic Task Usage Example

```c
#include <ti/devices/msp432p4xx/driverlib/driverlib.h>
#include <xdc/std.h>
#include <xdc/runtime/System.h>
#include <xdc/runtime/Error.h>
#include <ti/sysbios/BIOS.h>
#include <ti/sysbios/hal/Hwi.h>
#include <ti/sysbios/knl/Task.h>
#include <ti/sysbios/knl/Semaphore.h>
#include <ti/drivers/GPIO.h>
#include "Board.h"

#define TASKSTACKSIZE    512

 Error_Block eb;
 Hwi_Params hwiParams;
 Hwi_Handle myHwi;

 Task_Params taskParams;
 Task_Handle myTask1, myTask2;
 Char myTask1Stack[TASKSTACKSIZE], myTask2Stack[TASKSTACKSIZE];

 Semaphore_Params semParams;
 Semaphore_Handle mySem1, mySem2;

void MyTaskFxn1(UArg arg0, UArg arg1);
void MyTaskFxn2(UArg arg0, UArg arg1);
void MyHWIFxn(UArg arg);
void Timer_A_delay(void);

 int print = 0;
 int cnt = 0;

void main(void){

 Board_initGeneral();
 Board_initGPIO();

 Error_init(&eb);
 Hwi_Params_init(&hwiParams);
 hwiParams.arg = 0;
 hwiParams.enableInt = TRUE;
 myHwi = Hwi_create(26, MyHWIFxn, &hwiParams, &eb);
 if (myHwi == NULL) System_abort("Hwi create failed");

 Error_init(&eb);
 Task_Params_init(&taskParams);
 taskParams.stackSize = TASKSTACKSIZE;
 taskParams.stack = &myTask1Stack;
 taskParams.priority = 2;
 myTask1 = Task_create(MyTaskFxn1, &taskParams, &eb);
 if (myTask1 == NULL) System_abort("Task1 create failed");

 Error_init(&eb);
 Task_Params_init(&taskParams);
```

```
  taskParams.stackSize = TASKSTACKSIZE;
  taskParams.stack = &myTask2Stack;
  taskParams.priority = 1;
  myTask2 = Task_create(MyTaskFxn2, &taskParams, &eb);
  if (myTask2 == NULL) System_abort("Task2 create failed");

  Error_init(&eb);
  Semaphore_Params_init(&semParams);
  mySem1 = Semaphore_create(0, &semParams, &eb);
  if (mySem1 == NULL) System_abort("Semaphore create failed");

  Error_init(&eb);
  Semaphore_Params_init(&semParams);
  mySem2 = Semaphore_create(0, &semParams, &eb);
  if (mySem2 == NULL) System_abort("Semaphore create failed");

  Timer_A_delay();

  System_printf("Starting Task Example\n");
  System_flush();

  BIOS_start();
}

void MyHWIFxn(UArg arg){
 Timer_A_clearCaptureCompareInterrupt( TIMER_A1_BASE,
     TIMER_A_CAPTURECOMPARE_REGISTER_0 );
 Semaphore_post(mySem1);
}

void MyTaskFxn1(UArg arg0, UArg arg1){
 while(1){
  Semaphore_pend(mySem1, BIOS_WAIT_FOREVER);
  GPIO_toggle(Board_LED0);
  cnt ++;
  Semaphore_post(mySem2);}
}

void MyTaskFxn2(UArg arg0, UArg arg1){
 while(1){
  Semaphore_pend(mySem2, BIOS_WAIT_FOREVER);
  System_printf("LED Toggle cnt = %d \n", cnt);
  System_flush();}
}

void Timer_A_delay(void){
 ...
}
```

module thread called `myClock` and an idle loop. We associate these thread handles with functions `MyClockFxn` and `MyIdleFxn`, respectively. To note here, we make all these adjustments in the configuration file as given in Fig. 13.11.

We provide the C source code of our modified project in Listing 13.10. Here, the clock module generates system ticks every second and LED1 connected to pin P1.0 on

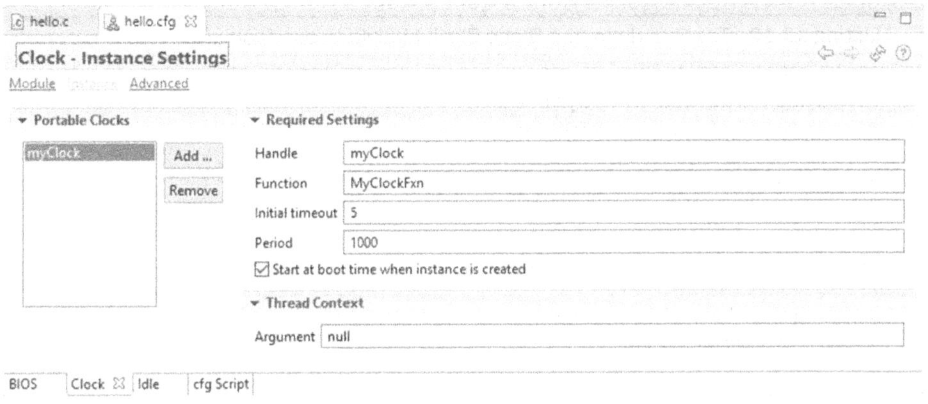

Real-time Operating System 337

FIGURE 13.11 Creating the clock module.

the MSP432 LaunchPad toggles accordingly. Here, we declare the clock module handle as `extern` to indicate that it is generated in the external configuration file.

In Listing 13.10, the clock module starts with the function `Clock_start(myClock)`. LED toggling operation is done in the `MyClockFxn`. The LED toggle count value is posted in the `MyIdleFxn`.

We can also create the clock module dynamically in a C code. We modify the previous clock module example this way and provide the C code in Listing 13.11. Here, `myClock` is the handle name for the clock module. `MyClockFxn` is the function triggered by the clock module. `clockParams` is the structure name that contains the clock module parameters. The timeout value is set as 5, which should not be zero since we let the clock module start instantly. `MyClockFxn` is triggered first after this timeout value. `arg` is the instance parameter that will be passed to `MyClockFxn` when it is triggered. `period` is the instance parameter to determine the triggering period of `MyClockFxn` function. `startFlag` is the instance parameter to start the clock ticks when the clock module runs. If this parameter is set as `FALSE`, clock is not started. Otherwise, it is started. Finally, `eb` is the error block to handle errors that may occur during clock module creation.

You can start or stop the clock dynamically during code execution using `Clock_start()` or `Clock_stop()` functions. Also, you can start or stop the timer-generating system tick using `Clock_tickStart()` or `Clock_tickStop()` functions. Finally, you can obtain the total number of clock ticks that have occurred since the starting of the instance by using `Clock_getTicks()` function.

13.3.2.2 The Timer Module

The timer module is used to create a timer and call the desired function with desired time specifications. This module is actually a special Hwi.

We can include the timer module to our TI-RTOS project statically using the configuration file mentioned in Sec. 13.2.4. To do so, open the `hello.cfg` file by right-clicking on it and select *Open With* → *XGCONF*. In the "BIOS - System Overview" window, click on the "Timer" box. In the opened window, tick the box titled "Add the portable Timer management module to my configuration". Then, select the "Instance" option on the top of the window. A new window should appear as in Fig. 13.12.

Listing 13.10 The C Code for the Static Clock Module Usage Example

```c
#include <xdc/std.h>
#include <xdc/runtime/System.h>
#include <ti/sysbios/BIOS.h>
#include <ti/sysbios/knl/Clock.h>
#include <ti/drivers/GPIO.h>
#include "Board.h"

extern Clock_Handle myClock;

void MyClockFxn(UArg arg);
void MyIdleFxn(void);

int print = 0;
int cnt = 0;

void main(void) {

 Board_initGeneral();
 Board_initGPIO();

 System_printf("Starting Clock Example\n");
 System_flush();

 Clock_start(myClock);

 BIOS_start();
}

void MyClockFxn(UArg arg) {
 GPIO_toggle(Board_LED0);
 cnt ++;
 print = 1;
}

void MyIdleFxn(void) {
 if(print == 1) {
  System_printf("LED Toggle cnt = %d \n", cnt);
  System_flush();
  print = 0;}
}
```

To create a new timer module (Hwi) thread in the window in Fig. 13.12, we should click on the "Add" button and set its properties in the "Required Settings" section. Here, we should provide a name to our thread handle, set the Timer ISR function name to be used with this handle, set the Timer Id, and set the period used in it. We can also adjust the properties of the timer module using options under "Additional Settings".

We repeat the previous LED toggle TI-RTOS project using the timer module. We provide the modified C code in Listing 13.12. Here, the timer module generates system ticks every second and LED1 connected to pin P1.0 on the MSP432 LaunchPad toggles accordingly. Here, we declare the timer module handle as extern to indicate that it is generated in the external configuration file.

Listing 13.11 The C Code for the Dynamic Clock Module Usage Example

```c
#include <xdc/std.h>
#include <xdc/runtime/System.h>
#include <xdc/runtime/Error.h>
#include <ti/sysbios/BIOS.h>
#include <ti/sysbios/knl/Clock.h>
#include <ti/drivers/GPIO.h>
#include "Board.h"

Clock_Params clockParams;
Clock_Handle myClock;
Error_Block eb;

void MyClockFxn(UArg arg);
void MyIdleFxn(void);

int print = 0;
int cnt = 0;

void main(void){

 Board_initGeneral();
 Board_initGPIO();

 Error_init(&eb);
 Clock_Params_init(&clockParams);
 clockParams.period = 1000;
 clockParams.startFlag = FALSE;
 clockParams.arg = 0;
 myClock = Clock_create(MyClockFxn, 1000, &clockParams, &eb);
 if (myClock == NULL) System_abort("Clock create failed");

 System_printf("Starting Clock Example\n");
 System_flush();

 Clock_start(myClock);

 BIOS_start();
}

void MyClockFxn(UArg arg){
 GPIO_toggle(Board_LED0);
 cnt ++;
 print = 1;
}

void MyIdleFxn(void){
 if(print == 1){
  System_printf("LED Toggle cnt = %d \n", cnt);
  System_flush();
  print = 0;}
}
```

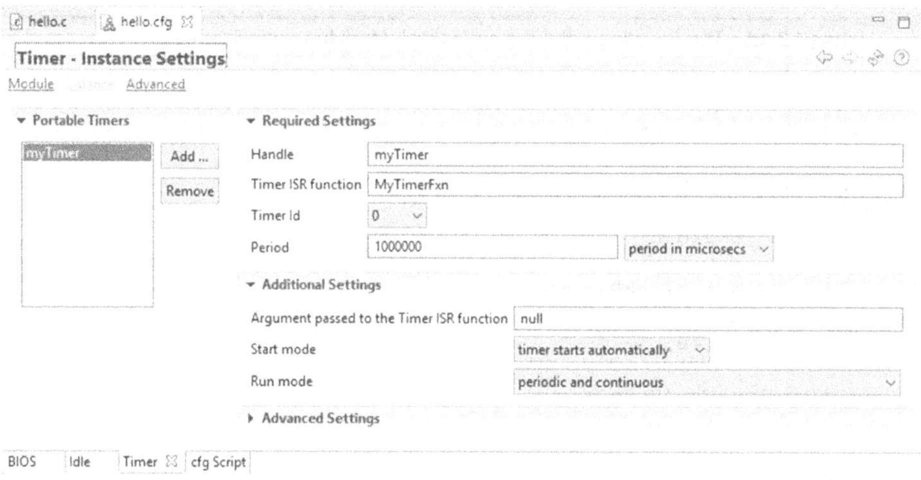

FIGURE 13.12 Creating the timer module.

In Listing 13.12, the timer module starts with the start of the kernel. LED toggling operation is done in `MyTimerFxn`. The LED toggle count value is posted in `MyIdleFxn`.

As in previous sections, we can create the timer module dynamically in a C code. We provide such an example in Listing 13.13. Here, `myTimer` is the handle name of the created timer module. `MyTimerFxn` is the function triggered by the timer module. `timerParams` is the structure name that contains the timer module instance parameters. `Timer_ANY` is a special definition to use any available hardware timer. `arg` is the instance parameter that will be passed to `MyTimerFxn` when it is triggered. `period` is the instance parameter to determine the triggering period of `MyTimerFxn` function. `startMode` is the instance parameter to determine how the timer module is started. If it is selected as `Timer_StartMode_AUTO`, the timer module starts automatically when the timer instance is created. If it is selected as `Timer_StartMode_USER`, the timer module can be started in the code by the user-defined functions. The `runMode` is the instance parameter to determine how many times the timer module will be run. If it is selected as `Timer_RunMode_ONESHOT`, the timer module is run once. Then, it stops. If it is selected as `Timer_RunMode_CONTINUOUS`, the timer module runs repeatedly with the given period. `periodType` is the instance parameter to determine period unit. If it is selected as `Timer_PeriodType_MICROSECS`, period unit is selected as microseconds. If it is selected as `Timer_PeriodType_COUNTS`, period unit is selected as timer counts. Finally, `eb` is the error block to handle errors that may occur during timer module creation.

There are two functions to start or stop the timer module dynamically during code execution. These are called `Timer_start()` and `Timer_stop()`. `Timer_start()` function clears counters, clears any pending interrupts, and enables the timer interrupt before starting the timer. `Timer_stop()` stops the timer and disables the timer interrupt.

13.3.2.3 Timestamp Module

The timestamp module can be used to add timestamps in your system. It can be easily used by adding the `#include <xdc/runtime/Timestamp.h>` line in the code. Then, it must be enabled statically using the configuration file mentioned in Sec. 13.2.4.

Listing 13.12 The C Code for the Static Timer Module Usage Example

```c
#include <xdc/std.h>
#include <xdc/runtime/System.h>
#include <ti/sysbios/BIOS.h>
#include <ti/sysbios/hal/Timer.h>
#include <ti/drivers/GPIO.h>
#include "Board.h"

 extern Timer_Handle myTimer;

void MyTimerFxn(UArg arg);
void MyIdleFxn(void);

 int print = 0;
 int cnt = 0;

void main(void){

 Board_initGeneral();
 Board_initGPIO();

 System_printf("Starting Clock Example\n");
 System_flush();

 BIOS_start();
}

void MyTimerFxn(UArg arg){
 GPIO_toggle(Board_LED0);
 cnt ++;
 print = 1;
}

void MyIdleFxn(void){
 if(print == 1){
  System_printf("LED Toggle cnt = %d \n", cnt);
  System_flush();
  print = 0;}
}
```

To do so, open the `hello.cfg` file by right-clicking on it and select *Open With → XGCONF*. In the "BIOS - System Overview" window, click on the "Timestamp" box. In the opened window, tick the box titled "Add Timestamp to my configuration". Afterwards, the `Timestamp_get32()` function can be used for obtaining 32-bit timestamp value. The `Timestamp_getFreq()` function can be used to obtain the timestamp timer frequency in Hz.

We can modify our previous TI-RTOS project to use the timestamp module. We provide the C code in Listing 13.14. In this project, we configure the clock module for triggering the `MyClockFxn` function with one-second interval to toggle LED1 connected to pin P1.0 on the MSP432 LaunchPad. We also use the idle loop to print the number of toggles and time duration between LED toggles obtained by the timestamp module in microseconds.

Listing 13.13 The C Code for the Dynamic Timer Module Usage Example

```c
#include <xdc/std.h>
#include <xdc/runtime/System.h>
#include <xdc/runtime/Error.h>
#include <ti/sysbios/BIOS.h>
#include <ti/sysbios/hal/Timer.h>
#include <ti/drivers/GPIO.h>
#include "Board.h"

 Timer_Params timerParams;
 Timer_Handle myTimer;
 Error_Block eb;

void MyTimerFxn(UArg arg);
void MyIdleFxn(void);

 int print = 0;
 int cnt = 0;

void main(void){

 Board_initGeneral();
 Board_initGPIO();

 Error_init(&eb);
 Timer_Params_init(&timerParams);
 timerParams.startMode = Timer_StartMode_AUTO;
 timerParams.runMode = Timer_RunMode_CONTINUOUS;
 timerParams.periodType = Timer_PeriodType_MICROSECS;
 timerParams.period = 1000000;
 timerParams.arg = 0;
 myTimer = Timer_create(Timer_ANY, MyTimerFxn, &timerParams, &eb);
 if (myTimer == NULL) System_abort("Timer create failed");

 System_printf("Starting Clock Example\n");
 System_flush();

 BIOS_start();
}
void MyTimerFxn(UArg arg){
 GPIO_toggle(Board_LED0);
 cnt ++;
 print = 1;
}
void MyIdleFxn(void){
 if(print == 1){
  System_printf("LED Toggle cnt = %d \n", cnt);
  System_flush();
  print = 0;}
}
```

Listing 13.14 The C Code for the Timestamp Module Usage Example

```c
#include <xdc/std.h>
#include <xdc/runtime/System.h>
#include <ti/sysbios/BIOS.h>
#include <ti/sysbios/knl/Clock.h>
#include <ti/drivers/GPIO.h>
#include <xdc/runtime/Timestamp.h>
#include <xdc/runtime/Types.h>
#include "Board.h"

 extern Clock_Handle myClock;

void MyClockFxn(UArg arg);
void MyIdleFxn(void);

 int print = 0;
 int cnt = 0;
 Types_FreqHz freq;
 unsigned int tstart,tstop;

void main(void){

 Board_initGeneral();
 Board_initGPIO();

 System_printf("Starting Timestamp Example\n");
 System_flush();

 Clock_start(myClock);
 tstart = Timestamp_get32();
 BIOS_start();
}
void MyClockFxn(UArg arg){
 tstop = Timestamp_get32();
 GPIO_toggle(Board_LED0);
 cnt ++;
 print = 1;
}
void MyIdleFxn(void){
 if(print == 1){
  Timestamp_getFreq(&freq);
  System_printf("LED Toggle after %u us cnt = %d \n", (tstop-tstart) /
      (freq.lo / 1000000), cnt);
  System_flush();
  tstart = tstop;
  print = 0;}
}
```

13.3.3 Synchronization

If more than one thread is using the same data, they should be synchronized. Synchronization may also be needed when organizing the thread running orders. This can be done by using semaphores, events, mailboxes, or queues. We will explain these methods in detail in this section.

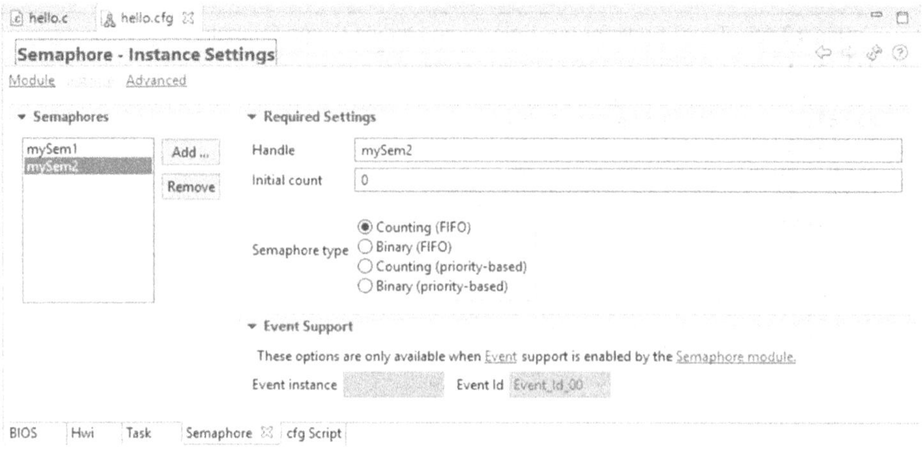

FIGURE 13.13 Creating a semaphore.

13.3.3.1 Semaphore

A semaphore is used to provide synchronization between tasks. The semaphore can be selected as binary or counting according to the number of tasks they handle. Binary semaphores are used for maximum of two tasks. Counting semaphores can coordinate count plus one tasks. Normally, tasks wait semaphores in first-in first-out (FIFO) order regardless of their priority level. However, priority semaphores can be used to run tasks in a desired priority order.

There are two functions defined to use semaphores as Semaphore_pend and Semaphore_post. The Semaphore_pend function is used for keeping a task in block state. The Semaphore_post function is used for signaling a semaphore and putting the task in ready state. The Semaphore_delete function should be used to delete a semaphore.

We can create a semaphore using the configuration file mentioned in Sec. 13.2.4. To do so, open the hello.cfg file by right-clicking on it and select *Open With → XGCONF*. In the "BIOS - System Overview" window, click on the "Semaphore" box. In the opened window, tick the box titled "Add the Semaphore management module to my configuration". Then, select the "Instance" option on the top of the window. A new window should appear as in Fig. 13.13.

To create a new semaphore in the window in Fig. 13.13, we should click on the "Add" button and set the properties of the semaphore in the "Required Settings" section. Here, we should provide a name to our semaphore and set its initial count value.

We had to use semaphores in our previous task example in Listing 13.8. There, we had two different tasks, one for toggling LED1 (connected to pin P1.0 on the MSP432 LaunchPad) with higher priority and the other for printing the number of LED toggles in the console window. Both tasks wait for a semaphore initially. We also configured the hardware timer for generating interrupts with one-second interval, triggering the MyHWIFxn function. In this function, we posted a semaphore and run the higher priority task. At the end of this task, we posted the semaphore again and run the lower priority task this time.

We used the dynamic semaphore generation part in our previous C code in Listing 13.9. There, mySem was the handle name of the created semaphore. semParams

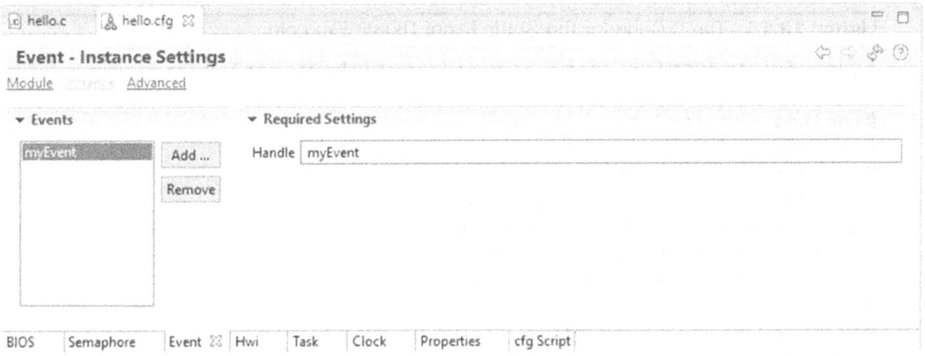

FIGURE 13.14 Creating an event.

was the structure name that contained semaphore instance parameters. mode was the instance parameter to determine the semaphore mode. It could be selected as Semaphore_Mode_BINARY, Semaphore_Mode_COUNTING, Semaphore_Mode_BINARY_PRIORITY, or Semaphore_Mode_COUNTING_PRIORITY.

13.3.3.2 Event

An event is used just like a semaphore. However, the user can define multiple conditions to return the thread from block state when using an event. There are two functions defined to use events. These are Event_pend and Event_post. The Event_pend function is used for keeping a task in block state for specified events. The Event_post function is used for signaling a selected event and putting the task in ready state.

We can create an event using the configuration file mentioned in Sec. 13.2.4. To do so, open the hello.cfg file by right-clicking on it and select *Open With → XGCONF*. In the "BIOS - System Overview" window, click on the "Event" box. In the opened window, tick the box titled "Add the Event module to my configuration". Then, select the "Instance" option on the top of the window. A new window should appear as in Fig. 13.14.

To create a new event in the window in Fig. 13.14, we should click on the "Add" button and set the properties of the event in the "Required Settings" section. Here, we should provide a handle name to our event.

We form a TI-RTOS project to show how events work. Within the project, we configure the Hwi to generate interrupts from the buttons S1 and S2 connected to pins P1.1 and P1.4 on the MSP432 LaunchPad. To prevent the switch bouncing problem, we use the clock module and two tasks. We provide the C code of the project in Listing 13.15.

In Listing 13.15, when one of the buttons is pressed, the clock module starts counting and the first task is kept in block state by a semaphore until the clock module finishes counting. Afterwards, the first task posts events according to the interrupt source and the second task toggles LED1 (connected to pin P1.0 on the MSP432 LaunchPad) or red color of LED2 (connected to pin P2.0 on the MSP432 LaunchPad) according to the posted event. We provide the Set_Buttons function used in the C code for the event example in Listing 13.16.

We also consider the event usage example in which all RTOS objects are created dynamically. We provide the corresponding C code in Listing 13.17. Here, myEvent is the handle name of the created event. The eb is the error block to handle errors that may occur during event creation.

Listing 13.15 The C Code for the Static Event Usage Example

```c
#include <ti/devices/msp432p4xx/driverlib/driverlib.h>

#include <xdc/std.h>
#include <xdc/runtime/System.h>
#include <ti/sysbios/BIOS.h>
#include <ti/sysbios/hal/Hwi.h>
#include <ti/sysbios/knl/Event.h>
#include <ti/sysbios/knl/Clock.h>
#include <ti/sysbios/knl/Semaphore.h>
#include <ti/sysbios/knl/Task.h>
#include <ti/drivers/GPIO.h>
#include "Board.h"

 extern Hwi_Handle myHwi;
 extern Event_Handle myEvent;
 extern Semaphore_Handle mySem;
 extern Task_Handle myTask1, myTask2;
 extern Clock_Handle myClock;

void MyHWIFxn(UArg arg);
void MyTaskFxn1(UArg arg0, UArg arg1);
void MyTaskFxn2(UArg arg0, UArg arg1);
void MyClockFxn(UArg arg);
void Set_Buttons(void);

 uint32_t status;

void main(void){

 Board_initGeneral();

 Set_Buttons();

 System_printf("Starting HWI Example\n");
 System_flush();

 BIOS_start();
}

void MyHWIFxn(UArg arg){
 status = MAP_GPIO_getEnabledInterruptStatus(GPIO_PORT_P1);

 Clock_start(myClock);

 MAP_GPIO_clearInterruptFlag(GPIO_PORT_P1, status);
}

void MyTaskFxn1(UArg arg0, UArg arg1){
 for (;;){
  Semaphore_pend(mySem, BIOS_WAIT_FOREVER);
  Clock_stop(myClock);

  if(status == GPIO_PIN1)Event_post(myEvent, Event_Id_00);
  if(status == GPIO_PIN4)Event_post(myEvent, Event_Id_01);}
```

```
}

void MyTaskFxn2(UArg arg0, UArg arg1){
 unsigned int events;

 for (;;){
  events = Event_pend(myEvent, Event_Id_NONE, Event_Id_00 +
     Event_Id_01, BIOS_WAIT_FOREVER);
  if (events & Event_Id_00)   MAP_GPIO_toggleOutputOnPin(GPIO_PORT_P1,
     GPIO_PIN0);
  if (events & Event_Id_01)   MAP_GPIO_toggleOutputOnPin(GPIO_PORT_P2,
     GPIO_PIN0);}
}

void MyClockFxn(UArg arg){
 Semaphore_post(mySem);
}

void Set_Buttons(void){
 ...
}
```

Listing 13.16 The Set_Buttons Function

```
void Set_Buttons(void){
 MAP_GPIO_setAsOutputPin(GPIO_PORT_P1, GPIO_PIN0);
 MAP_GPIO_setAsOutputPin(GPIO_PORT_P2, GPIO_PIN0);
 MAP_GPIO_setAsInputPinWithPullUpResistor(GPIO_PORT_P1, GPIO_PIN1|
    GPIO_PIN4);
 MAP_GPIO_clearInterruptFlag(GPIO_PORT_P1, GPIO_PIN1|GPIO_PIN4);
 MAP_GPIO_enableInterrupt(GPIO_PORT_P1, GPIO_PIN1|GPIO_PIN4);
 MAP_Interrupt_enableInterrupt(INT_PORT1);
}
```

Listing 13.17 The C Code for the Dynamic Event Usage Example

```
#include <ti/devices/msp432p4xx/driverlib/driverlib.h>
#include <xdc/std.h>
#include <xdc/runtime/System.h>
#include <xdc/runtime/Error.h>
#include <xdc/runtime/Types.h>
#include <ti/sysbios/BIOS.h>
#include <ti/sysbios/hal/Hwi.h>
#include <ti/sysbios/knl/Event.h>
#include <ti/sysbios/knl/Clock.h>
#include <ti/sysbios/knl/Semaphore.h>
#include <ti/sysbios/knl/Task.h>
#include <ti/drivers/GPIO.h>
#include "Board.h"

#define TASKSTACKSIZE    512

 Hwi_Params hwiParams;
 Hwi_Handle myHwi;
```

```
    Error_Block eb;
    Event_Handle myEvent;
    Semaphore_Params semParams;
    Semaphore_Handle mySem;
    Task_Params taskParams;
    Task_Handle myTask1, myTask2;
    Char myTask1Stack[TASKSTACKSIZE], myTask2Stack[TASKSTACKSIZE];
    Clock_Params clockParams;
    Clock_Handle myClock;
    Error_Block eb;

void MyHWIFxn(UArg arg);
void MyTaskFxn1(UArg arg0, UArg arg1);
void MyTaskFxn2(UArg arg0, UArg arg1);
void MyClockFxn(UArg arg);
void Set_Buttons(void);

 uint32_t status;

void main(void){

 Board_initGeneral();

 Set_Buttons();

 Error_init(&eb);
 Hwi_Params_init(&hwiParams);
 hwiParams.arg = 0;
 hwiParams.enableInt = TRUE;
 myHwi = Hwi_create(51, MyHWIFxn, &hwiParams, &eb);
 if (myHwi == NULL) System_abort("Hwi create failed");

 Error_init(&eb);
 myEvent = Event_create(NULL, &eb);
 if (myEvent == NULL) System_abort("Event create failed");

 Error_init(&eb);
 Semaphore_Params_init(&semParams);
 mySem = Semaphore_create(0, &semParams, &eb);
 if (mySem == NULL) System_abort("Sem create failed");

 Error_init(&eb);
 Clock_Params_init(&clockParams);
 clockParams.period = 10;
 clockParams.startFlag = FALSE;
 clockParams.arg = 0;
 myClock = Clock_create(MyClockFxn, 10, &clockParams, &eb);
 if (myClock == NULL) System_abort("Clock create failed");

 Error_init(&eb);
 Task_Params_init(&taskParams);
 taskParams.stackSize = TASKSTACKSIZE;
 taskParams.stack = &myTask1Stack;
 taskParams.priority = 1;
 myTask1 = Task_create(MyTaskFxn1, &taskParams, &eb);
```

```
  if (myTask1 == NULL) System_abort("Task1 create failed");

  Error_init(&eb);
  Task_Params_init(&taskParams);
  taskParams.stackSize = TASKSTACKSIZE;
  taskParams.stack = &myTask2Stack;
  taskParams.priority = 1;
  myTask2 = Task_create(MyTaskFxn2, &taskParams, &eb);
  if (myTask2 == NULL) System_abort("Task2 create failed");

  System_printf("Starting HWI Example\n");
  System_flush();

  BIOS_start();
}
void MyHWIFxn(UArg arg){
 status = MAP_GPIO_getEnabledInterruptStatus(GPIO_PORT_P1);
 Clock_start(myClock);
 MAP_GPIO_clearInterruptFlag(GPIO_PORT_P1, status);
}

void MyTaskFxn1(UArg arg0, UArg arg1){
 for (;;){
  Semaphore_pend(mySem, BIOS_WAIT_FOREVER);
  Clock_stop(myClock);

  if(status == GPIO_PIN1)Event_post(myEvent, Event_Id_00);
  if(status == GPIO_PIN4)Event_post(myEvent, Event_Id_01);}
}

void MyTaskFxn2(UArg arg0, UArg arg1){
 unsigned int events;
 for (;;){
  events = Event_pend(myEvent, Event_Id_NONE, Event_Id_00 +
      Event_Id_01, BIOS_WAIT_FOREVER);
  if (events & Event_Id_00)   MAP_GPIO_toggleOutputOnPin(GPIO_PORT_P1,
      GPIO_PIN0);
  if (events & Event_Id_01)   MAP_GPIO_toggleOutputOnPin(GPIO_PORT_P2,
      GPIO_PIN0);}
}

void MyClockFxn(UArg arg){
 Semaphore_post(mySem);
}

void Set_Buttons(void){
 ...
}
```

13.3.3.3 Mailbox

A mailbox is used to send messages between tasks. These messages are kept in buffers. The size and number of these buffers are specified when the mailbox is created.

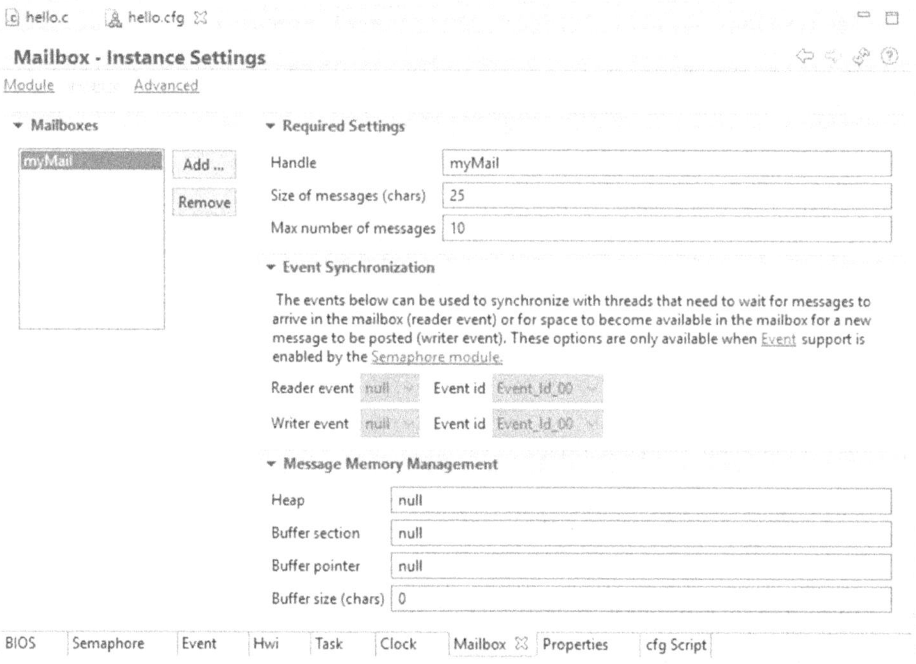

350 Chapter Thirteen

FIGURE 13.15 Creating a mailbox.

There are three functions defined to use a mailbox. The `Mailbox_post` function is used for sending the message buffer. The `Mailbox_pend` function is used for waiting until the message buffer is received. The `Mailbox_delete` function should be used to delete a mailbox.

We can create a mailbox using the configuration file mentioned in Sec. 13.2.4. To do so, open the `hello.cfg` file by right-clicking on it and select *Open With → XGCONF*. In the "BIOS - System Overview" window, click on the "Mailbox" box. In the opened window, tick the box titled "Add the Mailbox module to my configuration". Then, select the "Instance" option on the top of the window. A new window should appear as in Fig. 13.15.

To create a new mailbox in the window in Fig. 13.15, we should click on the "Add" button and set the properties of the mailbox in the "Required Settings" section. Here, we should provide a handle name to our mailbox. We should also set the size of messages in characters and the maximum number of messages. For our case, the size of messages is set as 25 since we have an array with 17 characters and two integers (each requiring four bytes or characters). The maximum number of messages is set as 10. In the same window, we can also make advanced settings under the "Event Synchronization" and "Message Memory Management" sections.

We next form a TI-RTOS project to show how the mailbox works. Within the project, we configure the Hwi to generate interrupts using the button S1 connected to pin P1.1 on the MSP432 LaunchPad. To prevent the switch bouncing problem, we use the clock module and two tasks. We provide the source code of the project in Listing 13.18.

In Listing 13.18, when the button S1 is pressed, the clock module starts counting. The first task is kept in block state by a semaphore until the clock module finishes counting.

Listing 13.18 The C Code for the Static Mailbox Usage Example

```c
#include <ti/devices/msp432p4xx/driverlib/driverlib.h>
#include <xdc/std.h>
#include <xdc/runtime/System.h>
#include <ti/sysbios/BIOS.h>
#include <ti/sysbios/hal/Hwi.h>
#include <ti/sysbios/knl/Mailbox.h>
#include <ti/sysbios/knl/Clock.h>
#include <ti/sysbios/knl/Semaphore.h>
#include <ti/sysbios/knl/Task.h>
#include <ti/drivers/GPIO.h>
#include "Board.h"
#include <string.h>

 extern Hwi_Handle myHwi;
 extern Mailbox_Handle myMail;

struct MsgObj{
 unsigned int id;
 unsigned int cnt;
 char buf[17];
} MsgObj;

 struct MsgObj msg;

 extern Semaphore_Handle mySem;
 extern Task_Handle myTask1, myTask2;
 extern Clock_Handle myClock;

void MyHWIFxn(UArg arg);
void MyTaskFxn1(UArg arg0, UArg arg1);
void MyTaskFxn2(UArg arg0, UArg arg1);
void MyClockFxn(UArg arg);
void Set_Button(void);

 unsigned int counter = 0;

void main(void){

 Board_initGeneral();

 Set_Button();

 System_printf("Starting Mailbox Example\n");
 System_flush();

 BIOS_start();
}

void MyHWIFxn(UArg arg){
 uint32_t status;
 status = MAP_GPIO_getEnabledInterruptStatus(GPIO_PORT_P1);
 Clock_start(myClock);
 MAP_GPIO_clearInterruptFlag(GPIO_PORT_P1, status);
}
```

```
void MyTaskFxn1(UArg arg0, UArg arg1){
 for (;;){
  Semaphore_pend(mySem, BIOS_WAIT_FOREVER);
  Clock_stop(myClock);

  counter ++;

  msg.id = 0;
  msg.cnt = counter;
  strcpy(msg.buf, "Led Toggle cnt =");
  MAP_GPIO_toggleOutputOnPin(GPIO_PORT_P1, GPIO_PIN0);
  Mailbox_post(myMail, &msg, BIOS_WAIT_FOREVER);}
}

void MyTaskFxn2(UArg arg0, UArg arg1){
 struct MsgObj msgread;
 int i;

 for (i=0; i < 10; i++) {
  Mailbox_pend(myMail, &msgread,  BIOS_WAIT_FOREVER);
  System_printf(msgread.buf);
  System_printf("'%d'.\n", msg.cnt);
  System_flush();}

  System_printf("Example done.\n");
  System_exit(0);
}

void MyClockFxn(UArg arg){
 Semaphore_post(mySem);
}

void Set_Button(void){
 ...
}
```

Listing 13.19 The Set_Button Function

```
void Set_Button(void){
 MAP_GPIO_setAsOutputPin(GPIO_PORT_P1, GPIO_PIN0);
 MAP_GPIO_setAsInputPinWithPullUpResistor(GPIO_PORT_P1, GPIO_PIN1);
 MAP_GPIO_clearInterruptFlag(GPIO_PORT_P1, GPIO_PIN1);
 MAP_GPIO_enableInterrupt(GPIO_PORT_P1, GPIO_PIN1);
 MAP_Interrupt_enableInterrupt(INT_PORT1);
}
```

Afterwards, the first task toggles LED1 (connected to pin P1.0 on the MSP432 LaunchPad) and posts the msg object. Then, the second task prints the message in the console window. This process is repeated 10 times and the system exits from BIOS. We provide the Set_Button function used in the C code for the mailbox example in Listing 13.19.

We also consider the mailbox usage example in which all RTOS objects are created dynamically. We provide the corresponding C code in Listing 13.20. Here, myMail is the

Listing 13.20 The C Code for the Dynamic Mailbox Usage Example

```c
#include <ti/devices/msp432p4xx/driverlib/driverlib.h>
#include <xdc/std.h>
#include <xdc/runtime/System.h>
#include <xdc/runtime/Error.h>
#include <ti/sysbios/BIOS.h>
#include <ti/sysbios/hal/Hwi.h>
#include <ti/sysbios/knl/Mailbox.h>
#include <ti/sysbios/knl/Clock.h>
#include <ti/sysbios/knl/Semaphore.h>
#include <ti/sysbios/knl/Task.h>
#include <ti/drivers/GPIO.h>
#include "Board.h"
#include <string.h>

#define TASKSTACKSIZE    512

Hwi_Params hwiParams;
Hwi_Handle myHwi;
Error_Block eb;
Mailbox_Params mailParams;
Mailbox_Handle myMail;

struct MsgObj{
 unsigned int id;
 unsigned int cnt;
 char buf[17];
} MsgObj;

 struct MsgObj msg;

Semaphore_Params semParams;
Semaphore_Handle mySem;
Task_Params taskParams;
Task_Handle myTask1, myTask2;
Char myTask1Stack[TASKSTACKSIZE], myTask2Stack[TASKSTACKSIZE];
Clock_Params clockParams;
Clock_Handle myClock;
Error_Block eb;

void MyHWIFxn(UArg arg);
void MyTaskFxn1(UArg arg0, UArg arg1);
void MyTaskFxn2(UArg arg0, UArg arg1);
void MyClockFxn(UArg arg);
void Set_Button(void);

 unsigned int counter = 0;

void main(void){

 Board_initGeneral();

 Set_Button();

 Error_init(&eb);
```

```c
    Hwi_Params_init(&hwiParams);
    hwiParams.arg = 0;
    hwiParams.enableInt = TRUE;
    myHwi = Hwi_create(51, MyHWIFxn, &hwiParams, &eb);
    if (myHwi == NULL) System_abort("Hwi create failed");

    Error_init(&eb);
    Semaphore_Params_init(&semParams);
    mySem = Semaphore_create(0, &semParams, &eb);
    if (mySem == NULL) System_abort("Sem create failed");

    Error_init(&eb);
    Mailbox_Params_init(&mailParams);
    myMail = Mailbox_create(sizeof(msg), 10, &mailParams, &eb);
    if (myMail == NULL) System_abort("Mailbox create failed");

    Error_init(&eb);
    Clock_Params_init(&clockParams);
    clockParams.period = 10;
    clockParams.startFlag = FALSE;
    clockParams.arg = 0;
    myClock = Clock_create(MyClockFxn, 10, &clockParams, &eb);
    if (myClock == NULL) System_abort("Clock create failed");

    Error_init(&eb);
    Task_Params_init(&taskParams);
    taskParams.stackSize = TASKSTACKSIZE;
    taskParams.stack = &myTask1Stack;
    taskParams.priority = 1;
    myTask1 = Task_create(MyTaskFxn1, &taskParams, &eb);
    if (myTask1 == NULL) System_abort("Task1 create failed");

    Error_init(&eb);
    Task_Params_init(&taskParams);
    taskParams.stackSize = TASKSTACKSIZE;
    taskParams.stack = &myTask2Stack;
    taskParams.priority = 1;
    myTask2 = Task_create(MyTaskFxn2, &taskParams, &eb);
    if (myTask2 == NULL) System_abort("Task2 create failed");

    System_printf("Starting Mailbox Example\n");
    System_flush();

    BIOS_start();
}

void MyHWIFxn(UArg arg){
    uint32_t status;
    status = MAP_GPIO_getEnabledInterruptStatus(GPIO_PORT_P1);
    Clock_start(myClock);
    MAP_GPIO_clearInterruptFlag(GPIO_PORT_P1, status);
}

void MyTaskFxn1(UArg arg0, UArg arg1){
    for (;;){
```

```
  Semaphore_pend(mySem, BIOS_WAIT_FOREVER);
  Clock_stop(myClock);

  counter ++;

  msg.id = 0;
  msg.cnt = counter;
  strcpy(msg.buf, "Led Toggle cnt =");
  MAP_GPIO_toggleOutputOnPin(GPIO_PORT_P1, GPIO_PIN0);
  Mailbox_post(myMail, &msg, BIOS_WAIT_FOREVER);}
}

void MyTaskFxn2(UArg arg0, UArg arg1){
 struct MsgObj msgread;
 int i;

 for (i=0; i < 10; i++) {
  Mailbox_pend(myMail, &msgread,  BIOS_WAIT_FOREVER);
  System_printf(msgread.buf);
  System_printf("'%d'.\n", msg.cnt);
  System_flush();}

 System_printf("Example done.\n");
 System_exit(0);
}

void MyClockFxn(UArg arg){
 Semaphore_post(mySem);
}

void Set_Button(void){
 ...
}
```

handle name of the created mailbox. `mailParams` is the structure name that contains mailbox instance parameters. `sizeof(msg)` is the size of the message object in bytes. 10 is the number of internal mailbox buffers. The `eb` is the error block to handle errors that may occur during mailbox creation.

13.3.3.4 Queue

A queue is used to create a list of objects. Threads can put or get queue elements from any point in this list. Hence, a queue does not have a size limit unlike mailbox. The queue also does not have a pend operation like mailbox.

There are four functions defined to use a queue. These are `Queue_enqueue`, `Queue_dequeue`, `Queue_insert`, and `Queue_remove`. The `Queue_enqueue` function is used for adding the queue element at the back of the list. The `Queue_dequeue` function is used for removing and returning the queue element from the beginning of the list. The `Queue_insert` function is used for inserting the queue element in front of the specified queue element. The `Queue_remove` function is used for removing the specified queue element.

We can create a queue object dynamically in a TI-RTOS project. We provide such a C code in Listing 13.21. Here, `myQueue` is the handle name of the created queue.

Listing 13.21 The C Code for the Dynamic Queue Usage Example

```c
#include <xdc/std.h>
#include <xdc/runtime/System.h>
#include <xdc/runtime/Error.h>
#include <ti/sysbios/BIOS.h>
#include <ti/sysbios/hal/Timer.h>
#include <ti/sysbios/knl/Clock.h>
#include <ti/sysbios/knl/Semaphore.h>
#include <ti/sysbios/knl/Task.h>
#include <ti/sysbios/knl/Queue.h>
#include "Board.h"
#include <string.h>

#define TASKSTACKSIZE    512

 Error_Block eb;
 Timer_Params timerParams;
 Timer_Handle myTimer;
 Clock_Params clockParams;
 Clock_Handle myClock;
 Semaphore_Params semParams;
 Semaphore_Handle mySem1, mySem2;
 Task_Params taskParams;
 Task_Handle myTask1, myTask2;
 Char myTask1Stack[TASKSTACKSIZE], myTask2Stack[TASKSTACKSIZE];
 Queue_Handle myQueue;

void MyTimerFxn(UArg arg);
void MyClockFxn(UArg arg);
void MyTaskFxn1(UArg arg0, UArg arg1);
void MyTaskFxn2(UArg arg0, UArg arg1);

typedef struct QueueObj{
 Queue_Elem elem;
 char Id[5];
 int Data;
} QueueObj;

 int cnt = 0;

 QueueObj send[40];

void main(void){

 Board_initGeneral();
 Board_initGPIO();

 Error_init(&eb);
 Timer_Params_init(&timerParams);
 timerParams.startMode = Timer_StartMode_USER;
 timerParams.runMode = Timer_RunMode_CONTINUOUS;
 timerParams.periodType = Timer_PeriodType_MICROSECS;
 timerParams.period = 3000;
 timerParams.arg = 5;
 myTimer = Timer_create(Timer_ANY, MyTimerFxn, &timerParams, &eb);
```

```
    if (myTimer == NULL) System_abort("Timer create failed");

    Error_init(&eb);
    Semaphore_Params_init(&semParams);
    mySem1 = Semaphore_create(0, &semParams, &eb);
    if (mySem1 == NULL) System_abort("Semaphore create failed");

    Error_init(&eb);
    Semaphore_Params_init(&semParams);
    mySem2 = Semaphore_create(0, &semParams, &eb);
    if (mySem2 == NULL) System_abort("Semaphore create failed");

    Error_init(&eb);
    Clock_Params_init(&clockParams);
    clockParams.period = 1;
    clockParams.startFlag = FALSE;
    clockParams.arg = 0;
    myClock = Clock_create(MyClockFxn, 1, &clockParams, &eb);
    if (myClock == NULL) System_abort("Clock create failed");

    Error_init(&eb);
    Task_Params_init(&taskParams);
    taskParams.stackSize = TASKSTACKSIZE;
    taskParams.stack = &myTask1Stack;
    taskParams.priority = 1;
    myTask1 = Task_create(MyTaskFxn1, &taskParams, &eb);
    if (myTask1 == NULL) System_abort("Task create failed");

    Error_init(&eb);
    Task_Params_init(&taskParams);
    taskParams.stackSize = TASKSTACKSIZE;
    taskParams.stack = &myTask2Stack;
    taskParams.priority = 15;
    myTask2 = Task_create(MyTaskFxn2, &taskParams, &eb);
    if (myTask2 == NULL) System_abort("Task create failed");

    myQueue = Queue_create(NULL, NULL);

    System_printf("\nStarting Queue Example\n");
    System_flush();

    Timer_start(myTimer);
    Clock_start(myClock);

    BIOS_start();
}

void MyClockFxn(UArg arg){
 Semaphore_post(mySem1);
}

void MyTimerFxn(UArg arg){
 if(cnt<arg){
  send[cnt].Data = cnt+1;
  strcpy(send[cnt].Id, "Timer");
```

```
      Queue_enqueue(myQueue, &send[cnt].elem);
      cnt++;}
    else{
      Clock_stop(myClock);
      send[cnt].Data = cnt+1;
      strcpy(send[cnt].Id, "Timer");
      Queue_enqueue(myQueue, &send[cnt].elem);
      Semaphore_post(mySem2);}
}

void MyTaskFxn1(UArg arg0, UArg arg1){
  for (;;){
    Semaphore_pend(mySem1, BIOS_WAIT_FOREVER);
    send[cnt].Data = cnt+1;
    strcpy(send[cnt].Id, "Task1");
    Queue_enqueue(myQueue, &send[cnt].elem);
    cnt++;}
}

void MyTaskFxn2(UArg arg0, UArg arg1){
  QueueObj* receive;
  Semaphore_pend(mySem2, BIOS_WAIT_FOREVER);
  Timer_stop(myTimer);
  while (!Queue_empty(myQueue)){
    receive = Queue_dequeue(myQueue);
    System_printf("Id:%s,Data:%d\n", receive->Id, receive->Data);}
  System_printf("Example done.\n");
  System_exit(0);
}
```

Queue_create function is used to create the queue. As can be seen in the code, the first element of the queue object must be the type of Queue_Elem.

13.3.4 Memory Management

There are two options for managing memory space in TI-RTOS. These are static and dynamic memory allocations. The memory for desired variables is allocated at the beginning of the program in static memory allocation. Hence, the memory space for these variables becomes fixed. Stacks are used for static memory allocation. We provided stack usage while explaining tasks in Sec. 13.3.1.4. The memory for desired variables is allocated during runtime in dynamic memory allocation. Hence, the memory space for dynamic variables is not fixed. Dynamically allocated memory space can also be freed on the run by the deallocation process. However, the user must be careful during these allocation/deallocation processes to prevent memory leakage.

Heaps are used for dynamic memory allocation. There are different heap types defined in TI-RTOS. We will only explain HeapMem and HeapBuf in this section. For further information on the remaining heap types, please see [44].

There are two functions, Memory_alloc and Memory_free, for dynamic memory allocation. The function Memory_alloc(IHeap_Handle heap, SizeT size, SizeT align, Error_Block *eb) is for allocating the desired memory space dynamically. Here, heap is the created and configured heap memory; size is the allocation size; align is the alignment of the block of memory; and eb is the pointer to error block. The function Memory_free(IHeap_Handle heap, Ptr block, SizeT

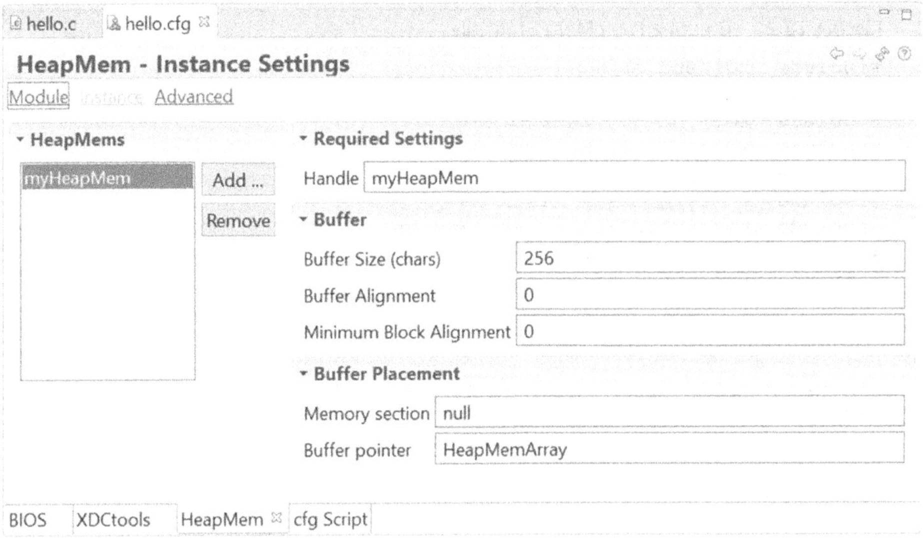

FIGURE 13.16 Creating HeapMem.

size) is for freeing the desired memory space dynamically. Here, block is the block of memory to free back to heap and size is the size of the block of memory to free. As can be seen, these two functions take an IHeap_Handle as their first argument. Therefore, the heaps created with HeapMem and HeapBuf must be casted to IHeap_Handle by HeapMem_Handle_upCast and HeapBuf_Handle_upCast functions, respectively.

13.3.4.1 Allocating Variable-Sized Memory Blocks

The HeapMem is used for allocating variable-sized memory blocks. It is ideal when the size of dynamic memory is not known until runtime. There are two major drawbacks in using HeapMem. These are fragmentation and nondeterministic performance. Allocation/freeing processes of variable-size blocks can result in fragmentation. Also, the memory manager traverses the memory to find a suitable block to allocate a new block in this fragmented memory. This may cause a nondeterministic performance.

We can create HeapMem using the configuration file in Sec. 13.2.4. To do so, open the hello.cfg file by right-clicking on it and select *Open With → XGCONF*. In the "BIOS - System Overview" window, click on the "Memory" box. In the opened window, tick the box titled "Add Memory management to my configuration". In this window, we should fill the Default Heap instance and Default Heap size (chars) boxes. For our case, the former will be myHeapMem and the latter will be 4096. Then, return to the "BIOS - System Overview" window, click on the "HeapMem" box. In the opened window, tick the box titled "Add HeapMem module to my configuration". Then, select the "Instance" option on the top of the window. A new window should appear as in Fig. 13.16.

To create a new HeapMem in the window in Fig. 13.16, we should click on the "Add" button and set the properties of the HeapMem in the "Required Settings" section. Here, we should provide a handle name to our HepMem. We should also set the Buffer Size, Buffer Alignment, and Minimum Block Alignment boxes. We should also fill the Memory section and Buffer pointer boxes under the Buffer Placement section.

We next form a TI-RTOS project to show how HeapMem works. Within the project, we configure the HeapMem to form four arrays. We calculate the average of each array.

Listing 13.22 Creating HeapMem Statically

```c
#include <xdc/std.h>
#include <xdc/runtime/IHeap.h>
#include <xdc/runtime/System.h>
#include <xdc/runtime/Memory.h>
#include <ti/sysbios/BIOS.h>
#include <ti/sysbios/knl/Task.h>
#include <ti/sysbios/heaps/HeapMem.h>
#include <xdc/runtime/Error.h>
#include "Board.h"

#define HEAPMEMSIZE 256

 extern HeapMem_Handle myHeapMem;
 extern Task_Handle myTask;
 extern char HeapMemArray[HEAPMEMSIZE];

void MyTaskFxn(UArg arg0, UArg arg1);

void main(){

 Board_initGeneral();

 System_printf("Memory HeapMem Example Started.\n");

 BIOS_start();
}

void MyTaskFxn(UArg arg0, UArg arg1){
...
}
```

Then, we free the memory space used. We provide the source code of the project in Listing 13.22. Within the code, we use the extern keyword to define variables generated in the configuration file. We provide the MyTaskFxn function used in the code in Listing 13.23.

We can create the HeapMem dynamically in a TI-RTOS project. We provide the modified form of our previous example in Listing 13.24. Here, MyHeapMem is the handle name of the created HeapMem. HeapMemParams is the structure name that contains HeapMem instance parameters. size parameter is used to define the size of the HeapMem memory. buf parameter is used to define the address of the HeapMem memory. The eb is the error block to handle errors that may occur during mailbox creation. We provide the MyTaskFxn function used in the code in Listing 13.23.

13.3.4.2 Allocating Fixed-Sized Memory Blocks

The HeapBuf is used for allocating fixed-size blocks of memory. When using HeapBuf, there is no fragmentation problem because of the fixed-size blocks. Also, the time of allocation/freeing processes is fixed. Hence, HeapBuf has a deterministic performance.

We can create a HeapBuf using the configuration file in Sec. 13.2.4. To do so, open the hello.cfg file by right-clicking on it and select *Open With → XGCONF*. In the "BIOS - System Overview" window, click on the "Memory" box. In the opened window, tick

Listing 13.23 The `MyTaskFxn` Function Used in HeapMem Creation

```c
void MyTaskFxn(UArg arg0, UArg arg1){
 IHeap_Handle heap = HeapMem_Handle_upCast(myHeapMem);

 char *ptr1, *ptr2, *ptr3, *ptr4;
 int sum1=0, sum2=0, sum3=0, sum4=0;
 int average1, average2, average3, average4;
 int i;

 ptr1 = (char *) Memory_alloc(heap, HEAPMEMSIZE/4, 0, NULL);
 ptr2 = (char *) Memory_alloc(heap, HEAPMEMSIZE/8, 0, NULL);
 ptr3 = (char *) Memory_alloc(heap, HEAPMEMSIZE/16, 0, NULL);
 ptr4 = (char *) Memory_alloc(heap, HEAPMEMSIZE/32, 0, NULL);

 for(i=0;i<HEAPMEMSIZE/4;i++)
  *(ptr1 + i) = i;
 for(i=0;i<HEAPMEMSIZE/4;i=i+2)
  *(ptr2 + i/2) = i+HEAPMEMSIZE/4;
 for(i=0;i<HEAPMEMSIZE/4;i=i+4)
  *(ptr3 + i/4) = i+2*HEAPMEMSIZE/4;
 for(i=0;i<HEAPMEMSIZE/4;i=i+8)
  *(ptr4 + i/8) = i+3*HEAPMEMSIZE/4;

 for(i=0;i<HEAPMEMSIZE/4;i++)
  sum1 += *(ptr1 + i);
 for(i=0;i<HEAPMEMSIZE/8;i++)
  sum2 += *(ptr2 + i);
 for(i=0;i<HEAPMEMSIZE/16;i++)
  sum3 += *(ptr3 + i);
 for(i=0;i<HEAPMEMSIZE/32;i++)
  sum4 += *(ptr4 + i);

 average1 = sum1/(HEAPMEMSIZE/4);
 average2 = sum2/(HEAPMEMSIZE/8);
 average3 = sum3/(HEAPMEMSIZE/16);
 average4 = sum4/(HEAPMEMSIZE/32);

 Memory_free(heap, ptr1, HEAPMEMSIZE/4);
 Memory_free(heap, ptr2, HEAPMEMSIZE/8);
 Memory_free(heap, ptr3, HEAPMEMSIZE/16);
 Memory_free(heap, ptr4, HEAPMEMSIZE/32);

 System_printf("Average of 1st block:%d\n", average1);
 System_printf("Average of 2nd block:%d\n", average2);
 System_printf("Average of 3rd block:%d\n", average3);
 System_printf("Average of 4th block:%d\n", average4);

 System_exit(0);
}
```

the box titled "Add Memory management to my configuration". In this window, we should fill the Default Heap instance and Default Heap size (chars) boxes. For our case, the former will be `heap0` and the latter will be 4096. Then, return to the "BIOS - System Overview" window, click on the "HeapMem" box. Fill the entries of this window as in the previous static HeapMem example. Finally, return to the "BIOS - System Overview" window, click on the "HeapBuf" box. In the opened window, tick the box titled "Add

Listing 13.24 Creating HeapMem Dynamically

```c
#include <xdc/std.h>
#include <xdc/runtime/IHeap.h>
#include <xdc/runtime/System.h>
#include <xdc/runtime/Memory.h>
#include <ti/sysbios/BIOS.h>
#include <ti/sysbios/knl/Task.h>
#include <ti/sysbios/heaps/HeapMem.h>
#include <xdc/runtime/Error.h>
#include "Board.h"

#define TASKSTACKSIZE   512
#define HEAPMEMSIZE     256

 Error_Block eb;
 Task_Params taskParams;
 Task_Handle myTask;
 Char myTaskStack[TASKSTACKSIZE];
 HeapMem_Params HeapMemParams;
 HeapMem_Struct HeapMemStruct;
 HeapMem_Handle myHeapMem;
 static char HeapMemArray[HEAPMEMSIZE];

void MyTaskFxn(UArg arg0, UArg arg1);

void main(){

 Board_initGeneral();

 Error_init(&eb);
 Task_Params_init(&taskParams);
 taskParams.stackSize = TASKSTACKSIZE;
 taskParams.stack = &myTaskStack;
 taskParams.priority = 1;
 myTask = Task_create(MyTaskFxn, &taskParams, &eb);
 if (myTask == NULL) System_abort("Task create failed");

 Error_init(&eb);
 HeapMem_Params_init(&HeapMemParams);
 HeapMemParams.size = 256;
 HeapMemParams.buf = (Ptr)HeapMemArray;
 myHeapMem = HeapMem_create(&HeapMemParams , &eb);
 if (myHeapMem == NULL) System_abort("HeapMem create failed");

 System_printf("Memory HeapMem Example Started.\n");

 BIOS_start();
}

void MyTaskFxn(UArg arg0, UArg arg1){
 ...
}
```

HeapBuf module to my configuration". Then, select the "Instance" option on the top of the window. A new window should appear as in Fig. 13.17.

To create a new HeapBuf in the window in Fig. 13.17, we should click on the "Add" button and set the properties of the HeapBuf in the "Required Settings" section. Here, we should provide a handle name to our HeapBuf. We should also set the Block size,

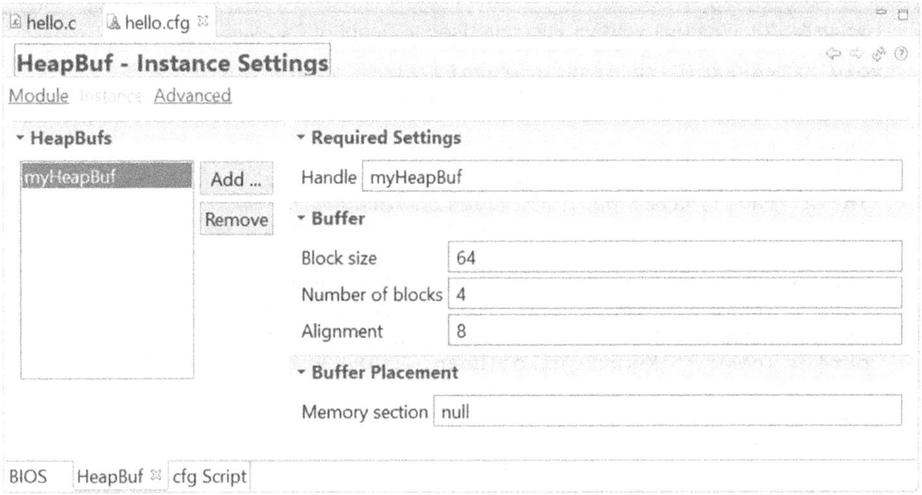

Real-time Operating System

FIGURE 13.17 Creating a HeapBuf.

Listing 13.25 Creating HeapBuf Statically

```c
#include <xdc/std.h>
#include <xdc/runtime/IHeap.h>
#include <xdc/runtime/System.h>
#include <xdc/runtime/Memory.h>
#include <ti/sysbios/BIOS.h>
#include <ti/sysbios/knl/Task.h>
#include <ti/sysbios/heaps/HeapBuf.h>
#include <xdc/runtime/Error.h>
#include "Board.h"

#define HEAPBUFSIZE 256

 extern HeapBuf_Handle myHeapBuf;
 extern Task_Handle myTask;

void MyTaskFxn(UArg arg0, UArg arg1);

void main(){

 Board_initGeneral();

 System_printf("Memory HeapBuf Example Started.\n");

 BIOS_start();
}

void MyTaskFxn(UArg arg0, UArg arg1){
 ...
}
```

Listing 13.26 The `MyTaskFxn` Function Used in HeapBuf Creation

```
void MyTaskFxn(UArg arg0, UArg arg1){
IHeap_Handle heap = HeapBuf_Handle_upCast(myHeapBuf);

char *ptr1, *ptr2, *ptr3, *ptr4;
int sum1=0, sum2=0, sum3=0, sum4=0;
int average1, average2, average3, average4;
int i;

ptr1 = (char *) Memory_alloc(heap, HEAPBUFSIZE/4, 0, NULL);
ptr2 = (char *) Memory_alloc(heap, HEAPBUFSIZE/4, 0, NULL);
ptr3 = (char *) Memory_alloc(heap, HEAPBUFSIZE/4, 0, NULL);
ptr4 = (char *) Memory_alloc(heap, HEAPBUFSIZE/4, 0, NULL);

for(i=0;i<HEAPBUFSIZE/4;i++){
 *(ptr1 + i) = i;
 *(ptr2 + i) = i+HEAPBUFSIZE/4;
 *(ptr3 + i) = i+2*HEAPBUFSIZE/4;
 *(ptr4 + i) = i+3*HEAPBUFSIZE/4;}

for(i=0;i<HEAPBUFSIZE/4;i++){
 sum1 += *(ptr1 + i);
 sum2 += *(ptr2 + i);
 sum3 += *(ptr3 + i);
 sum4 += *(ptr4 + i);}

average1 = sum1/(HEAPBUFSIZE/4);
average2 = sum2/(HEAPBUFSIZE/4);
average3 = sum3/(HEAPBUFSIZE/4);
average4 = sum4/(HEAPBUFSIZE/4);

Memory_free(heap, ptr1, HEAPBUFSIZE/4);
Memory_free(heap, ptr2, HEAPBUFSIZE/4);
Memory_free(heap, ptr3, HEAPBUFSIZE/4);
Memory_free(heap, ptr4, HEAPBUFSIZE/4);

System_printf("Average of 1st block:%d\n", average1);
System_printf("Average of 2nd block:%d\n", average2);
System_printf("Average of 3rd block:%d\n", average3);
System_printf("Average of 4th block:%d\n", average4);

System_exit(0);
}
```

Number of blocks, and Alignment boxes. We can also fill the Memory section box under Buffer Placement.

We next form a TI-RTOS project to show how HeapBuf works. Within the project, we configure the HeapBuf to form four arrays as in our previous HeapMem example. We calculate the average of each array. Then, we free the memory space used. We provide the source code of the project in Listing 13.25. Within the code, we use the extern keyword to define variables generated in the configuration file. We provide the `MyTaskFxn` function used in the code in Listing 13.26.

Listing 13.27 Creating HeapBuf Dynamically

```c
#include <xdc/std.h>
#include <xdc/runtime/IHeap.h>
#include <xdc/runtime/System.h>
#include <xdc/runtime/Memory.h>
#include <ti/sysbios/BIOS.h>
#include <ti/sysbios/knl/Task.h>
#include <ti/sysbios/heaps/HeapBuf.h>
#include <xdc/runtime/Error.h>
#include "Board.h"

#define TASKSTACKSIZE   512
#define HEAPBUFSIZE     256

 Error_Block eb;
 Task_Params taskParams;
 Task_Handle myTask;
 Char myTaskStack[TASKSTACKSIZE];
 HeapBuf_Params HeapBufParams;
 HeapBuf_Struct HeapBufStruct;
 HeapBuf_Handle myHeapBuf;
 static char HeapBufArray[HEAPBUFSIZE];

void MyTaskFxn(UArg arg0, UArg arg1);

void main(){

 Board_initGeneral();

 Error_init(&eb);
 Task_Params_init(&taskParams);
 taskParams.stackSize = TASKSTACKSIZE;
 taskParams.stack = &myTaskStack;
 taskParams.priority = 2;
 myTask = Task_create(MyTaskFxn, &taskParams, &eb);
 if (myTask == NULL) System_abort("Task create failed");

 Error_init(&eb);
 HeapBuf_Params_init(&HeapBufParams);
 HeapBufParams.blockSize = HEAPBUFSIZE/4;
 HeapBufParams.numBlocks = 4;
 HeapBufParams.buf = (Ptr)HeapBufArray;
 HeapBufParams.bufSize = HEAPBUFSIZE;
 myHeapBuf = HeapBuf_create(&HeapBufParams , &eb);
 if (myHeapBuf == NULL) System_abort("HeapBuf create failed");

 System_printf("Memory HeapBuf Example Started.\n");

 BIOS_start();
}

void MyTaskFxn(UArg arg0, UArg arg1){
 ...
}
```

We can also create the HeapBuf dynamically in TI-RTOS project. We provide the modified form of our previous example in Listing 13.27. Here, `MyHeapBuf` is the handle name of the created HeapBuf. `HeapBufParams` is the structure name that contains HeapBuf instance parameters. `blockSize` parameter is used to define the size of one block of HeapBuf memory. The `numBlocks` parameter is used to define the number of blocks of the HeapBuf memory. The `bufSize` parameter is used to define the size of the HeapBuf memory. `buf` parameter is used to define the address of the HeapBuf memory. The `eb` is the error block to handle errors that may occur during mailbox creation. We have provided the `MyTaskFxn` function used in the code in Listing 13.26.

13.4 TI-RTOS Drivers

We have reached MSP432 peripheral devices either in register-level or using DriverLib functions up to now. TI-RTOS offers special driver functions to reach these peripheral devices. These functions ensure that they do not interfere with RTOS while working. In this section, we will only explain the driver functions on the peripheral devices of interest. For further information on the remaining driver functions, please see [44].

Before going further, we should mention one important point. To use driver functions explained in this section, the `Board.h` header file should be added to the source code of the TI-RTOS project. Besides, the initialization function `Board_initGeneral()` must be called at the top of the main function in the code.

13.4.1 GPIO

We can use GPIO functions under TI-RTOS by including the header file `GPIO.h` to our C code. Then, we should use the `Board_initGPIO()` at the top of the main function. Afterwards, we can use the GPIO functions given below in our TI-RTOS project.

```
Board_BUTTON0 => Onboard S1 button
Board_BUTTON1 => Onboard S2 button
Board_LED0    => Onboard LED1
Board_LED1    => Onboard red color of LED2
Board_LED_ON  => Used for turning on the specific LED
Board_LED_OFF => Used for turning off the specific LED

unsigned int GPIO_read(unsigned int index)
/*
Returns the value of GPIO pin.
index: GPIO index of specific pin.
*/

void GPIO_write(unsigned int index, unsigned int value)
/*
Sets the value of GPIO pin.
index: GPIO index of specific pin.
value: the value to be written.
*/

void GPIO_toggle(unsigned int index)
/*
Toggles the current state of GPIO pin.
index: GPIO index of specific pin.
*/
```

```
void GPIO_enableInt(unsigned int index)
/*
Enables the interrupt for selected GPIO pin.
index: GPIO index of specific pin.
*/

void GPIO_disableInt(unsigned int index)
/*
Disables the interrupt for selected GPIO pin.
index: GPIO index of specific pin.
*/

void GPIO_clearInt(unsigned int index)
/*
Clears the interrupt flag for selected GPIO pin.
index: GPIO index of specific pin.
*/

void GPIO_setCallback(unsigned int index, GPIO_CallbackFxn callback)
/*
Connects a callback function to a GPIO pin interrupt.
index: GPIO index of specific pin.
callback: address of the callback function.
*/

void (*GPIO_CallbackFxn)(unsigned int index)
/*
Callback function for GPIO interrupt.
index: GPIO index of specific pin.
*/
```

13.4.2 ADC

We can use ADC functions under TI-RTOS by including the header file ADC.h to our C code. Then, we should use the Board_initADC() at the top of the main function. Afterwards, we can use the ADC functions given below in our TI-RTOS project:

```
Board_ADC0 => Use ADC0
Board_ADC1 => Use ADC1
ADC_Handle => Handle for ADC peripheral.
ADC_Params => It is the struct for ADC parameters.

void ADC_Params_init(ADC_Params *params)
/*
Initializes the ADC_Params structure to default values.
params: pointer to the parameter structure.
*/

ADC_Handle ADC_open(uint_fast16_t index, ADC_Params *params)
/*
Opens the given ADC instance and sets ADC parameters.
index: ADC index.
params: pointer to the parameter structure.
*/
```

```
void ADC_close(ADC_Handle handle)
/*
Closes the ADC instance specified by the ADC handle.
handle: ADC handle.
*/

int_fast16_t ADC_convert(ADC_Handle handle, uint16_t *value)
/*
Performs the ADC conversion.
handle: ADC handle.
value: pointer for the conversion result.
*/
```

13.4.3 PWM

We can use PWM functions under TI-RTOS by including the header file PWM.h to our C code. Then, we should use the Board_initPWM() at the top of the main function. Afterwards, we can use the PWM functions given below in our TI-RTOS project:

```
Board_PWM0 => Use TA1_1
Board_PWM1 => Use TA1_2
PWM_Handle => Handle for PWM peripheral.

PWM_Params
/*
Structure for PWM parameters. Its elements are listed below.
periodUnits: Unit of PWM period. It can be PWM_PERIOD_US (us),
    PWM_PERIOD_HZ (Hz) or PWM_PERIOD_COUNTS (timer cycles). Default
    value is PWM_PERIOD_HZ.
dutyUnits: Unit of PWM duty cycle. It can be PWM_DUTY_US (us),
    PWM_DUTY_FRACTION (fraction of period) or PWM_DUTY_COUNTS (timer
    cycles). Default value is PWM_DUTY_FRACTION.
periodValue: PWM period. Default value is 1e6 corresponding to 1 MHz.
dutyValue: PWM duty cycle. Default value is 0.
idleLevel: Pin output state when PWM is stopped. It can be PWM_IDLE_LOW
    (0) or PWM_IDLE_HIGH (1). Default value is PWM_IDLE_LOW.
*/

void PWM_Params_init(PWM_Params *params)
/*
Initializes the PWM_Params structure to default values.
params: pointer to the parameter structure.
*/

PWM_Handle PWM_open(unsigned int index, PWM_Params *params)
/*
Opens the given PWM instance and sets PWM parameters.
index: PWM index.
params: pointer to the parameter structure.
*/

void PWM_close(PWM_Handle handle);
/*
Closes the PWM instance specified by the PWM handle.
handle: PWM handle.
*/
```

```
int PWM_setDuty(PWM_Handle handle, uint32_t duty)
/*
Sets the duty cycle of the specified PWM handle.
handle: PWM handle.
duty: PWM duty cycle value.
*/

int PWM_setPeriod(PWM_Handle handle, uint32_t period)
/*
Sets the period of the specified PWM handle.
handle: PWM handle.
period: PWM period value.
*/

void PWM_start(PWM_Handle handle);
/*
Starts the specified PWM handle with current settings.
handle: PWM handle.
*/

void PWM_stop(PWM_Handle handle)
/*
Stops the specified PWM handle.
handle: PWM handle.
*/
```

13.4.4 UART

We can use UART functions under TI-RTOS by including the header file UART.h to our C code. Then, we should use the Board_initUART() at the top of the main function. Afterwards, we can use the UART functions given below in our TI-RTOS project:

```
Board_UART0 => Use UART A0
Board_UART1 => Use UART A2
UART_Handle => Handle for UART peripheral.

UART_Params
/*
Structure for UART parameters. Its elements are listed below.
readMode: mode for all read calls. It can be UART_MODE_BLOCKING or
    UART_MODE_CALLBACK. In UART_MODE_BLOCKING, RTOS uses a semaphore to
     block task execution until all
the data in buffer has been read. In UART_MODE_CALLBACK, RTOS does not
    block task execution. Instead, a callback function is called when
    the transfer is finished. Default value is UART_MODE_BLOCKING.
writeMode: mode for all write calls. It can be UART_MODE_BLOCKING or
    UART_MODE_CALLBACK. In UART_MODE_BLOCKING, RTOS uses a semaphore to
     block task execution until all
the data in buffer has been written. In UART_MODE_CALLBACK, RTOS does
    not block task execution. Instead, a callback function is called
    when the transfer is finished. Default value is UART_MODE_BLOCKING.
readTimeout: timeout for read semaphore. Default value is
    UART_WAIT_FOREVER.
writeTimeout: timeout for write semaphore. Default value is
    UART_WAIT_FOREVER.
```

```
readCallback: pointer to read callback. Default value is NULL.
writeCallback: pointer to write callback. Default value is NULL.
readReturnMode: receive return mode. It can be UART_RETURN_FULL
    (Unblock/callback when buffer is full) or UART_RETURN_NEWLINE
    (Unblock/callback when newline character is received). Default
        value is UART_RETURN_NEWLINE.
readDataMode: type of data being read. It can be UART_DATA_BINARY or
    UART_DATA_TEXT. Default value is UART_DATA_TEXT.
writeDataMode: type of data being written. It can be UART_DATA_BINARY
    or UART_DATA_TEXT. Default value is UART_DATA_TEXT.
readEcho: echo received data back. It can be UART_ECHO_OFF (no echo) or
      UART_ECHO_ON (received data is echoed). Default value is
    UART_ECHO_ON.
baudRate: baud rate for UART. Default value is 115200.
dataLength: data length for UART. It can be UART_LEN_x, where x can get
    values 5, 6, 7 or 8. Default value is UART_LEN_8.
stopBits: stop bits for UART. It can be UART_STOP_ONE (one stop bit) or
    UART_STOP_TWO (two stop bits). Default value is UART_STOP_ONE.
parityType: parity bit type for UART. It can be UART_PAR_NONE (no
    parity), UART_PAR_EVEN (even parity), UART_PAR_ODD (odd parity),
    UART_PAR_ZERO (parity always zero) or UART_PAR_ONE (parity always
    one). Default value is UART_PAR_NONE.
*/

void UART_Params_init(UART_Params *params)
/*
Iinitializes the UART_Params structure to default values.
params: pointer to the parameter structure.
*/

UART_Handle UART_open(unsigned int index, UART_Params *params)
/*
Opens the given UART instance and sets UART parameters.
index: UART index.
params: pointer to the parameter structure.
*/

void UART_close(UART_Handle handle)
/*
Closes the UART instance specified by the UART handle.
handle: UART handle.
*/

int UART_write(UART_Handle handle, const void *buffer, size_t size)
/*
Writes data to a UART with interrupts enabled.
handle: UART handle.
buffer: pointer to the buffer containing data to be written to the
    UART.
size: number of bytes in the buffer.
*/

int UART_read(UART_Handle handle, void *buffer, size_t size)
/*
Reads data from a UART with interrupt enabled.
handle: UART handle.
buffer: pointer to an empty buffer in which received data should be
    written to.
size: number of bytes in the buffer.
```

```
void (*UART_Callback) (UART_Handle, void *buf, size_t count)
/*
Callback function for UART interrupt.
handle: UART handle.
buf: pointer to read or write buffer.
count: number of elements read or written.
*/
```

13.4.5 SPI

We can use SPI functions under TI-RTOS by including the header file SPI.h to our C code. Then, we should use the Board_initSPI() at the top of the main function. Afterwards, we can use the SPI functions given below in our TI-RTOS project:

```
Board_SPI0 => Use SPI B0
Board_SPI1 => Use SPI B2
SPI_Handle => Handle for SPI peripheral.

SPI_Params
/*
Structure for SPI parameters. Its elements are listed below.
transferMode: blocking or Callback mode. It can be SPI_MODE_BLOCKING or
    SPI_MODE_CALLBACK. In SPI_MODE_BLOCKING, SPI_transfer will block
    task execution until the data transfer has completed. In
    SPI_MODE_CALLBACK, SPI_transfer does not block task execution and
    calls the callback function. Default value is SPI_MODE_BLOCKING.
transferTimeout: transfer timeout in system ticks. Default value is
    SPI_WAIT_FOREVER.
transferCallbackFxn: pointer to callback function. Default value is
    NULL.
mode: master or Slave mode. It can be SPI_MASTER or SPI_SLAVE. Default
    value is SPI_MASTER.
bitRate: SPI bit rate in Hz. Default value is 1000000.
dataSize: SPI data frame size in bits. Default value is 8.
frameFormat: SPI frame format. It can be SPI_POL0_PHA0(Polarity 0
    Phase 0), SPI_POL0_PHA1(Polarity 0 Phase 1), SPI_POL1_PHA0(
        Polarity 1 Phase 0), SPI_POL1_PHA1(Polarity 1 Phase 1), SPI_TI
        (TI mode) or SPI_MW (micro-wire mode). Default value is
        SPI_POL0_PHA0.
*/

void SPI_Params_init(SPI_Params *params)
/*
Initializes the SPI_Params structure to default values.
params: pointer to the parameter structure.
*/

SPI_Handle SPI_open(unsigned int index, SPI_Params *params)
/*
Opens the given SPI instance and sets the SPI parameters.
index: SPI index.
params: pointer to the parameter structure.
*/

void SPI_close(SPI_Handle handle)
```

```
/*
Closes the SPI instance specified by the SPI handle.
handle: SPI handle.
*/

bool SPI_transfer(SPI_Handle handle, SPI_Transaction *transaction)
/*
Performs the SPI data transfer.
handle: SPI handle.
transaction: structure including transfer size, transmit buffer pointer
    and receive buffer pointer.
*/

void (*SPI_CallbackFxn) (SPI_Handle handle, SPI_Transaction *
    transaction)
/*
Callback function for SPI interrupt.
handle: SPI handle.
transaction: structure including transfer size, transmit buffer pointer
    and receive buffer pointer.
*/
```

13.4.6 I²C

We can use I²C functions under TI-RTOS by including the header file `I2C.h` to our C code. Then, we should use the `Board_initI2C()` at the top of the main function. Afterwards, we can use the I²C functions given below in our TI-RTOS project.

```
Board_I2C0 => Use I2C0
I2C_Handle => Handle for I2C peripheral.

I2C_Params
/*
Structure for I2C parameters. Its elements are listed below.
transferMode: blocking or callback mode. It can be I2C_MODE_BLOCKING or
    I2C_MODE_CALLBACK. In I2C_MODE_BLOCKING, I2C_transfer will block
    task execution until the data transfer has completed. In
    I2C_MODE_CALLBACK, I2C_transfer does not block task execution and
    calls the callback function. default value is I2C_MODE_BLOCKING.
transferCallbackFxn: pointer to callback function. Default value is
    NULL.
bitRate: I2C bus bit rate. It can be I2C_100kHz or I2C_400kHz. Default
    value is I2C_100kHz.
*/

void I2C_Params_init(I2C_Params *params)
/*
Initializes the I2C_Params structure to default values.
params: pointer to the parameter structure.
*/

I2C_Handle I2C_open(unsigned int index, I2C_Params *params)
/*
Opens the given I2C instance and sets I2C parameters.
index: I2C index.
params: pointer to the parameter structure.
*/
```

```
void I2C_close(I2C_Handle handle)
/*
Closes the I2C instance specified by the I2C handle.
handle: I2C handle.
*/

bool I2C_transfer(I2C_Handle handle, I2C_Transaction *transaction)
/*
Performs the I2C data transfer with an I2C slave peripheral.
handle: I2C handle.
transaction: structure including number of bytes to be written to the
    slave, number of bytes to be read from the slave, transmit buffer
    pointer, receive buffer pointer and address of the I2C slave
    device.
*/

void (*I2C_CallbackFxn)(I2C_Handle handle, I2C_Transaction
    *transaction, bool transferStatus)
/*
Callback function for I2C interrupt.
handle: I2C handle.
transaction: structure including number of bytes to be written to the
    slave, number of bytes to be read from the slave, transmit buffer
    pointer, receive buffer pointer and address of the I2C slave
    device.
transferStatus: result of the I2C data transfer.
*/
```

13.5 TI-RTOS Instrumentation

The instrumentation tool can be used to analyze and debug the TI-RTOS project during execution. Hence, it will be very useful to understand and pinpoint any malfunction and system resource usage during runtime. We should first add the diagnostics tools to the RTOS project to use the debugging option under instrumentation. In this section, we will focus on the CPU Load option within diagnostics tools. Then, we will look at other debugging features provided by CCS.

13.5.1 Adding the CPU Load Option to the TI-RTOS Project

We can add the CPU Load option under diagnostics option to our TI-RTOS project using the configuration file mentioned in Sec. 13.2.4. To do so, open the `hello.cfg` file by right-clicking on it and select *Open With → XGCONF*. In the "BIOS - System Overview" window, click on the "CPU Load" box. In the opened window, tick the box titled "Add the CPU load monitoring module to my configuration". The window should appear as in Fig. 13.18. Then, fill in the boxes under Threads to Monitor and Monitor Options sections in this window.

13.5.2 RTOS Object Viewer

RTOS object viewer (ROV) can be used to analyze the state of the kernel. It can be opened by *Tools → RTOS Object Viewer (ROV)* after debugging the project. For more information

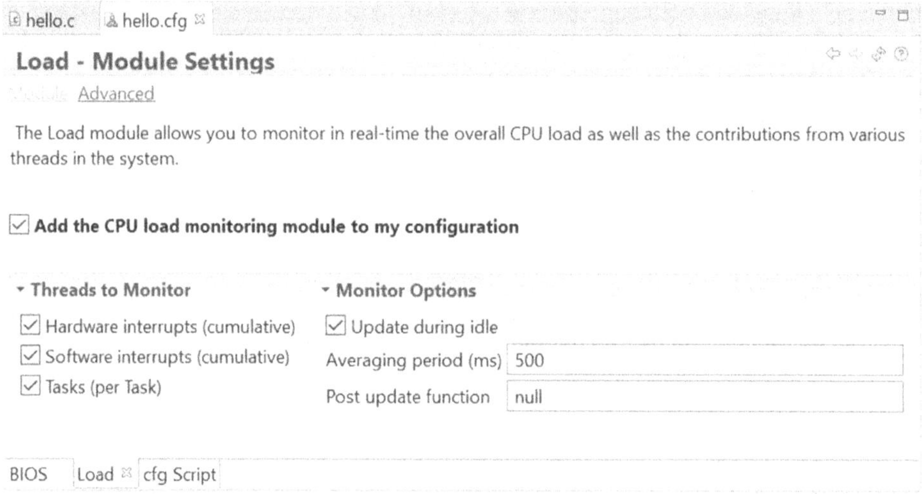

FIGURE 13.18 Adding the CPU Load option to the TI-RTOS project.

on ROV, please see `tirtos_basics.html` file under the `simplelink\_academy` in Resource Explorer.

13.5.3 Runtime Object Viewer

Runtime object viewer is an extended version of the ROV. It can be opened by *Tools → Runtime Object Viewer* after debugging the project. For more information on the runtime object viewer, please see `tirtos_basics.html` file under the `simplelink\_academy` folder in Resource Explorer.

13.5.4 RTOS Analyzer

RTOS analyzer can be used to visually observe the RTOS working status. It can be opened by *Tools → RTOS Analyzer* after debugging the project. In this section, we will only focus on the Execution Graph under the RTOS analyzer. For more information on other options in the RTOS analyzer, please see `tirtos_basics.html` file under the `simplelink\_academy` folder in Resource Explorer.

The execution graph under the RTOS analyzer deserves special consideration since it visually shows how different threads are handled by the TI-RTOS in time. We will use two projects to show how the RTOS analyzer can be of help. In our first project, we have two tasks to toggle LED1 (connected to pin P1.0 on the MSP432 LaunchPad) and red color of LED2 (connected to pin P2.0 on the MSP432 LaunchPad) with different speeds. In our second project, we have one task and one hardware interrupt.

Let's start with our first TI-RTOS project having two tasks. We provide the C code of the project in Listing 13.28. Within this code the task 1 has lower priority and task 2 has higher priority. Within task 1 we turn on LED1, wait in a loop for a while, turn off LED1, and go to blocked state for four seconds. Similarly, within task 2 we turn on the red color of LED2, wait in a loop as one-tenth of the wait time in task 1, turn off the red color of LED2, and go to blocked state for 0.4 seconds. We do not have any operations in the idle loop. It is solely used to observe that the system goes to idle loop when both

Listing 13.28 The C Code for the RTOS Analyzer Example with Two Tasks

```c
#include <xdc/std.h>
#include <xdc/runtime/System.h>
#include <xdc/runtime/Error.h>
#include <ti/sysbios/BIOS.h>
#include <ti/sysbios/knl/Clock.h>
#include <ti/drivers/GPIO.h>
#include <ti/sysbios/knl/Task.h>
#include "Board.h"

#define TASKSTACKSIZE 512

Error_Block eb;
Task_Params taskParams;
Task_Handle myTask1, myTask2;
char myTask1Stack[TASKSTACKSIZE], myTask2Stack[TASKSTACKSIZE];

void MyTaskFxn1(UArg arg0, UArg arg1);
void MyTaskFxn2(UArg arg0, UArg arg1);
void MyIdleFxn(void);

int cnt = 0;
int i,j;

void main(void){

 Board_initGeneral();
 Board_initGPIO();

 Error_init(&eb);
 Task_Params_init(&taskParams);
 taskParams.stackSize = TASKSTACKSIZE;
 taskParams.stack = &myTask1Stack;
 taskParams.priority = 1;
 myTask1 = Task_create(MyTaskFxn1, &taskParams, &eb);
 if (myTask1 == NULL) System_abort("Task1 create failed");

 Error_init(&eb);
 Task_Params_init(&taskParams);
 taskParams.stackSize = TASKSTACKSIZE;
 taskParams.stack = &myTask2Stack;
 taskParams.priority = 15;
 myTask2 = Task_create(MyTaskFxn2, &taskParams, &eb);
 if (myTask2 == NULL) System_abort("Task2 create failed");

 System_printf("Starting Instrumentation Example\n");
 System_flush();

 BIOS_start();
}

void MyTaskFxn1(UArg arg0, UArg arg1){
 while(1){
  GPIO_write(Board_GPIO_LED0, Board_GPIO_LED_OFF);
  for(j=0;j<5000000;j++);
```

```
      GPIO_write(Board_GPIO_LED0, Board_GPIO_LED_ON);
      Task_sleep(4000 * (1000 / Clock_tickPeriod));}
}

void MyTaskFxn2(UArg arg0, UArg arg1){
 while(1){
   GPIO_write(Board_GPIO_LED1, Board_GPIO_LED_OFF);
   for(i=0;i<500000;i++);
   GPIO_write(Board_GPIO_LED1, Board_GPIO_LED_ON);
   Task_sleep(400 * (1000 / Clock_tickPeriod));}
}

void MyIdleFxn(void){}
```

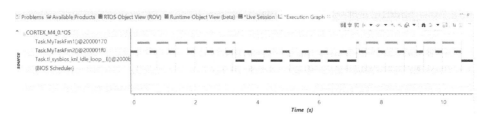

FIGURE 13.19 Execution graph for the TI-RTOS project with two tasks.

tasks are in blocked state. To use the debugging tools, we should add the below code lines to our project's config file:

```
var LoggingSetup = xdc.useModule('ti.uia.sysbios.LoggingSetup');
LoggingSetup.sysbiosLoggerSize = 9192;
LoggingSetup.loadLogging = false;
```

We should run the C code in Listing 13.28 for a couple of seconds and pause execution. To observe the Execution Graph under the RTOS analyzer example, select it from *Tools → RTOS Analyzer → Execution Analysis*. A new window should appear asking for the Analysis Settings. Here, tick the *Execution Graph* box. As we press the Start button in this window, a new graph opens as in Fig. 13.19. We should zoom out to observe all graph properties. As can be seen in this figure, the system starts executing task 2 since it has a higher priority. Afterwards, task 2 goes to blocked state with the function `Task_sleep`. Meanwhile, task 1 starts working. However, before the operations in task 1 has ended task 2 becomes active again. It preempts task 1 and continues its execution. This scenario is repeated a couple of times till task 1 goes to the blocked state for four seconds with the function `Task_sleep`. When both task 1 and task 2 are in the blocked state, the idle loop becomes active.

Our second TI-RTOS project has one task and one hardware interrupt. We provide the C code of the project in Listing 13.29. Within this code, task 1 waits in a loop for a while and goes to the blocked state for five milliseconds. The timer module generates interrupts periodically every millisecond. Within the function associated with the timer module, only the variable `i` is incremented. When the task is in blocked state and the timer does not generate an interrupt, the system goes to the idle loop. Within the idle loop, the system checks whether the variable `i` has reached the value 20. If this is the

Listing 13.29 The C Code for the RTOS Analyzer Example with One Task and One Hardware Interrupt

```c
#include <xdc/std.h>
#include <xdc/runtime/System.h>
#include <xdc/runtime/Error.h>
#include <ti/sysbios/BIOS.h>
#include <ti/sysbios/knl/Clock.h>
#include <ti/drivers/GPIO.h>
#include <ti/sysbios/knl/Task.h>
#include <ti/sysbios/hal/Timer.h>
#include "Board.h"

#define TASKSTACKSIZE   512

 Error_Block eb;
 Task_Params taskParams;
 Task_Handle myTask1;
 char myTask1Stack[TASKSTACKSIZE];
 Timer_Params timerParams;
 Timer_Handle myTimer;

void MyTaskFxn1(UArg arg0, UArg arg1);
void MyTimerFxn(UArg arg);
void MyIdleFxn(void);

 int cnt = 0;
 int i,j;

void main(void){

 Board_initGeneral();
 Board_initGPIO();

 Error_init(&eb);
 Timer_Params_init(&timerParams);
 timerParams.startMode = Timer_StartMode_AUTO;
 timerParams.runMode = Timer_RunMode_CONTINUOUS;
 timerParams.periodType = Timer_PeriodType_MICROSECS;
 timerParams.period = 1000;
 timerParams.arg = 0;
 myTimer = Timer_create(Timer_ANY, MyTimerFxn, &timerParams, &eb);
 if (myTimer == NULL) System_abort("Timer create failed");

 Error_init(&eb);
 Task_Params_init(&taskParams);
 taskParams.stackSize = TASKSTACKSIZE;
 taskParams.stack = &myTask1Stack;
 taskParams.priority = 1;
 myTask1 = Task_create(MyTaskFxn1, &taskParams, &eb);
 if (myTask1 == NULL) System_abort("Task1 create failed");

 System_printf("Starting Instrumentation Example\n");
 System_flush();

 BIOS_start();
}
```

378 Chapter Thirteen

```
void MyTaskFxn1(UArg arg0, UArg arg1){
 while(1){
  for(j=0;j<5000;j++);
  Task_sleep(5 * (1000 / Clock_tickPeriod));}
}

void MyTimerFxn(UArg arg){
 i++;
}

void MyIdleFxn(void){
 if(i==20) System_exit(0);
}
```

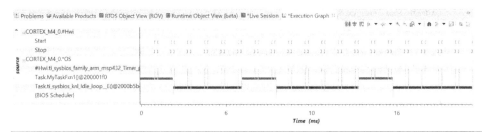

FIGURE 13.20 Execution graph for the TI-RTOS project with one task and one hardware interrupt.

case, the system stops working. To use the debugging tools, we should add the below code lines to our project's config file.

```
var LoggingSetup = xdc.useModule('ti.uia.sysbios.LoggingSetup');
LoggingSetup.sysbiosLoggerSize = 9192;
LoggingSetup.loadLogging = false;
Load.hwiEnabled = true;
Load.swiEnabled = true;
```

We should run the C code in Listing 13.29 for a couple of seconds and pause execution. To observe the Execution Graph under the RTOS analyzer example, select it from *Tools → RTOS Analyzer → Execution Analysis*. A new window should appear asking for the Analysis Settings. Here, we should tick the *Execution Graph* box. As we press the Start button in this window, a new graph opens as in Fig. 13.20. As can be seen in this figure, the task is executed when it is not in the blocked state with the function `Task_sleep` and the timer does not generate an interrupt. When the timer generates an interrupt, the task is preempted and the function for the timer module is executed. When the task is in blocked state and the timer module does not generate an interrupt, the system executes the idle loop. As the timer module generates an interrupt, the idle loop is preempted and the function for the timer module is executed.

Listing 13.30 Energia TI-RTOS Code

```c
#include <xdc/runtime/Error.h>
#include <ti/sysbios/BIOS.h>
#include <ti/sysbios/knl/Clock.h>
#include <xdc/runtime/Timestamp.h>
#include <xdc/runtime/Types.h>

 Clock_Params clockParams;
 Clock_Handle myClock;
 Error_Block eb;

int print = 0;
int cnt = 0;

 Types_FreqHz freq;
 unsigned int tstart,tstop;

 uint8_t status = 0;
 const uint8_t myLED = YELLOW_LED;

void MyClockFxn(UArg arg){
 tstop = Timestamp_get32();
 status = 1 - status;
 digitalWrite(myLED, status);
 cnt ++;
 print = 1;
}

void setup(){
 Serial.begin(115200);
 Error_init(&eb);
 Clock_Params_init(&clockParams);

 clockParams.period = 1000;
 clockParams.startFlag = FALSE;
 clockParams.arg = 0;
 myClock = Clock_create(MyClockFxn, 1000, &clockParams, &eb);
 if (myClock == NULL) System_abort("Clock create failed");

 Serial.println("Starting Timestamp Example");

 Clock_start(myClock);
 tstart = Timestamp_get32();
}

void loop(){
 if(print == 1){
  Timestamp_getFreq(&freq);

  Serial.print("Led Toggle after ");
  Serial.print((tstop-tstart) / (freq.lo / 1000000));
  Serial.print(" us cnt = ");
  Serial.println(cnt);

  tstart = tstop;
  print = 0;}
}
```

13.6 Energia and TI-RTOS

The Energia versions 15 and above support multitasking (MT). Moreover, these versions are built on TI-RTOS. In other words, a new Energia project is directly constructed as a TI-RTOS project. Therefore, it is suitable to be used by MSP432 RTOS applications.

Each tab represents a different task in the Energia MT project. There is a `setup()` and `loop()` function for each task. However, these functions should have different names for each task such as `setup1()`, `loop1()`, and `setup2()`, `loop2()`. Communication between these tasks are done via global variables declared in the first task. However, we cannot assign priorities to tasks in this setup. This limits the usage of advanced RTOS concepts. On the other hand, this structure is very useful if the user wants to benefit from the basic RTOS properties.

Energia allows the usage of advanced RTOS concepts by defining them dynamically as explained in previous sections. The main differences in Energia are as follows. Energia does not need the `BIOS_start()` function. Besides, we cannot use the `System_printf()` function to observe outputs on the console window. Instead, we can use the `Serial.print()` and `Serial.println()` functions to observe outputs via UART interface.

We next provide a sample Energia code in Listing 13.30. This code uses the clock and timestamp modules of TI-RTOS. Every time the clock function is called, LED1 connected to pin P1.0 on the MSP432 LaunchPad toggles. Duration between these toggling actions are calculated by the timestamp module. In an idle loop, the number of toggles and the time passed between each toggle can be observed on the terminal via UART interface.

13.7 Applications

TI-RTOS can be used in implementing applications introduced thus far. We leave it to the reader to form the CCS projects accordingly. In this section, we will provide a sample application to specifically show the TI-RTOS working principles. We did not use any external hardware besides the MSP432 LaunchPad. Hence, our sample application can be implemented without any difficulty.

In our sample TI-RTOS application, we will design a system that starts the ADC conversion, takes desired number of samples, averages them to obtain the final ADC result, and stops the ADC conversion periodically. The ADC data will be obtained from pin P5.5. Also, if the button S1 connected to pin P1.4 on the MSP432 LaunchPad is pressed, the last final ADC result will be sent to host PC via UART interface. Here, we will use TI-RTOS drivers to control ADC14, UART, and GPIO modules. When the system starts, the user will be asked to enter a number between 1 and 512 that will be the size of the dynamic heap memory to be created. This dynamic heap memory will be used to keep obtained ADC samples. The timer module will activate the ADC14 module and start clock module periodically every five seconds. The clock module is used with 1-millisecond period to obtain an ADC sample and send it to a task using mailbox to be saved in dynamic heap memory. When the dynamic heap memory is full, the ADC14 module will be deactivated and clock module will be stopped until the next timer trigger. Afterwards, the sample values at the dynamic heap memory will be averaged and the result will be written to the variable `ADC_result`. During this process, an Hwi will be triggered whenever the button S1 is pressed. This Hwi will post a Swi to send the `ADC_result` value to the host PC via UART interface.

13.8 Summary

RTOS provides a standardized environment to manage complex operations on the microcontroller with efficient use of its resources. To benefit from RTOS in our projects, we explored it in this chapter. We specifically picked TI-RTOS since it was supported by CCS. We provided sample projects on all TI-RTOS kernel components including threads, synchronization, timing, and memory management. While providing the sample projects, we considered both static and dynamic kernel component generation. Then, we introduced TI-RTOS drivers to use the peripheral devices in RTOS projects safely. Afterwards, we focused on instrumentation and diagnostics tools offered by CCS. These tools can be used to understand the working principles of RTOS better. Finally, we provided a sample application to show the usefulness of RTOS. The concepts introduced in this chapter can be used in developing complex embedded systems in an effective manner. Therefore, we kindly ask the reader to master them.

13.9 Exercises

TI-RTOS can also be used in solving almost all problems till this chapter. Therefore, we kindly ask the reader to revisit them and propose a solution from an RTOS perspective. We believe this will improve the understanding of RTOS concepts better. In this chapter, we only provided conceptual problems emphasizing some properties of RTOS.

13.1 What is the main difference between a Swi and task? Can the task be used instead of Swi in an RTOS application?

13.2 Can we use a global variable instead of a semaphore? What is the disadvantage of this approach?

13.3 What is the difference between a semaphore and an event? Is there a case such that the event usage is mandatory?

13.4 TI-RTOS provides the FatFS as a part of middleware and drivers. Construct a project to use it in reaching and modifying SD card contents.

13.5 TI provides a multi-threaded RTOS thermostat application under SimpleLink SDK (in Resource Explorer). First, analyze it to understand its working principles. Then, implement it if possible.

CHAPTER 14

Advanced Applications

MSP432 can be used in implementing advanced applications. We provide 20 such applications based on real-life problems in this chapter. The aim here is showing the usefulness of the microcontroller in our daily lives. We provide the circuit layout and the equipment list for each application. Hence, the reader can implement them easily. We did not provide the CCS project for the applications expecting the reader can solve them based on the gained experience till this chapter. Related to this, we would like to remind that register level, DriverLib, or Energia can be used in developing the CCS projects. Besides, an RTOS-based approach can be followed in forming the applications.

14.1 Garage Gate Control System

The goal of this application is to learn how to use the digital and wireless communication, mixed signal systems, and timer modules of the MSP432 microcontroller. As a real-life application, we examine a garage gate control system. In this section, we provide the equipment list, circuit layout, and system design specifications.

14.1.1 Equipment List and Layout

Following is a list of equipments to be used in the garage gate control system:

- One servo motor
- One GP2Y0A60SZLF proximity sensor
- One HM-10 Bluetooth module

In this application, we will use a mini servo motor to simulate the garage gate. A servo motor is a device that rotates its shaft to a specific angular position according to its input. Its rotation is limited with π rad. Moreover, the servo motor cannot rotate continuously unless otherwise specified. Servo motors are controlled by PWM signal such that a 1.5-millisecond pulse width (with the period 50 Hz) will make the motor turn to the $\pi/2$ rad position (often called the neutral position). If the pulse width is shorter than 1.5 milliseconds, then the motor will turn the shaft between $\pi/2$ and 0 rad accordingly. One-millisecond pulse width corresponds to rotation of 0 rad. If the pulse width is longer than 1.5 milliseconds, then the motor will turn the shaft between $\pi/2$ and π rad accordingly. Two-millisecond pulse width corresponds to rotation of π rad. Following

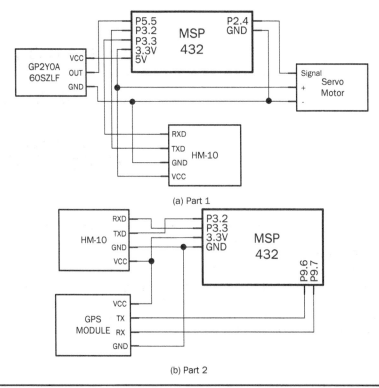

FIGURE 14.1 Layout of the garage gate control system.

is an additional equipment list for the garage gate control system when the wireless communication mode is added to it:

- One HM-10 Bluetooth module
- One Neo-6M GPS module

To note here, a second MSP432 is also needed to form a wireless network between the car and garage gate controller. The layout of garage gate control system is shown in Fig. 14.1.

14.1.2 System Design Specifications

In this application, we will design a garage gate control system using a Bluetooth module, servo motor, and proximity sensor. The user can open the garage gate by using his or her cell phone via Bluetooth communication. Here, a simple Android application developed under MIT App Inventor may be sufficient on the cell phone [45]. Once the user approaches the gate, he or she sends the character O via cell phone to the Bluetooth module. Then, the gate opens and waits for 12 seconds to close unless the car is still passing through. Here, the proximity sensor is used to detect whether the car is passing through the gate or not. Also, the servo motor is used to open and close the garage gate. Initially, the servo motor stays at 0 rad to indicate that the gate is closed. When the gate is opened, it goes to π rad.

We can place another module into the car that detects the user location and sends the character O via the Bluetooth module to the garage gate control system automatically. The user location can be detected using the GPS module. When the user is within range of garage with his or her car, the Bluetooth module tries to connect to the garage gate control system. Once the connection is established between the two Bluetooth modules, then the character O is sent.

14.2 Vending Machine

The goal of this application is to learn how to use the digital I/O, digital and wireless communication, mixed signal systems, and timer modules of the MSP432 microcontroller. As a real-life application, we examine a vending machine. In this section, we provide the equipment list, circuit layout, and system design specifications.

14.2.1 Equipment List and Layout

Following is a list of equipments to be used in the vending machine. To note here, we can also use a 2×16 LCD module instead of the four-digit seven-segment display. More information on the LCD module can be found in [1]:

- One stepper motor
- One ULN2003 motor driver
- Two push buttons
- Two 100-ηF capacitors
- Four-digit seven-segment display
- Four 330-Ω resistors
- One GSM click module

In this application, we will use a stepper motor to drop the selected product into delivery basket. A stepper motor is a device that rotates in smaller steps rather than a continuous 2π rad rotation. More information on the stepper motor can be found in [1]. The layout of the vending machine is shown in Fig. 14.2.

14.2.2 System Design Specifications

In this application, we will design a vending machine containing three products as bottled water (costs $ 1.25), soft drink (costs $ 0.75), and energy drink (costs $ 1.75). The button S1 connected to pin P1.1 on the MSP432 LaunchPad is used to switch between products. The red, green, and blue colors of LED2 connected to pins P2.0, P2.1, and P2.2 on the MSP432 LaunchPad are associated with the available products. The vending machine will start its operation with the bottled water selection.

We simulate the money entrance with two external buttons. The button connected to pin P1.5 on the MSP432 LaunchPad is used for $0.25 coin entrance. The button connected to pin P1.6 on the MSP432 LaunchPad is used for $1.00 coin entrance. The total money entered to the vending machine is displayed on the four-digit seven-segment display in the format XX.XX. The button S2 connected to pin P1.4 on the MSP432 LaunchPad will be used for the final buy command. If the selected product is in stock, its cost will be deduced from the entered money and stock will be decreased by one. Then, the stepper motor will turn 10, 15, or 20 rotations to the right if the bought product is water,

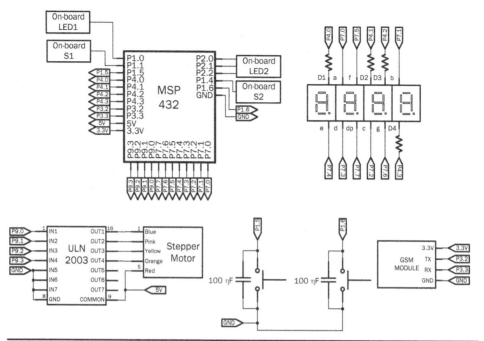

FIGURE 14.2 Layout of the vending machine.

soft drink, or energy drink, respectively. If the selected product is out of stock, related warning message will be sent to the vending machine company via SMS message using the GSM click module. If the entered money is not enough for the selected product, LED1 connected to pin P1.0 on the MSP432 LaunchPad will turn on for three seconds for warning.

14.3 Digital Clock

The goal of this application is to learn how to use the digital I/O, mixed signal systems, and timer modules of the MSP432 microcontroller. As a real-life application, we examine a digital clock. In this section, we provide the equipment list, circuit layout, and system design specifications.

14.3.1 Equipment List and Layout

Following is a list of equipments to be used in the digital clock. To note here, we can also use a 2 × 16 LCD module instead of the four-digit seven-segment display. The layout of the digital clock is shown in Fig. 14.3:

- Four-digit seven-segment display
- Five 330-Ω resistors
- One rotary encoder
- Three 100-ηF capacitors
- Five 10-kΩ resistors
- One piezo buzzer

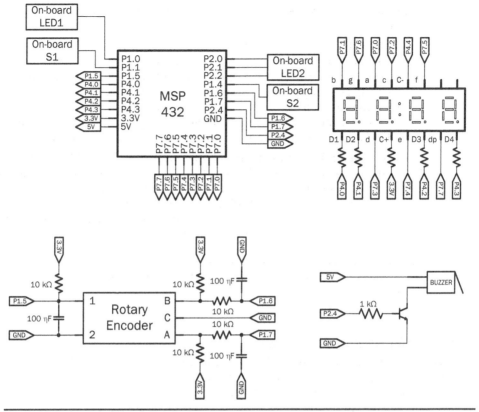

FIGURE 14.3 Layout of the digital clock.

- One BC547 transistor
- One 1-kΩ resistor

14.3.2 System Design Specifications

In this application, we will design a digital clock with three different modes as clock, alarm, and chronometer. The button S1 connected to pin P1.1 on the MSP432 LaunchPad is used to switch between these modes. The red, green, and blue colors of LED2 connected to pins P2.0, P2.1, and P2.2 on the MSP432 LaunchPad will display the active mode. Our clock will start with the initial value 00:00. The three modes of the clock are explained in detail below:

- **Clock:** In this mode, the leftmost two digits of the seven-segment display show the hour value. The rightmost two digits of the seven-segment display show the minute value. Also the colon-dots of the seven-segment display will toggle every second. The user can set the clock time first by pressing the button S2 connected to pin P1.4 on the MSP432 LaunchPad. The user should select the desired digit using the rotary encoder. Afterwards, he or she should press the push button of the rotary encoder. Next, this digit can be modified using the rotary encoder. The value will be set by pressing the push button of rotary encoder again. After the desired time value is set, the clock can be started again by pressing the button S2.

- **Alarm:** In this mode, the alarm can be set using the rotary encoder. To do so, the user should first select the desired digit using the rotary encoder and press its push button. Afterwards, this digit can be modified using the rotary encoder. The value will be set by pressing the push button of the rotary encoder again. When the alarm time comes, LED1 connected to pin P1.0 on the MSP432 LaunchPad will toggle every second until the button S2 is pressed. Also the alarm buzzer will turn on with two-second interval until the button S2 is pressed.

- **Chronometer:** In this mode, the chronometer will start with the initial value 00:00 by pressing the push button of the rotary encoder. Here, the leftmost two digits of the seven-segment display show the minute value. The rightmost two digits of the seven-segment display show the seconds value. If the push button of the rotary encoder is pressed again, the chronometer pauses. The user can reset the chronometer by pressing the button S2.

All three modes can work at the same time. Therefore, if the user wants to use the chronometer, he or she can do so without disturbing the digital clock and alarm operations.

14.4 Audio Spectrum Analyzer

The aim of this application is to learn how to use the digital I/O and mixed signal systems modules of the MSP432 microcontroller. As a real-life application, we examine an audio spectrum analyzer. In this section, we provide the equipment list, circuit layout, and system design specifications.

14.4.1 Equipment List and Layout

Following is a list of equipments to be used in the audio spectrum analyzer. The layout of the application is shown in Fig. 14.4:

- One mini electret microphone
- One TLC272 dual operational amplifier
- One 100-kΩ resistor
- Three 10-kΩ resistors
- One 1-kΩ resistor
- Eight 220-Ω resistors
- One 100-ηF capacitor
- One 8 × 8 LED matrix

14.4.2 System Design Specifications

In this application, we will design an audio spectrum analyzer using a microphone and 8 × 8 LED matrix. When the button S1 connected to pin P1.1 on the MSP432 LaunchPad is pressed, the system starts capturing audio data coming from the microphone. The audio spectrum will be displayed on the 8 × 8 LED matrix. To obtain the audio spectrum, the

Advanced Applications 389

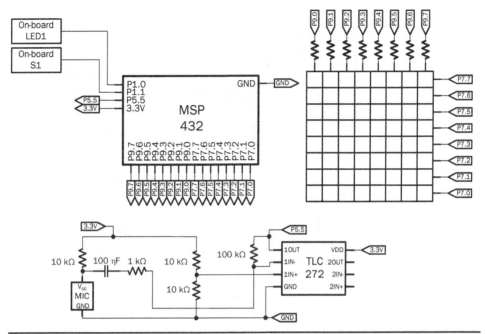

FIGURE 14.4 Layout of the audio spectrum analyzer.

fast Fourier transform (FFT) operation must be performed on the signal. FFT simply provides the frequency content of a given signal. Please see a digital signal processing book to grasp how FFT works in detail. In our application, each column in the 8×8 LED matrix shows different frequency components as 1, 1.5, 2, 3, 4, 5, 6, and 7.5 kHz, respectively. Each row in a column shows the magnitude of the associated frequency component. When the button S1 is pressed again, the system stops capturing the audio signal. LED1 connected to pin P1.0 on the MSP432 LaunchPad will be used to indicate the on/off state of the system.

14.5 Air Freshener Dispenser

The goal of this application is to learn how to use the digital I/O and timer modules of the MSP432 microcontroller. As a real-life application, we examine an air freshener dispenser. In this section, we provide the equipment list, circuit layout, and system design specifications.

14.5.1 Equipment List and Layout

Following is a list of the equipment to be used in the air freshener dispenser application. The layout of this application is shown in Fig. 14.5:

- One LED
- One 220-Ω resistor
- One push button
- One 100-ηF capacitor

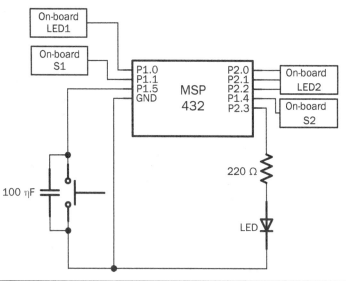

Figure 14.5 Layout of the air freshener dispenser.

14.5.2 System Design Specifications

The design steps of the air freshener dispenser system are as follows. In the first part of the application, an air freshener dispenser with three different programs will be implemented. These programs are called short, medium, and long. They correspond to the spraying of fresh odor in 5-, 10-, and 15-second periods. These should be in minutes in an actual system. Also, the spraying operation should be done by a mechanism in the actual system. We simulate this operation by turning on LED1 connected to pin P1.0 on the MSP432 LaunchPad for one second. In our system, the button S1 connected to pin P1.1 on the MSP432 LaunchPad will be used to switch between programs. The red, green, and blue colors of LED2 connected to pins P2.0, P2.1, and P2.2 on the MSP432 LaunchPad will be associated with the programs. Therefore, selecting each program will turn on the associated color of LED2. Also, the button S2 connected to pin P1.4 on the MSP432 LaunchPad will be used as an instant spray button. When it is pressed, the system will spray the odor and reset the counting process. In this part, one of the three programs must be selected as the initial starting program.

In the second part, an on/off button connected to pin P1.5 and a warning LED connected to pin P2.3 will be added to the system. When the system is in off state, all LEDs are turned off and the system goes into an appropriate low-power mode. All buttons except the on/off button will be unavailable in this state. When the system is turned on by this button, the warning LED will turn on. This LED should blink every second during operation. Initially, the system must be in the off state. The watchdog timer can be used for the restarting process in the second part of the application.

14.6 Obstacle Avoiding Tank

The goal of this application is to learn how to use the digital I/O, mixed signal systems, and timer modules of the MSP432 microcontroller. As a real-life application, we examine

Advanced Applications 391

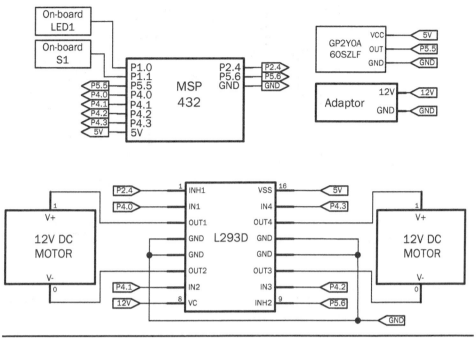

FIGURE 14.6 Layout of the obstacle-avoiding tank.

an obstacle avoiding tank. In this section, we provide the equipment list, circuit layout, and system design specifications.

14.6.1 Equipment List and Layout

Following is a list of equipments to be used in the obstacle avoiding tank application. The layout of this application is shown in Fig. 14.6:

- One 12-V DC adaptor
- Two 12-V DC motors
- One L293D motor driver IC
- One GP2Y0A60SZLF proximity sensor

14.6.2 System Design Specifications

In this application, we will design an obstacle-avoiding tank with a proximity sensor and two DC motors. The tank should check the obstacle distance every 0.2 seconds. The tank will move forward if there is no obstacle closer than 15 centimeters. This step is carried out by rotating the DC motors in the same direction with a suitable PWM signal. The tank should move backwards diagonally if there is an obstacle closer than 15 centimeters. This is achieved by stopping one of the DC motors and rotating the other in the reverse direction with a suitable PWM signal. During this phase, the tank should check the obstacle distance continuously. If there is no obstacle closer than 15 centimeters, the tank should continue to move in that direction. Then, it should return checking the obstacle distance every 0.2 seconds. The system can be turned off and on by

392 Chapter Fourteen

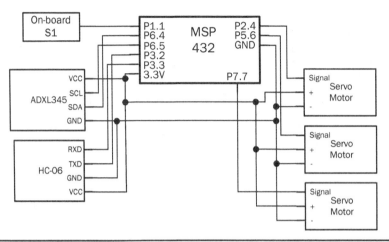

FIGURE 14.7 Layout of the robot arm control system.

the button S1 connected to pin P1.1 on the MSP432 LaunchPad. This operation must be accomplished by a suitable low-power mode. LED1 connected to pin P1.0 on the MSP432 LaunchPad should be turned on if the system is working.

14.7 Robot Arm Control System

The goal of this application is to learn how to use the digital I/O, mixed signal systems, and digital and wireless communication modules of the MSP432 microcontroller. As a real-life application, we construct a robot arm control system. In this section, we provide the equipment list, circuit layout, and system design specifications.

14.7.1 Equipment List and Layout

Following is a list of equipments to be used in the robot arm control system. We can also include the HC-06 Bluetooth module when the wireless communication mode is added to the application. The layout of the robot arm control system is shown in Fig. 14.7:

- Three servo motors
- One ADXL345 three-axis accelerometer module

14.7.2 System Design Specifications

In this application, we will design a robot arm control system using three servo motors and an ADXL345 three-axis accelerometer module. Servo motors will be used to mimic up-down, left-right, and forward-backward movements of the robot arm. Initially, all servo motors stay at $\pi/2$ rad. According to data coming from the accelerometer, the position of the servo motors will be changed between 0 and π rad. Also, servo motors can be reset to their initial condition by pressing the button S1 connected to pin P1.1 on the MSP432 LaunchPad.

We can improve our robot arm control system further by adding the wireless communication module introduced in Sec. 10.3. Then, we can control our robot arm from a remote location. To do so, we should add the HC-06 Bluetooth module to the system.

The user can control the robot arm by using his or her cell phone via Bluetooth communication. Here, a simple Android application developed under MIT App Inventor may be sufficient on the cell phone [45]. The characters u, d, l, r, f, b and rotation angle value can be sent from the cell phone to mimic the up-down, left-right, and forward-backward movements of the arm. Also, servo motors can be reset to their initial condition by sending the character i.

14.8 Intelligent Washing Machine

The goal of this application is to learn how to use the digital I/O, digital and wireless communication, and mixed signal systems modules of the MSP432 microcontroller. As a real-life application, we examine an intelligent washing machine. In this section, we provide the equipment list, circuit layout, and system design specifications.

14.8.1 Equipment List and Layout

Following is a list of equipments to be used in the intelligent washing machine application. We can also include the GSM click and two-level sensor modules when the wireless communication mode is added to the application. The layout of the intelligent washing machine is shown in Fig. 14.8:

- Three 100-ηF capacitors
- One stepper motor
- One ULN2003 motor driver
- Three push buttons
- One IR transmitter LED
- One IR receiver LED
- One LED
- One 10-kΩ resistor
- Two 220-Ω resistors

14.8.2 System Design Specifications

The intelligent washing machine will be controlled by five push buttons. Two of them are S1 and S2 connected to pins P1.1 and P1.4 on the MSP432 LaunchPad for main on/off and rotation speed control, respectively. The remaining three buttons (connected to pins P3.5, P3.6, and P3.7) are for program selection as follows:

- **Prewash:** 30 rotations in one direction, then 30 rotations in the other direction.
- **Normal wash:** 100 rotations in one direction, then 100 rotations in the other direction.
- **Final spin:** 50 rotations in one direction, but faster than prewash and normal wash.

The normal wash program can be improved by adding intelligence to it. To do so, we can include an IR transmitter and receiver LED pair. The IR transmitter LED emits IR light when fed with voltage. The IR receiver LED produces voltage when it absorbs

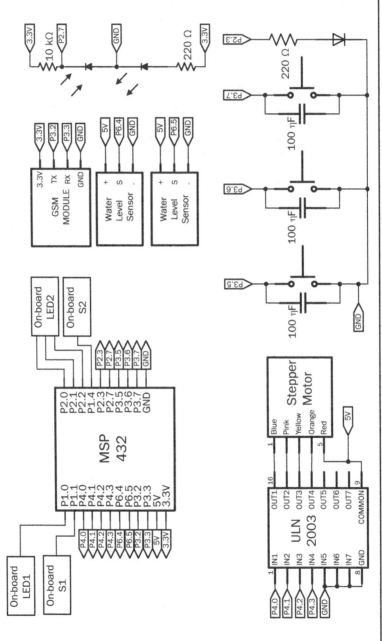

Figure 14.8 Layout of the intelligent washing machine.

IR light. We can form a structure by them such that when the water passing through is dirty, no light transmission occurs. Hence, the output of the receiver LED can be taken as logic level 1. When the water passing through these LEDs is clean, light transmission occurs. Hence, the output of the receiver LED can be taken as logic level 0. Therefore, when the water is dirty, normal wash program is repeated again. This program ends when the water becomes clean.

When the button S1 is pressed, the system will be activated. To indicate this, LED1 connected to pin P1.0 on the MSP432 LaunchPad will turn on. In this state, all programs (prewash, normal wash, and final spin) can be performed. Each program can be selected by a specific button. The red, green, and blue colors of LED2 connected to pins P2.0, P2.1, and P2.2 on the MSP432 LaunchPad will turn on according to the program selection. The button S2 will be used to adjust the rotation speed to slow or fast. Depending on the selection, the LED connected to pin P2.3 will be either turned on or off. When the button S1 is pressed again, the system will be deactivated. To indicate this, LED1 will turn off.

We can improve our intelligent washing machine further. Therefore, we can add sensor modules to detect detergent and softener levels. We can also add the GSM click module to send SMS message to the market if the detergent or softener level gets low.

14.9 Non-Touch Paper Towel Dispenser

The aim of this application is to learn how to use the digital I/O, digital and wireless communication, mixed signal systems, and timer modules of the MSP432 microcontroller. As a real-life application, we examine a non-touch paper towel dispenser. In this section, we provide the equipment list, circuit layout, and system design specifications.

14.9.1 Equipment List and Layout

Following is a list of equipments to be used in the non-touch paper towel dispenser application:

- One 12-V DC adaptor
- One 12-V DC motor
- One L293D motor driver IC
- One IR sensor

The additional equipment list for the non-touch paper towel dispenser application when the wireless communication mode is added to it is as follows. Note that a second MSP432 is also needed to create a wireless network. The layout of the non-touch paper towel dispenser is shown in Fig. 14.9:

- One IR sensor module
- Two Xbee modules
- One GSM click module

14.9.2 System Design Specifications

In the first part of the application, we will design a non-touch towel dispenser using an IR sensor and LED1 connected to pin P1.0 on the MSP432 LaunchPad. When the user crosses his or her hand by the IR sensor, this will indicate that the paper towel is needed.

396 Chapter Fourteen

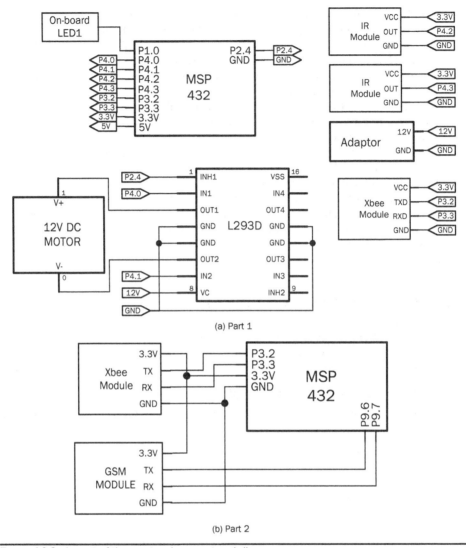

FIGURE 14.9 Layout of the non-touch paper towel dispenser.

This should generate a timer interrupt. Then, LED1 will turn on for four seconds to indicate that the paper towel is fed. During this time, no other paper towel request is accepted. When the waiting time is over, LED1 will turn off. The system will wait for a new paper towel request.

In the second part of the application, we will repeat the first part using a DC motor instead of LED1. To do so, we should set the PWM frequency to 5 kHz. The duty cycle of the PWM signal should be 50%. The DC motor will rotate for four seconds to simulate the feeding of the paper towel. Again, no other paper towel request is accepted during this time. After the waiting time is over, the motor will stop.

In the third part of the application, we will add the wireless communication option to our non-touch paper towel dispenser. To do so, we will add an IR sensor module

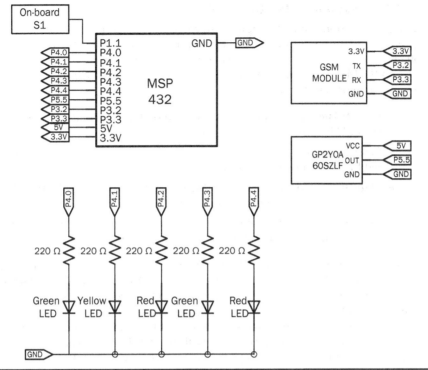

FIGURE 14.10 Layout of the traffic lights system.

to detect whether the paper is below a critical level. If this is the case, our system will connect to the remote main controller via ZigBee protocol by sending the character P. When the main controller receives this request, it sends an SMS message using the GSM click module to the market requesting new towel.

14.10 Traffic Lights

The aim of this application is to learn how to use the digital I/O, digital and wireless communication, mixed signal systems, and timer modules of the MSP432 microcontroller. As a real-life application, we examine a traffic lights system. In this section, we provide the equipment list, circuit layout, and system design specifications.

14.10.1 Equipment List and Layout

Following is a list of equipments to be used in the traffic lights system. We can also include the GSM click module and proximity sensor when the wireless communication mode is added to the application. The layout of the traffic lights system is shown in Fig. 14.10:

- Five LEDs (two green, two red, one yellow)
- Five 220-Ω resistors

398 Chapter Fourteen

14.10.2 System Design Specifications

In this application, we will design a traffic lights system for a street with crosswalk. The green, yellow, and red LEDs connected to pins P4.0, P4.1, and P4.2 are for the cars. The other green and red LEDs connected to pins P4.3 and P4.4 and button S1 connected to pin P1.1 on the MSP432 LaunchPad are for pedestrians. When the button S1 is not pressed, the system works in a loop as follows:

- **State 1:** The green LED for cars is turned on for 90 seconds. During this time, the red LED is turned on for pedestrians.
- **State 2:** The yellow LED for cars is turned on for 5 seconds. During this time, the red LED is turned on for pedestrians.
- **State 3:** The red LED for cars is turned on for 20 seconds. During this time, the green LED is turned on for pedestrians.
- **State 4:** The red and yellow LEDs for cars are turned on for 5 seconds. During this time, the red LED is turned on for pedestrians.

If a pedestrian presses the button S1 in State 1 after 60 seconds, then the system will jump to State 2. If the button S1 is pressed before 60 seconds, then the system will wait until 60 seconds have passed. Then, it will jump to State 2. If the system is in State 3 or State 4, the button S1 will not be activated and cannot be used.

We can also add a traffic rule violation module to our system. When the traffic light is red for the cars, the proximity sensor will start working. If a car pass is detected during this time, our module will report this violation to a nearby traffic police station via sending an SMS message using the GSM click module.

14.11 Car Parking Sensor System

The goal of this application is to learn how to use the digital I/O and mixed signal systems modules of the MSP432 microcontroller. As a real-life application, we examine a car parking sensor system. In this section, we provide the equipment list, circuit layout, and system design specifications.

14.11.1 Equipment List and Layout

Following is a list of equipments to be used in the car parking sensor system. The layout of this application is shown in Fig. 14.11:

- One GP2Y0A60SZLF proximity sensor
- One buzzer
- One BC547 transistor
- One 1-kΩ resistor

14.11.2 System Design Specifications

In this application, we will design a car parking sensor system using an analog proximity sensor. The system will start working when the proximity sensor reads a value corresponding to one meter. Then, LED1 connected to pin P1.0 on the MSP432 LaunchPad will turn on. As the distance falls lower than 50 centimeters, the buzzer starts working.

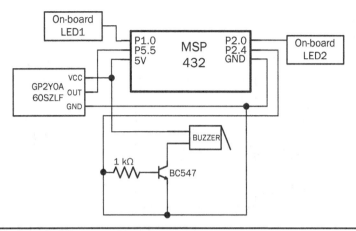

FIGURE 14.11 Layout of the car parking sensor system.

LED1 will turn off and red color of LED2 connected to pin P2.0 on the MSP432 Launch-Pad will turn on. Frequency of the sound produced by the buzzer will increase with respect to proximity such that when the car is 10 centimeters close to the obstacle, the buzzer will have the highest frequency value.

14.12 Body Weight Scale

The aim of this application is to learn how to use the digital I/O, mixed signal systems, digital and wireless communication, and timer modules of the MSP432 microcontroller. As a real-life application, we examine a body weight scale. In this section, we provide the equipment list, circuit layout, and system design specifications.

14.12.1 Equipment List and Layout

Following is a list of equipments to be used in the body weight scale application. We can include the HC-06 Bluetooth module when the wireless communication mode is added to the application. To note here, we can also use a 2×16 LCD module instead of the four-digit seven-segment display. The layout of the body weight scale is shown in Fig. 14.12:

- One load cell
- One HX711 load cell amplifier
- Four-digit seven-segment display
- Four 330-Ω resistors

14.12.2 System Design Specifications

The aim of this application is to build a body weight scale used in our homes. Basically, we need a load cell and amplifier to boost its signal. The load cell converts the applied force on it to electrical voltage. The weight of the user will be displayed on the four-digit seven-segment display in the format XXX.X. To avoid false measurements, the system

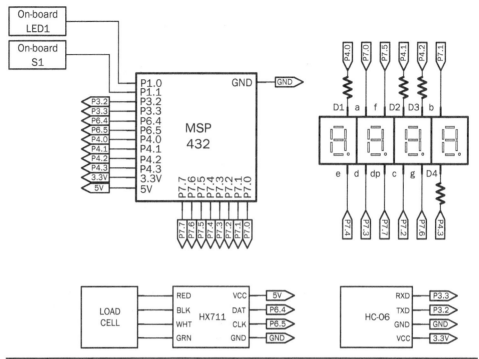

Figure 14.12 Layout of the body weight scale.

should wait for the sensor to get stabilized. If the measurement stays within 0.1-kg interval for 5 seconds, then the average value will be displayed on the display. This system can be turned on by pressing the button S1 connected to pin P1.1 on the MSP432 LaunchPad. LED1 connected to pin P1.0 on the MSP432 LaunchPad will be used to indicate the on/off state. The system will start working when it detects a weight on it. After making the weight measurement, it will turn off within four seconds.

We can improve the body weight scale by adding a wireless connectivity option. Therefore, we can add the HC-06 Bluetooth module to the system to send the weight log to the user's cell phone via Bluetooth communication. A simple Android application developed under MIT App Inventor may be sufficient on the cell phone side [45]. The RTC can be used to keep time on the body weight scale. When the weight measurement is complete, it can be saved in memory with the time value. When user connects to the body weight scale via his/her cell phone, all values can be transmitted to the phone via Bluetooth communication.

14.13 Intelligent Billboard

The goal of this application is to learn how to use the digital I/O, mixed signal systems, digital and wireless communication, and timer modules of the MSP432 microcontroller. As a real-life application, we examine an intelligent billboard. In this section, we provide the equipment list, circuit layout, and system design specifications.

Advanced Applications

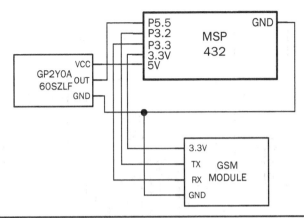

FIGURE 14.13 Layout of the intelligent billboard.

14.13.1 Equipment List and Layout

Following is a list of equipments to be used in the intelligent billboard. The layout of this application is shown in Fig. 14.13:

- One GP2Y0A60SZLF proximity sensor
- One GSM click module

14.13.2 System Design Specifications

In this application, we will design an intelligent billboard system using an analog proximity sensor and GSM module. Nowadays, companies want to measure the impact of an advertisement published on a billboard. One way of doing this is extracting statistics based on who viewed or paid attention to the billboard while the advertisement is on. We can develop a prototype system for this purpose. Our system is composed of a proximity sensor and GSM module. The proximity sensor will be directed to the people passing by. When someone gets closer to the billboard and stays within a predefined distance for five seconds, a counter in the system will increase. Since the billboard is an autonomous device located in a public area, we have to integrate GSM capability to send the count results to the company at the end of each day.

14.14 Elevator Cabin Control System

The goal of this application is to learn how to use the digital I/O, mixed signal systems, and timer modules of the MSP432 microcontroller. As a real-life application, we examine an elevator cabin control system. In this section, we provide the equipment list, circuit layout, and system design specifications.

14.14.1 Equipment List and Layout

Following is a list of equipments to be used in the elevator cabin control system. To note here, we can also use a 2 × 16 LCD module instead of the four-digit seven-segment display. The layout of the elevator cabin control system is shown in Fig. 14.14:

- One DIP switch with eight positions
- One seven-segment display

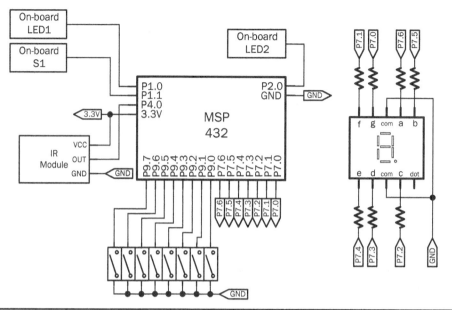

FIGURE 14.14 Layout of the elevator cabin control system.

- Seven 220-Ω resistors
- One IR sensor module

14.14.2 System Design Specifications

In this application, we will design a prototype elevator cabin control system. Our elevator works in a building with five floors. We will use the first three switches on the DIP switch (one to three) to identify which floor we are calling the cabin from. We also need an elevator call button. The button S1 connected to pin P1.1 on the MSP432 LaunchPad can be used for this purpose. We also need five buttons inside the cabin to indicate the target floor. We will use the next five switches (four to eight) for this purpose. For example, after the elevator door is closed and DIP switch eight goes to logic level 1, the elevator should go to the fifth floor.

The system works as follows. The cabin starts at first floor. If someone at this floor presses the call button, door of the cabin will open. We will use the proximity sensor to avoid closing of the door when someone is entering or leaving the cabin. If someone from another floor presses the call button, the cabin moves to that floor. Assume that the travel time between each floor is three seconds. When the cabin reaches the target floor, its door opens and stays in that state for 10 seconds. Since this is a prototype system, we assumed one user at a time. Therefore, scheduling issues within the elevator control are avoided. However, we suggest the reader to think about this possibility as well for a more advanced elevator cabin control system.

The seven-segment display will show which floor the cabin is at. LED1 connected to pin P1.0 on the MSP432 LaunchPad shows whether the elevator is busy or not. The red color of LED2 connected to pin P2.0 on the MSP432 LaunchPad indicates whether the cabin door is open or closed.

Advanced Applications 403

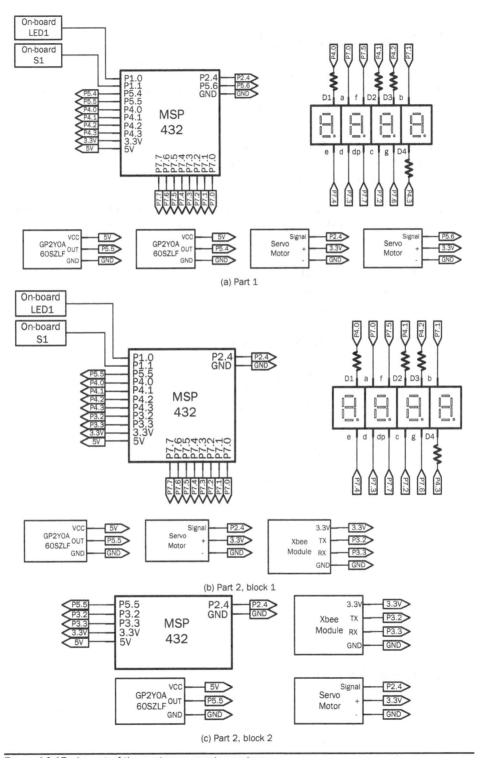

FIGURE 14.15 Layout of the customer counter system.

14.15 Customer Counter System

The aim of this application is to learn how to use the digital I/O, mixed signal systems, digital and wireless communication, and timer modules of the MSP432 microcontroller. As a real-life application, we examine a customer counter system. In this section, we provide the equipment list, circuit layout, and system design specifications.

14.15.1 Equipment List and Layout

Following is a list of equipments to be used in the customer counter system. We can also include two Xbee modules when the wireless communication mode is added to the application. To create a wireless network a second MSP432 is also required. To note here, we can also use a 2×16 LCD module instead of the four-digit seven-segment display. The layout of the customer counter system is shown in Fig. 14.15:

- Two servo motors
- Two GP2Y0A60SZLF proximity sensors
- Four-digit seven-segment display
- Four 330-Ω resistors

14.15.2 System Design Specifications

In this application, we will design a customer counter system for counting customers in a shopping mall with designated doors for entrance and leaving. To do so, we should place a proximity sensor to each door. Hence, we can detect whether a customer passing through the gate is entering or leaving. Movement of the doors will be represented by servo motors. If a door is opened, related servo motor will stay in 0 rad. If the door is closed, related servo motor will stay in π rad. If a customer pass is detected, the door will be opened. Then, it will stay in this state for 10 seconds if another pass is not detected during this time. The customer counter is reset as the mall opens by pressing the button S1 connected to pin P1.1 on the MSP432 LaunchPad. The count value is increased by one for each entering customer. It is decreased by one for each leaving customer. The total number of customers in the mall can be shown on the four-digit seven-segment display. As the shopping mall closes, security check can be done via the count value. If no one is left in the mall, gates will be locked by pressing the button S1. This can be simulated by turning on LED1 connected to pin P1.0 on the MSP432 LaunchPad.

The entrance and leaving doors may be located far away. Then, we should have two microcontrollers communicating with the Xbee modules introduced in Sec. 10.4. In this setup, one microcontroller will act as a master having the display unit on it. The other microcontroller will act as slave.

14.16 Frequency Meter

The aim of this application is to learn how to use the digital I/O, mixed signal systems, and timer modules of the MSP432 microcontroller. As a real-life application, we examine a frequency meter. In this section, we provide the equipment list, circuit layout, and system design specifications.

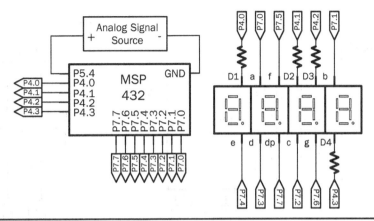

FIGURE 14.16 Layout of the frequency meter.

14.16.1 Equipment List and Layout

Following is a list of equipments to be used in the frequency meter. To note here, we can also use a 2 × 16 LCD module instead of the four-digit seven-segment display. The layout of the frequency meter is shown in Fig. 14.16:

- Four-digit seven-segment display
- Four 330-Ω resistors

14.16.2 System Design Specifications

The frequency meter is a device to measure frequency (repetition rate per second) of an analog signal. Assume that analog signal oscillates between 0 and 3 V. Hence, the DC value of the signal will be 1.5 V. We can use the ADC14 module to detect 1.5-V crossings of the periodic signal. Total number of 1.5-V crossings within one second can be used to calculate the frequency of the signal. The measured signal can be noisy. To avoid any miscalculations, the window comparator can be used under the ADC14 module. We can display the measured frequency on the four-digit seven-segment display. Hence, frequency values between 0 and 9999 Hz can be measured by the system.

14.17 Pedometer

The aim of this application is to learn how to use the digital I/O, digital and wireless communication, and timer modules of the MSP432 microcontroller. As a real-life application, we examine a pedometer. In this section, we provide the equipment list, circuit layout, and system design specifications.

14.17.1 Equipment List and Layout

Following is a list of equipments to be used in the pedometer. The layout of this application is shown in Fig. 14.17:

- One ADXL345 three-axis accelerometer module
- One HC-06 Bluetooth module

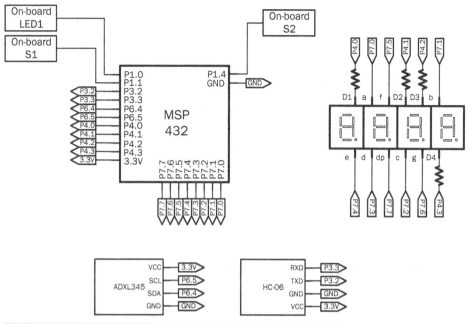

FIGURE 14.17 Layout of the pedometer.

- Four-digit seven-segment display
- Four 330-Ω resistors

14.17.2 System Design Specifications

The pedometer is a device that counts steps when you carry it on. We can design a pedometer using an ADXL345 three-axis accelerometer module. This module communicates over I²C interface providing 16 bits of data per axis. Once you get the acceleration data, you have to work on it to extract steps. There is a good article explaining how to do this [46]. Therefore, we strongly suggest applying the method there. Once you extract the steps, you can count them and let the user know when total number of steps reach a limit (let's say per day). LED1 connected to pin P1.0 on the MSP432 LaunchPad will turn on in this case. Also, the number of steps will be shown on the four-digit seven-segment display. This value can be reset by pressing the button S1 connected to pin P1.1 on the MSP432 LaunchPad. The user can also send the total number of steps to his or her cell phone through the Bluetooth module when the button S2 connected to pin P1.4 on the MSP432 LaunchPad is pressed.

14.18 Digital Camera

The aim of this application is to learn how to use the digital I/O and digital communication modules of the MSP432 microcontroller. As a real-life application, we examine a digital camera. In this section, we provide the equipment list, circuit layout, and system design specifications.

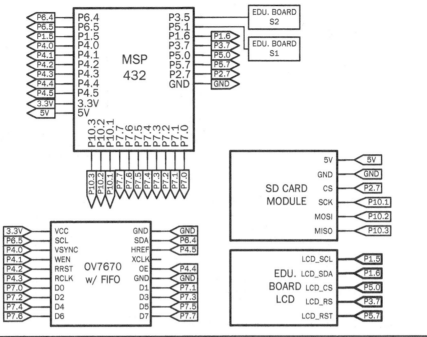

FIGURE 14.18 Layout of the digital camera.

14.18.1 Equipment List and Layout

Following is a list of equipments to be used in the digital camera. The layout of this application is shown in Fig. 14.18:

- One OV7670 camera module with AL422 FIFO
- TI BOOSTXL-EDUMKII Educational BoosterPack
- One SD card module
- One SD card

14.18.2 System Design Specifications

In this application, we will design a complete digital camera system using an OV7670 camera, LCD, and SD card modules. When the button S1 connected to pin P5.1 on the MSP432 LaunchPad is pressed, the image will be captured and saved to the SD card. This image will be displayed on the LCD screen of the TI BOOSTXL-EDUMKII Educational BoosterPack when the button S2 connected to pin P3.5 on the MSP432 LaunchPad is pressed. If the button S1 is pressed again, the next image will be captured and saved to the SD card with a different name. To note here, buttons S1 and S2 are on the TI BOOSTXL-EDUMKII Educational BoosterPack for this application.

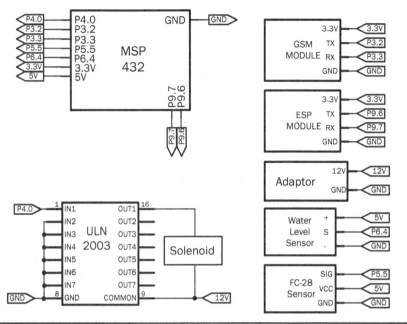

Figure 14.19 Layout of the irrigation system.

14.19 Irrigation System

The aim of this application is to learn how to use the digital I/O, digital and wireless communication, mixed signal systems, and timer modules of the MSP432 microcontroller. As a real-life application, we examine an irrigation system. In this section, we provide the equipment list, circuit layout, and system design specifications.

14.19.1 Equipment List and Layout

Following is a list of equipments to be used in the irrigation system. The layout of this application is shown in Fig. 14.19:

- One 12-V DC adaptor
- One ULN2003 motor driver
- One 12-V plastic water solenoid valve
- One FC-28 moisture sensor
- One water-level sensor
- One GSM click module
- One ESP-01 Wi-Fi module

14.19.2 System Design Specifications

In this application, we will design an irrigation system using moisture and water-level sensors. The moisture sensor will be used to measure the moisture level of the soil every

10 seconds. When the moisture level falls below 20%, the water solenoid valve will be opened to irrigate the soil. When the moisture level is above 70%, the water solenoid valve will be closed. We will use the ESP-01 module to get weather data as in Sec. 10.2. Hence, the irrigation system will not work when the weather condition is not suitable (such as a rainy day or cold weather). The water tank will be used as water source in the irrigation system. The water level in this tank will be checked every 10 seconds (using the comparator module) by the water-level sensor. If the water level falls below 20%, an SMS message will be sent to the user by the GSM click module. The system must wait in a suitable low-power mode when its modules are not working.

14.20 Speech Recognizer

We would like to mention the speech recognizer library introduced by TI. This library can be used in several applications. Hence, commands can be given to the system by speaking to it. Since TI provides the complete application, we did not include it here. We kindly ask the reader to visit the product website to learn how to use it in applications [47].

References

1. Ünsalan, C. and Gürhan, H.D. (2013). *Programmable Microcontrollers with Applications: MSP430 LaunchPad with CCS and Grace*, McGraw-Hill.

2. Texas Instruments. (2017). *MSP432P401R SimpleLink Microcontroller LaunchPad Development Kit (MSPEXP432P401R) User's Guide*, slau597c edn.

3. Texas Instruments. (2017). *MSP432P401R, MSP432P401M SimpleLink Mixed-Signal Microcontrollers*, slas826f edn.

4. Texas Instruments. (2017). *MSP432P4xx SimpleLink Microcontrollers Technical Reference Manual*, slau356f edn.

5. Texas Instruments. (2015). *Code Composer Studio v6.1 for MSP432 User's Guide*, slau575b edn.

6. Instruments, T., http://processors.wiki.ti.com/index.php/Download_CCS. Accessed: May 25, 2017.

7. Instruments, T., https://dev.ti.com/about. Accessed: May 25, 2017.

8. Deitel, P. and Deitel, H. (2015). *C How to Program*, Prentice Hall, 8th edn.

9. Texas Instruments. (2016). *ARM Assembly Language Tools v16.12.0.STS User's Guide*, spnu118q edn.

10. Texas Instruments. (2011). *Cortex-M3/M4F Instruction Set Technical User's Manual*, umcoreism-7703 edn.

11. Texas Instruments. (2015). *MSP432 Peripheral Driver Library User's Guide*.

12. Energia, http://energia.nu/guide/#gettingstarted. Accessed: March 23, 2017.

13. Yiu, J. (2014). *The Definitive Guide to ARM Cortex-M3 and Cortex-M4 Processors*, Newness, 3rd edn.

14. Oppenheim, A.V. and Schafer, R.W. (2009). *Discrete-Time Signal Processing*, Prentice Hall, 3rd edn.

15. Lathi, B.P. and Ding, Z. (2009). *Modern Digital and Analog Communication Systems*, Oxford University Press, 4th edn.

16. Gaspar, P.D., Santo, A.E., and Riberio, B. (2009). *MSP430 Teaching Rom*, Texas Instruments.

17. Microchip. (2002). *Differential ADC Biasing Techniques, Tips and Tricks*, an842 edn.

18. Ünsalan, C. and Tar, B. (2017). *Digital System Design with FPGA: Implementation using Verilog and VHDL*, McGraw-Hill.

19. NXP. (2012). *UM10204 I2C-bus Specification and User Manual*, rev. 5 edn.
20. ROHM Semiconductor. (2010). *Digital 16-bit Serial Output Type Ambient Light Sensor IC BH1750FVI*, 10046th edn.
21. Microchip. (2011). *MCP9808 Maximum Accuracy Digital Temperature Sensor*, ds25095a edn.
22. IEEE, http://standards.ieee.org/about/get/802/802.11.html. Accessed: May 11, 2017.
23. Tanenbaum, A.S. and Wetherall, D.J. (2010). *Computer Networks*, Pearson, 5th edn.
24. Microsoft, https://technet.microsoft.com/en-us/library/cc958821.aspx. Accessed: May 11, 2017.
25. Espressif, https://www.espressif.com/en/products/hardware/esp8266ex/overview. Accessed: May 11, 2017.
26. Espressif. (2017). *ESP8266 AT Instruction Set*, v2.1.0 edn.
27. Map, O.W., http://openweathermap.org/api. Accessed: May 25, 2017.
28. Map, O.W., http://openweathermap.org/appid. Accessed: May 25, 2017.
29. Map, O.W., http://openweathermap.org/weather-conditions. Accessed: May 25, 2017.
30. IEEE, http://www.ieee802.org/15/pub/TG1.html. Accessed: May 11, 2017.
31. SIG, B., https://www.bluetooth.com/. Accessed: May 11, 2017.
32. Guangzhou HC Information Technology Co. Ltd. (2011). *Product Data Sheet*, rev 1 edn.
33. JNHuaMao Technology Company. (2013). *Bluetooth 4.0 BLE Module Datasheet*, v507 edn.
34. Alliance, Z., http://www.zigbee.org/. Accessed: May 11, 2017.
35. International, D., http://docs.digi.com/display/XBeeArduinoCodingPlatform/AT+commands. Accessed: May 11, 2017.
36. Mikroelektronika, https://shop.mikroe.com/click/wireless-connectivity/gsm. Accessed: May 11, 2017.
37. Teilt, http://www.telit.com. Accessed: May 11, 2017.
38. Telit. (2015). *AT Commands Reference Guide*, 80000th edn.
39. Sturgess, B.N. and Carey, F.T. (1995). *The Surveying Handbook*, Springer, chap. 12: Trilateration, 2nd edn.
40. NMEA, http://www.gpsinformation.org/dale/nmea.htm. Accessed: May 11, 2017.
41. u-blox. (2011). *NEO-6 u-blox 6 GPS Modules Data Sheet*, gps.g6-hw-09005-e edn.
42. Lethaby, N. (2013). Why use a real-time operating system in mcu applications, *White paper*, Texas Instruments.
43. Instruments, T., https://training.ti.com/ti-rtos-workshop-series. Accessed: May 25, 2017.
44. Texas Instruments. (2016). *TI-RTOS 2.20 Users Guide*, spruhd4m edn.
45. Inventor, M.A., http://appinventor.mit.edu/. Accessed: May 25, 2017.
46. Zhao, N. (2010). Full-featured pedometer design realized with 3-axis digital accelerometer. *Analog Dialogue*, **44** (06):1–5.
47. Instruments, T., http://www.ti.com/tool/msp-speech-recognizer. Accessed: May 25, 2017.

Index

Index

A

ACLK, 108–109, 110
active high/low input, 74
active mode, 113–114, 115–116
ADC in TI-RTOS projects, 367–368
ADC14
  applications, 192–196, 196–197
  coding examples, 180–188, 310–313
  described, 167–169
  registers, 170–175
  usage via DriverLib functions, 175–180, 181, 184–188
  usage via Energia function, 180, 188
  window comparator, 170, 172, 174–175, 187–188
ADC14CLRIFGR0, 174
ADC14CLRIFGR1, 174–175
ADC14CTL0, 170–171
ADC14CTL1, 171–172
ADC14HI0, 172
ADC14HI1, 172
ADC14IER0, 174
ADC14IER1, 174
ADC14IFGR0, 174
ADC14IFGR1, 174
ADC14IV, 175
ADC14LO0, 172
ADC14LO1, 172
ADC14MCTLx, 172
ADC14MEMx, 173–174
advanced applications
  air freshener dispenser, 389–390
  audio spectrum analyzer, 388–389
  car parking sensor system, 398–399
  customer counter system, 403–404
  digital camera, 406–407
  digital clock, 386–388
  elevator cabin control system, 401–402
  frequency meter, 404–405
  garage gate control system, 383–384
  intelligent billboard, 400–401
  intelligent washing machine, 393–395
  irrigation system, 408–409
  non-touch paper towel dispenser, 395–397
  obstacle avoiding tank, 390–392
  pedometer, 405–406
  robot arm control system, 392–393
  speech recognizer, 409
  traffic lights, 397–398
  vending machine, 385–386
advanced encryption standard accelerator module, 296–298
advertising mode, 265
AES256, 296–298
air freshener dispenser, 389–390
.align directive, 50–51
AM (active mode), 113–114, 115–116
American Standard Code for Information Interchange, 32–33
analog signals, 151
analog-to-digital conversion. *see* ADC14
application program status register, 50, 51
applications. *see also* advanced applications; applications by topic
  baby monitor system, 313
  blinds control system, 196–197
  code updater, 299
  critical household items tracking, 269–271
  door control, 279–281
  energy saver system, 148
  flame detector, 248–253
  garage gate control, 286–288
  heater control, 281–282
  home alarm system, 79–81
  home entrance system, 99–104, 298–299
  intelligent refrigerator module, 267–269
  intelligent umbrella holder, 258–261
  leakage control system, 192–196
  leakage sensor values display, 261–264, 265, 266
  washing machine with sensors, 272–274, 283–285
  window sensor, 275–278, 279
applications by topic
  ADC14, 192–196, 196–197

applications by topic (*Cont.*):
   Bluetooth BR/EDR, 267–269
   Bluetooth LE, 269–271, 272–274
   digital communication, 248–253
   digital I/O, 79–81
   DMA module, 313
   flash controller, 298–299
   GPS, 286–288
   GSM, 281–285
   LPMs, timers, and timer interrupts, 148
   MPU, 298–299
   port interrupts, 99–104
   PWM, 196–197
   RFID, 279–281
   TI-RTOS, 380
   using multiple modules. *see* advanced applications
   WLAN, 258–264
   ZigBee, 275–278, 279
APSR, 50, 51
architecture, 2–5
arithmetic operations
   in assembly, 54–57, 61–62
   in C, 34–37
ARM Cortex-M4F architecture, 1. *see also* instruction set
ARM PL230 microDMA controller. *see* DMA module
arrays, 39–41
ASCII, 32–33
assembly programming
   anatomy of a program, 50–51
   creating projects, 13
   inline, 64–66
   instructions. *see* instruction set
Assembly Step Into button, 14
Assembly Step Over button, 14
asynchronous communication. *see* UART
audio spectrum analyzer, 388–389
auto-request cycle mode, 303–304
auxiliary clock, 108–109, 110

B

baby monitor system, 313
backscatter coupling, 279
backup domain clock, 108–109, 110
bandgap reference voltage generator, 152, 153
BASEPRI register, 50
baud rate generation, 205–207
BC417, 266
BCLK, 108–109, 110
billboard application, 400–401
BIOS (TI-RTOS kernel), 316–317, 319–320, 321. *see also* TI-RTOS memory management; TI-RTOS synchronization; TI-RTOS threads; TI-RTOS timing

bit field operations, 58, 63–64
bit-banding, 4, 54
BITCLK, 200–201
BITCLK16, 200–201, 205
bitwise logic operations, 37–38
blinds control system, 196–197
Bluetooth
   BR/EDR request listener application, 267–269
   defined, 265–266
   LE beacon application, 269–271
   LE communication application, 272–274
body weight scale, 399–400
branch instructions, 57, 58, 62–63
BRCLK, 200–201, 205, 206
breakpoints
   in CCS, 14–15
   in CCS Cloud, 20–21
burst mode, 152
button bouncing, 74–75

C

C programming
   advanced mathematical operations, 45–46
   approaches, 5–6
   arithmetic operations, 34–37
   arrays and pointers, 39–41
   casting, 43–44
   constants, 42–43
   control structures, 38–39
   creating header files, 11–12
   creating projects, 10–11
   data types, 29–32
   inline assembly, 64–66
   logic operations, 37–38
   memory access, 44–45
   memory management, 27–29, 30
   sample data type usages, 33–34
   structures, 41–42
capacitive coupling, 279
capture blocks, 130, 132, 133
capture mode inputs, 191
capture operation, 134
capture/compare interrupt flag, 133, 134, 135, 138
car parking sensor system, 398–399
casting, 43–44
CC2541, 266
CCIFG, 133, 134, 135, 138
CCS Cloud. *see also* Code Composer Studio (CCS)
   compared with other platforms, 23
   working with, 19–21
   working with Energia sketches in, 23
central processing unit, 2, 49–50
CExCTL0, 157–158
CExCTL1, 158–159
CExCTL2, 159–160

Index   417

CExCTL3, 160
CExINT, 160–161
CExIV, 161
char data type, 32
clock module, 333, 336–337, 338, 339
clock system
   clocks and sources, 108–109
   faults, 109
   purpose of, 108
   registers, 109–111
   usage via DriverLib functions, 111–113
clocks
   eUSCI, 200–221, 204
   I^2C mode, 234
   real-time, 145–148
   SPI mode, 222, 223
   UART mode, 205
Code Composer Studio (CCS). *see also* CCS Cloud
   compared with other platforms, 23
   downloading and installing, 9
   introduction, 5
   observing hardware, 15–16
   perspectives, 10
   running and debugging programs, 13–15
   starting, 9–10
   terminal window, 16–18, 19
   working with Energia sketches in, 23
   working with projects, 10–13, 18
code updater application, 299
coding examples. *see under specific topics*
comments, 51
communication. *see* eUSCI; wireless communication
comparator modules (COMP_E0 and COMP_E1)
   coding examples, 163–166
   described, 155–157
   registers, 157–161
   usage via DriverLib functions, 161–162, 164–166
compare blocks, 130, 132, 133
compare mode output types, 189–190
compare operation, 134
condition checks in C, 38
condition codes in assembly, 51, 52
condition flags, 50, 51
const declaration, 42–43
continuous mode
   REF_A, 152
   Timer_A, 129, 132, 133
control instructions in assembly, 57, 58, 62–63
CONTROL register, 50
control structures in C, 38–39
conversion modes, 171
core registers, 49–50
core voltage levels, 113–115
CPU, 2, 49–50
CPU Load option, 373, 374
CS. *see* clock system

CSCLKEN, 111
CSCLRIFG, 111
CSCTL0, 109
CSCTL1, 109–110
CSIFG, 111
CSKEY, 109
customer counter system, 403–404
cycle operating modes, 303–304

D

data moving and processing instructions, 52–53, 60
data types, 29–32, 33–34
DC-DC switching voltage regulator, 113–114
DCO, 108, 109
DCORSEL bits, 109
Debug perspective, 10
Debug window, 13–14
deep sleep mode, 114, 117
define statement, 42–43
delay, adding, 59
destination register, instruction syntax, 51
digital camera application, 406–407
digital clock application, 386–388
digital communication. *see* eUSCI
digital I/O
   application, 79–81
   coding examples, 75–79
   hardware issues, 74–75
   pin layout for, 69–70
   registers, 70–72
   via DriverLib functions, 72–73
   via Energia functions, 73–74
digital signals, 151
digitally controlled oscillator, 108, 109
digital-to-analog conversion. *see* pulse width modulation (PWM)
direct memory access module. *see* DMA
direction register, 71
directives, assembly, 50–51
Disassembly window, 16, 18
division by zero, 36
DMA
   coding examples, 307–313
   described, 301–304
   usage via DriverLib functions, 304–307
door control application, 279–281
double data type, 32, 33
DriverLib functions
   ADC14 usage via, 175–180, 181, 184–188
   adding to projects, 13
   AES256 usage via, 296–297
   clock system usage via, 111–113
   comparator modules usage via, 161–162, 164–166
   digital I/O via, 72–73, 76–78
   DMA module usage via, 304–307

DriverLib functions (*Cont.*):
  flash controller usage via, 290–292, 293
  I²C mode usage via, 239–243, 249–251
  interrupts via, 88, 90
  introduction, 6
  MPU usage via, 294
  port interrupts via, 90–91, 94–98
  power control module usage via, 117–118
  PWM usage via, 191, 192, 194
  real-time clock usage via, 145–148
  reference module usage via, 153–154, 155, 156
  SPI mode usage via, 226–227, 229, 231–233
  system timer usage via, 124–125, 126, 127
  Timer_A usage via, 135–138, 139–144
  Timer32 usage via, 128–129, 131
  UART mode usage via, 208–209, 215–219
  watchdog timer usage via, 119–120, 121–122, 123
drivers. *see* TI-RTOS drivers
dynamic memory allocation, 358, 360, 362, 365–366

E

Edit perspective, 10
elevator cabin control system, 401–402
EM4100, 279
.end directive, 50–51
Energia
  ADC14 function, 180, 188
  compared with other platforms, 23
  digital I/O functions, 73–74, 78–79
  introduction, 6
  port interrupt functions, 91–92, 98–99
  PWM function, 192, 195
  TI-RTOS and, 379–380
  UART mode functions, 210, 220, 221
  using CSS and CSS cloud for sketches, 23
  working with, 21–23, 24
energy saver system, 148
enhanced universal serial communication interface. *see* eUSCI
EPSR, 50
ESP-01, 258–259, 261–262
ESP8266, 257–258
eUSCI. *see also* wireless communication
  application, 248–253
  clocks, 200–221
  introduction, 5, 199
  modes. *see* I²C; SPI; UART
  pin usage, 201, 202
  properties common to modes, 201
  registers, 199–200, 201, 204
event-driven scheduling, 316
events, 345–349
exceptions. *see also* interrupts
  defined, 83
  registers for, 50

execution program status register, 50
Expressions window, 15

F

FatFS, 318
FAULTMASK register, 50
faults, clock system, 109
FFD, 275
.field directive, 50
fixed-size memory blocks, 360–366
flame detector, 248–253
flash memory and controller (FLCTL)
  application, 298–299
  coding example, 292, 293
  described, 289–290
  introduction, 2, 4
  usage via DriverLib functions, 290–292, 293
float data type, 32, 33
floating-point numbers
  in C, 31–32
  instructions in assembly, 57–58
fractional numbers, 31–32
free-running mode, 125, 127
frequency meter, 404–405
full function devices, 275
functions (subroutines), 63

G

garage gate control system
  advanced, 383–384
  using GPS, 286–288
GL865-QUAD GSM IC, 281
.global directive, 50–51
global positioning system, 285–287
global system for mobile communication, 281–285
global variables, 15, 27–29
GPIO, 366–367
GPS, 285–287
GSM, 281–285

H

handler mode, 83. *see also* exceptions; interrupts
hardware
  digital I/O issues, 74–75
  observing, 15–16
hardware breakpoints, 14
hardware interrupts, 322–324, 325, 326
HC-05/06, 266, 267–268
header files, creating, 11–12
HeapBuf, 359, 360–366
HeapMem, 359–360, 361, 362
heater control application, 281–282
hexadecimal numbers, 33

high frequency (oversampling) mode, 205
high-frequency external crystal clock source (HFXT), 108, 109, 110–111
high-speed sub-main clock, 108–109, 110
HM-10, 266, 269, 271, 272–274
home alarm system, 79–81
home entrance system
   using flash controller and MPU, 298–299
   using port interrupts, 99–104
household items tracking application, 269–271
HSMCLK, 108–109, 110
HTTP, 257

I

I^2C. *see also* eUSCI
   coding examples, 243–248, 249–251
   described, 233–234
   registers, 234–239
   in TI-RTOS projects, 372–373
   usage via DriverLib functions, 239–243, 249–251
ICER[x], 88, 89
idle loop, 320–322
IEEE 754 standard, 31–32
IEEE 802.15.1, 265
IEEE 802.15.4, 275
inductive coupling, 279
infinite loops, 39
inline assembly, 64–66
input and reference voltage comparison. *see* comparator modules (COMP_E0 and COMP_E1)
input ports, 4
input register, 71
instruction set
   anatomy of an instruction, 51–52
   arithmetic and logic operations, 54–57, 61–62
   branch and control, 57, 62–63
   floating-point, 57–58
   general data moving and processing, 52–53, 60
   memory access, 53–54
   other instructions, 58–59, 63–64
integer numbers, 29–31
inter-integrated circuit. *see* I^2C
interrupt latency, 85
interrupt program status register, 50
interrupts. *see also* port interrupts
   ADC14 module, 174–175
   COMP_E module, 160–161
   definition and scenarios, 83–85
   DMA, 304
   hardware, 322–324, 325, 326
   I^2C mode, 238–239
   MPU, 292, 293
   multiple requests, 86–87
   NVIC module, 85–90

   single requests, 85–88
   software, 324–325, 327–331
   SPI mode, 225–226
   UART mode, 207–208
   via DriverLib functions, 88, 90
invalid cycle mode, 303
IoT, 255. *see also* wireless communication
IP[y], 88
IPSR, 50
irrigation system, 408–409
ISER[x], 88, 89
ISM bands, 255
ISR, 83. *see also* interrupts
ITU radio regulations, 255

K

kernel
   RTOS, 315–316
   TI-RTOS, 316–317, 319–320, 321. *see also* TI-RTOS memory management; TI-RTOS synchronization; TI-RTOS threads; TI-RTOS timing

L

labels, 51
languages. *see* assembly programming; C programming
late arrival, 87
LDO, 113–115
leakage control system, 192–196
leakage sensor values display application, 261–264, 265, 266
level-sensitive interrupts, 85
LFXT, 108, 109, 110–111
link register, 50
Link with Active Debug Context button, 16
little endian form
   for double data type, 32
   for long long data type, 29
   as shown on memory browser, 33, 35
local variables, 15, 16, 27–29
logic operations
   in assembly, 56, 61
   in C, 37–38
long long data type, 29, 33
loops
   in assembly, 62–63
   in C, 38–39
low-dropout linear voltage regulator, 113–115
low-frequency crystal, 108, 109, 110–111
low-power mode (LPM)
   ADC14, 172
   power control module, 113–117
LR (R14), 50

M

mailboxes, 349–355
masking interrupts, 84
master clock, 108–109, 110
master mode
   Bluetooth, 265–266
   I^2C, 234–235, 236–238, 239
   SPI, 222, 224–225
   ZigBee, 275
master receiver mode, 238
master transmitter mode, 237–238
mathematical operations, advanced, 45–46
math.h, 45–46
MCLK, 108–109, 110
memory
   in assembly, accessing, 53–54, 60
   in C, reaching specific addresses, 44–45
   direct access. *see* DMA
   encryption, 296–298
   flash. *see* flash memory and controller (FLCTL)
   introduction, 2, 4
   observing in CCS, 16, 18
   protection. *see* memory protection unit
memory management
   in C, 27–29, 30
   TI-RTOS. *see* TI-RTOS memory management
memory mapped I/O, 4
memory protection unit
   application, 298–299
   coding example, 295–296
   described, 292–293
   usage via DriverLib functions, 294
memory-to-memory transfer example, 307, 308
memory-to-peripheral transfer example, 307–310
mesh topology application, ZigBee, 275–278, 279
microDMA controller. *see* DMA
mixed signal processing modules, 4
mnemonics, 51
module oscillator (MODOSC), 108
MPU. *see* memory protection unit
MSP432 LaunchPad. *see also specific topics*
   introduction, 1–2
   linking with terminal program, 16–17
MSP432 microcontroller. *see also specific topics*
   architecture of, 2–5
   core registers, 49–50
   introduction, 1
   observing in CCS, 15–16
MSP432P401R, 1. *see also specific topics*

N

National Marine Electronics Association, 286
negative numbers, 30–31, 33
Neo-6M, 286
nested vectored interrupt controller module, 85–90
nesting interrupts, 84, 87
NMEA data, 286
non-maskable interrupts, 84, 86–87
non-touch paper towel dispenser, 395–397
NOP instruction, 59, 64
NVIC, 85–90

O

obstacle avoiding tank, 390–392
one-shot timer mode, 126, 127
open systems interconnection model, 256–257
openweather.com client application, 258–261
OSI model, 256–257
output modes (OUTMOD_x), 189–190
output ports, 4
output register, 71
overflows
   in assembly, 56–57
   in C, 35–36
oversampling mode, 205

P

packing instructions, 59, 63–64
paper towel dispenser, 395–397
parity bit, , 204
PC (R15), 50
PCM, 113–118
PCMCTL0, 115–116
PCMCTL1, 116–117
pedometer, 405–406
periodic timer mode, 125–126, 127
peripheral-to-memory transfer examples, 307–313
perspectives, 10
pin map, 5
ping-pong cycle mode, 304
pins
   digital I/O, 69–70
   eUSCI, 201, 202
   I^2C, 233
   introduction, 4, 5
   SPI mode, 221–222, 224–225
   UART mode, 203
pointers, 39–41
pop pre-emption, 87
port interrupts
   application, 99–104
   coding examples, 92–99
   registers, 90
   via DriverLib functions, 90–91
   via Energia functions, 91–92
ports
   discovering, on Windows 10, 16
   introduction, 4
   pin layout for digital I/O, 69–70

Index

positive numbers, 30–31, 33
power control module, 113–118
power modes
    ADC14, 172
    power control module, 113–117
power supply system, 107–108
preemptive, RTOS, 316
PRIMASK register, 50, 88
priorities
    interrupt, 84, 86–87
    thread. *see* TI-RTOS threads
privileged access, 292–293
program counter register, 50
program status register, 50
programming. *see* assembly programming; C programming
projects
    adding DriverLib functions to, 13
    closing, in CCS, 18
    creating, in CCS, 10–11, 13
    creating, in CCS Cloud, 20
    creating, in Energia, 21
PSR, 50
pull-up/down resistor register, 71
pulse interrupt signals, 85
pulse width modulation (PWM)
    application, 196–197
    coding examples, 192
    described, 188–189
    registers, 189–191
    TIMER_A and, 134
    in TI-RTOS projects, 368–369
    usage via DriverLib functions, 191, 192, 194
    usage via Energia function, 192, 195
PxDIR, 71
PxIE, 90
PxIES, 90
PxIFG, 90
PxIN, 71
PxOUT, 71
PxREN, 71
PxSEL, 72
PxSEL2, 72

Q

queues, 355–358

R

radio signals, 255–256
RAM encryption, 296–298
RAM protection. *see* memory protection unit
real-life applications. *see* applications
real-time clock, 145–148
real-time operating system. *see* RTOS
receive operation
    in I^{2}C mode, 236–237
    in SPI mode, 224–225
    in UART mode, 206–207, 208
reduced function devices, 275
REF_A, 152–156
REFCTL0, 153
reference module and voltage, 152–156
REFO, 108, 109, 111
refrigerator module application, 267–269
register-level programming
    ADC14 examples, 180–181, 182–183
    comparator module examples, 163–164
    digital I/O example, 75–76
    I^{2}C mode examples, 243, 244–247
    introduction, 5–6
    port interrupt examples, 92–94
    PWM example, 192, 193
    reference module example, 154–155
    SPI mode examples, 227–229, 230
    system timer example, 125, 126
    Timer_A examples, 138–139, 140, 141
    Timer32 example, 129, 130
    UART mode examples, 211–215
    watchdog timer example, 120–121, 122
registers
    ADC14, 170–175
    clock system, 109–111
    comparator modules, 157–161
    core, 49–50
    for digital I/O, 70–72
    eUSCI, 199–200, 201, 204
    I^{2}C, 234–239
    NVIC, 87–88
    observing in CCS, 16, 17
    port interrupt, 90
    power control module, 115–117
    PWM, 189–191
    reference module, 153
    SPI, 222–226
    system timer, 123–124
    Timer32, 126–128
    UART, 204–208
    watchdog timer, 118
Reset CPU button, 14
resets, 118
Restart button, 14
Resume button, 13
RF spectrum, 255
RFD, 275
RFID, 278–281
RMD6300 RFID card reader, 279–280
robot arm control system, 392–393
ROM, 4
RTC_C, 145–148

422 Index

RTOS. *see also* TI-RTOS
   analyzer, 374–378
   described, 315–316
   object viewer, 373–374
Runtime Object Viewer, 374

S

sampled mode, 152, 153, 159
SAR, 167, 168. *see also* ADC14
saturating arithmetic instructions, 56–57, 61–62
scanning mode, 265
select registers, 72
semaphores, 344–345
serial peripheral interface. *see* SPI
Show Source button, 16
shutdown mode, 114, 117
signed data types, 30–31
SimpleLink MSP432 SDK, 13
slave mode
   Bluetooth, 265–266
   I^2C, 234–235, 236–238, 239
   SPI, 222, 225
   ZigBee, 275
slave receiver mode, 237
slave transmitter mode, 237
__sleep() function, 117
sleep mode, 114, 117
smart cards, 279
SMCLK, 108–109, 110
software breakpoints, 14
software interrupts, 324–325, 327–331
SP (R13), 49–50
speech recognizer, 409
SPI. *see also* eUSCI
   coding examples, 227–233
   described, 220–222
   registers, 222–226
   in TI-RTOS projects, 371–372
   usage via DriverLib functions, 226–227, 229, 231–233
SRAM, 4
stack pointer register, 49–50
stacks
   in C, 28
   TI-RTOS static memory allocation, 358
static memory allocation, 358, 360, 363
static mode, 152, 153, 159
STCSR, 123–124
STCVR, 123
Step Into button, 14
Step Over button, 14
Step Return button, 14
stop mode, 129
structures in C, 41–42
STRVR, 123

sub-main clock, 108–109, 110
subroutines, 63
successive approximation register, 167, 168. *see also* ADC14
Suspend button, 14
switch bouncing, 74–75
synchronization. *see* TI-RTOS synchronization
synchronous communication. *see* I^2C; SPI
SYS/BIOS (TI-RTOS kernel), 316–317, 319–320, 321. *see also* TI-RTOS memory management; TI-RTOS synchronization; TI-RTOS threads; TI-RTOS timing
system oscillator (SYSOSC), 108, 109
system timer (SysTick), 122–127

T

T32CONTROLx, 127
T32LOADx, 127
TACCRx, 189, 191
tail chaining, 87
tasks (threads), 331–333, 334–336
TAxCCRy, 133, 134
TAxCCTLy, 134, 135
TAxCTL, 131, 133
TAxEX0, 135
TAxIV, 135
TAxR, 131, 133, 134, 189
TCP/IP, 257
Telit, 281
temperature sensor, 152, 153
terminal window, 16–17, 19
.text directive, 50–51
thread mode, 83
threads, RTOS, 315–316. *see also* TI-RTOS threads
TI SimpleLink SDK, 13
time sharing, 316
timer modules
   introduction, 4
   TI-RTOS, 337–338, 340, 341–342
Timer_A
   coding examples, 138–144
   described, 129–130, 132
   output modes, 189–190
   registers, 131–135
   usage via DriverLib functions, 135–138, 139–144
Timer32
   coding examples, 129
   described, 125–126
   registers, 126–128
   usage via DriverLib functions, 128–129, 131
timestamp module, 340–341, 343
timing. *see* TI-RTOS timing
TI-RTOS. *see also other TI-RTOS entries*
   application, 380
   creating projects, 318–319

Index   423

TI-RTOS (Cont.):
  Energia and, 379–380
  overview, 316–318
TI-RTOS drivers
  ADC, 367–368
  GPIO, 366–367
  I²C, 372–373
  overview, 317, 318, 366
  PWM, 368–369
  SPI, 371–372
  UART, 369–371
TI-RTOS instrumentation
  main discussion, 373–378
  overview, 317, 318
TI-RTOS kernel, 316–317, 319–320, 321. *see also*
    TI-RTOS memory management; TI-RTOS
    synchronization; TI-RTOS threads; TI-RTOS
    timing
TI-RTOS memory management
  allocating fixed-size memory blocks, 360–366
  allocating variable-sized blocks, 359–360, 361, 362
  overview, 358–359
TI-RTOS middleware, 317, 318
TI-RTOS synchronization
  events, 345–349
  mailboxes, 349–355
  queues, 355–358
  semaphores, 344–345
TI-RTOS threads
  hardware interrupt, 322–324, 325, 326
  idle loop, 320–322
  overview, 320
  software interrupt, 324–325, 327–331
  task, 331–333, 334–336
TI-RTOS timing
  clock module, 333, 336–337, 338, 339
  timer module, 337–338, 340, 341–342
  timestamp, 340–341, 343
traffic lights, 397–398
transmission control protocol/Internet protocol, 257
transmit operation
  in I²C mode, 236–237
  in SPI mode, 224–225
  in UART mode, 206–207, 208
trilateration, 286
two's complement form, 30–31

━━━ **U** ━━━

UART. *see also* eUSCI
  coding examples, 210–220
  described, 201, 203–204
  registers, 204–208
  in TI-RTOS projects, 369–371
  usage via DriverLib functions, 208–209, 215–219
  usage via Energia functions, 210, 220, 221

UCAxCTLW0, 204, 206, 223
UCAxIE, 207, 225
UCAxIFG, 207–208, 225
UCAxIV, 208, 225
UCAxMCTLW, 205
UCAxRXBUF, 199, 224–225, 226
UCAxSTATW, 204–205, 224
UCAxTXBUF, 199, 206, 224–225, 226
UCBxBRW, 201
UCBxCTLW0, 234
UCBxCTLW1, 235
UCBxI2COA0, 236, 237
UCBxI2CSA, 236, 237
UCBxIE, 238–239
UCBxIFG, 239
UCBxIV, 239, 240
UCBxRXBUF, 199, 237, 238
UCBxSTATW, 235
UCBxTBCNT, 235–236
UCBxTXBUF, 199, 237–238
UCxCLK, 222, 225
umbrella holder application, 258–261
United Nations ITU, 255
universal asynchronous receiver/transmitter. *see*
    UART
unpacking instructions, 59, 63–64
unprivileged access, 292–293
unsigned data types, 30–31
up mode, 129, 132, 133
up/down mode, 129, 132, 133, 138

━━━ **V** ━━━

variables
  adding watch expressions, in CCS, 15
  adding watch expressions, in CCS Cloud, 21
variable-sized memory blocks, 359–360, 361, 362
vending machine, 385–386
VLO, 108
voltage
  input and reference comparison. *see* comparator
    modules (COMP_E0 and COMP_E1)
  reference, 152–156
voltage levels, 113–115
voltage regulators, 113

━━━ **W** ━━━

washing machine applications
  advanced, 393–395
  using Bluetooth LE, 272–274
  using GSM, 283–285
watch expressions
  adding, in CCS, 15
  adding, in CCS Cloud, 21
watchdog timer, 118–123

watchpoints, 14
WDT_A, 118–123
WDTCTL, 118
weather client application, 258–261
Web client application, 258–261
Web server application, 261–264, 265, 266
Wi-Fi, 256. *see also* wireless local area network (WLAN)
window comparator, 170, 172, 174–175, 187–188
window sensor application, 275–278, 279
wireless communication. *see also* Bluetooth; wireless local area network (WLAN)
   background on, 255–256
   GPS, 285–287
   GSM, 281–285
   RFID, 278–281
   ZigBee, 274–278
wireless local area network (WLAN)
   defined, 256–257
   Web client application, 258–261
   Web server application, 261–264, 265, 266
wrapping mode, 127

X

Xbee, 275–277

Z

ZigBee, 274–278